Aging, Sex, and DNA Repair

Aging, Sex, and DNA Repair

Carol Bernstein

Department of Microbiology and Immunology
College of Medicine
University of Arizona
Tucson, Arizona

Harris Bernstein

Department of Microbiology and Immunology
College of Medicine
University of Arizona
Tucson, Arizona

Academic Press, Inc.
Harcourt Brace Jovanovich, Publishers
San Diego   New York   Boston
London   Sydney   Tokyo   Toronto

ACADEMIC PRESS, INC.
San Diego, California 92101

United Kingdom Edition published by
Academic Press Limited
24–28 Oval Road, London NW1 7DX

Library of Congress Cataloging-in-Publication Data

Bernstein, Carol, DATE
    Aging, sex, and DNA repair / Carol Bernstein and Harris Bernstein.
        p.   cm.
    Includes index.
    ISBN   0-12-092860-4
    1. Aging--Molecular aspects.   2. DNA repair.   3. DNA damage.
4. Sex (Biology)   I. Bernstein Harris.   II. Title.
    [DNLM: 1. Aging--genetics.   2. DNA Damage.   3. DNA Repair.
4. Sex.   QH 467 B531a]
QP86.B34   1991
574.3'72--dc20
DNLM/DLC
for Library of Congress   90-14467
                                                                        CIP

PRINTED IN THE UNITED STATES OF AMERICA
91   92   93   94   9   8   7   6   5   4   3   2   1

To Our Children, Beryl, Golda, and Benjamin
and
in Memory of Fred Hopf

Contents

Preface   xi

1. Introduction: Major Ideas and Historical Perspectives
   I. Definitions of Aging and Sex   1
   II. DNA Damage as the Basic Cause of Aging
       and Sex   2
   III. Historical Background   4
       References   11

2. DNA Damage
   I. DNA Damages as Physically Distinct Alterations in
      Polynucleotide Chains   15
   II. Kinds of DNA Damage Prevalent in Nature   18
       References   25

3. Immediate Consequences of DNA Damage
   I. Transcription Termination   27
   II. Impairment of Replication   29
   III. Decreased Cell Survival   32
       References   34

4. Accumulation of DNA Damage in Somatic Cells
   I. Cell Types Based on Rate of Division   37
   II. Low DNA Repair in Postmitotic Muscle and
       Neuronal Tissue   38
   III. Accumulation of DNA Damage in Adult Muscle and
        Neuronal Tissue   46
   IV. Accumulation of DNA Damage in Liver   52
   V. Accumulation of DNA Damage in Other Tissues   59

VI. Conclusions   60
      References   60

5. **The DNA Damage Hypothesis of Aging Applied to Mammals**
     I. Theoretical Considerations in Support of DNA Damage as the Cause of Aging in Mammals   66
     II. Experimental Evidence in Support of the Hypothesis that DNA Damage is the Cause of Aging in Mammals   68
     III. Mechanisms of Compensating for DNA Damage   82
     IV. Summary of the Stages of Aging Leading to Death in Mammals   84
      References   85

6. **Oxidative DNA Damage as a Potential Major Cause of Aging**
     I. The Nature and Cause of Oxidative DNA Damage   91
     II. Accumulation of Oxidative DNA Damages with Age   95
     III. Inverse Correlation between Incidence of Oxidative DNA Damage and Life Span   96
     IV. Premature Aging in Down's Syndrome Correlated with Increased Oxidative Damage   99
     V. Conclusions   102
      References   103

7. **Additional Evidence Bearing on the DNA Damage Hypothesis of Aging**
     I. Positive Correlation of Life Span with DNA Repair Capacity   108
     II. Experimental Acceleration of Aging by DNA-Damaging Agents   116
     III. Genetic Syndromes in Humans with Features of Premature Aging   121
     IV. Conclusions   139
      References   139

8. **Aging in Nonmammalian Organisms, with Comparisons to Aspects of Aging in Mammals**
     I. Aging and DNA Damage in Single-Celled Organisms   152

II. Aging in Multicellular Organisms   157
III. Similarities of Aging in Nonmammalian Organisms and
    in Mammals   164
IV. Aging of Mammals Viewed in a Broad Evolutionary
    Context   167
    References   168

9. DNA Repair with Emphasis on Single-Strand Damages
    I. Repair Depends on Redundant Information   173
    II. Repair of Single-Strand DNA Damages   175
    III. Relationship of DNA Repair Levels to Aging   189
    IV. Repair of Double-Strand Damages   196
    References   198

10. DNA Repair with Emphasis on Recombinational Repair
    I. Viruses   209
    II. Bacteria   213
    III. Eukaryotes   219
    IV. Overview   226
    References   227

11. Meiosis and Meiotic Recombination
    I. Events of Meiosis and of Meiotic
        Recombination   235
    II. Evidence Bearing on the Double-Strand Break Repair
        Model of Meiotic Recombination   240
    III. Implications   245
    References   246

12. Meiosis Viewed as an Adaptation for DNA Repair
    I. DNA Repair during Meiosis   248
    II. Premeiotic Replication   252
    III. Role of Mitotic Recombination Compared With
        Meiotic Recombination   255
    IV. Organisms Not Using Meiosis or Having Reduced
        Meiotic Recombination in Males   262
    V. Immortality of the Germ Line: Aging of the
        Somatic Line   265
    VI. Summary   265
    References   267

13. The Selective Advantage of Outcrossing
    I. Prevalence of Sexual Reproduction   272
    II. Sex Has Two Basic Features: Recombination
       and Outcrossing   274
    III. Outcrossing   277
    IV. Conclusions   287
    References   288

14. Evolutionary Aspects of Sex
    I. The Likely Continuous Evolutionary History of Sex:
      Evidence that Sex Arose prior to the Divergence
      of Prokaryotes and Eukaryotes   290
    II. Possible Origin of Sex in the Earliest Cells   293
    III. Selective Forces Leading to the Origin of
      Enzyme-Mediated Recombinational Repair   297
    IV. Advantage of Sex in Unicellular
      Haploid Organisms   299
    V. Emergence of Diploid Organisms   300
    VI. Sex and the Emergence of Species   303
    VII. Summary and Conclusions   308
    References   309

15. Other Theories of Aging and Sex
    I. Aging Theories   314
    II. Theories of Sex   327
    References   339

16. Overview, Future Directions, and Implications
    I. General Overview   346
    II. Medical Implications   360
    III. Future Research Directions   366
    IV. General Implications   369
    References   370

Index   373

Preface

Both aging and sexual reproduction (sex) reflect a universal property of life, vulnerability of the genetic material to damage. This idea, and the evidence bearing on it, are the subject of this book. In the scientific literature, aging and sex are usually dealt with as separate subjects. However, because we hypothesize that both phenomena are related consequences of a single cause—genome damage—we consider it logical to treat them together. We begin with established properties of DNA and show, first, that aging is a direct consequence of DNA damage and, second, that sex is a consequence of the need to transmit damage-free genetic information to progeny. Organisms cope with DNA damage by special repair processes. Understanding these processes is crucial to understanding aging and sex. The scientific literature in each of the individual areas of aging, sex, and DNA repair is vast, and so we focus only on those studies which are relevant to our general thesis. Within this framework, however, we try to be complete in presenting evidence for and against our hypothesis.

Over the years we have worked collaboratively on the ideas presented here and discussed them with many individuals. We have enjoyed extensive fruitful collaborations with our colleagues Helen Gensler, Fred Hopf, and Rick Michod. We have also enjoyed collaborations with Henry Byerly, Greg Byers, and Jennifer Hall on specific aspects. In addition, we have benefited greatly from the intellectual stimulation of our current and past Ph.D. graduate students, Steve Abedon, Marian Baldy, Jim Cornett, Peh-yean Cheah, Davis Chen, John Delaney, Kathleen Fisher, Eric Floor, George Holmes,

Paul Hyman, Risa Kandell, David McCarthy, Pat McCreary, Anne Menkens, Robin Miskimins, Siraj Mufti, Eileen Nonn, John Obringer, Paul Scotti, Pete Snustad, Lois Wilson, and Dan Yarosh. We especially thank Steve Abedon for his thoughtful and detailed comments on the entire manuscript.

We are also grateful to Maria Felix, Mary Ann Nelson and Donna Vandenberg for their careful and patient typing of the manuscript and its numerous revisions.

Chapter One

Introduction: Major Ideas and Historical Perspectives

I. *Definitions of Aging and Sex*

Aging can be defined as inherent, progressive impairments of function (Sonneborn, 1978). It occurs in multicellular species as well as in many unicellular organisms.

Some meanings of sex, such as genital union and gender, refer to particular aspects of the sexual process in particular species. Such aspects are not universal. Sexual reproduction does, however, have two features that can be regarded as defining characteristics. The first is recombination, which is the exchange of genetic material between pairs of chromosomes. The second is outcrossing, which means that the two chromosomes exchanging information originate from two different parents. Other features of sex are present almost universally, so that for most organisms sex has the following four steps: (1) two haploid genomes, each from a separate parent, come together in a shared cytoplasm; (2) the chromosomes comprising the two haploid genomes pair so that homologous sequences are aligned; (3) exchange of genetic material occurs between the pairs; and (4) progeny containing the new recombinant chromosomes are formed.

In Section II below, we outline the subjects developed in each of the succeeding chapters. In Section III, we review the historical background of the key ideas on aging and sex that are the basis of this book. We also review the historical background of the major conflicting ideas.

II. DNA Damage as the Basic Cause of Aging and Sex

Genome damage is a universal problem for life. The genetic material, DNA (RNA in some viruses), is unstable under physiological conditions. DNA is further assaulted by reactive chemicals that are found in the intracellular environment. Both instability and assault by reactive chemicals cause DNA to become damaged. As noted by Haynes (1988), DNA is composed of rather ordinary molecular subunits, which are not endowed with any peculiar kind of quantum mechanical stability. He considered that its very "chemical vulgarity" makes DNA subject to all the "chemical horrors" that might befall any such molecule in a warm aqueous medium. Evidence that DNA damages are common in nature is reviewed in Chapter 2.

When DNA damages occur, they are either repaired or they remain in the genome of a cell. Unrepaired DNA damages can result in loss of genetic information and interference with transcription and replication (see Chapter 3). Thus, unrepaired DNA damages are deleterious and often lethal. Aging, we will argue, is mainly due to the accumulation of unrepaired DNA damages in somatic cells. On the other hand, we will argue that the primary function of sexual reproduction is to increase the repair of damages in germ cell DNA so that it can be passed on undamaged to progeny. Thus, we propose that aging and sexual reproduction are two sides of the same coin: Aging reflects the accumulation of DNA damage, and sex reflects the removal of DNA damage. These two ideas are referred to, respectively, as the DNA damage hypothesis of aging and the DNA repair hypothesis of sex.

Most of the work on aging has been done with mammals; therefore, most of our discussion on aging will focus on that in mammals. In Chapter 4 we present evidence that, in some mammalian cell types, DNA damage accumulates with time. In Chapter 5 we present evidence that the accumulation of DNA damage is associated with a decline in gene expression as well as a decline in cellular, tissue, and organ function. This implies a cause and effect relationship between DNA damage and the functional declines that define aging. In Chapter 6 we present evidence that oxidative free radicals, produced as by-products of metabolism, are a general source of DNA damages, which are important in mammalian aging. In Chapter 7, we show that, among different mammalian species, life span correlates with

DNA repair capacity, suggesting that DNA repair is a determinant of longevity. Additionally, in Chapter 7 we present evidence that irradiation of mammals with X-rays or treatment with DNA-damaging chemicals shortens life span and causes some features of premature aging. Furthermore, we review evidence that genetic syndromes in humans with higher than normal levels of DNA damage have features of premature aging. The evidence from Chapter 7, taken together, implies that an increased rate of accumulation of DNA damage causes an increased rate of aging. In Chapter 8 we compare aging in mammals with aging in other species. We conclude that the strategy used by mammals to cope with DNA damage is but one of several options in nature. The range of DNA repair processes available to overcome DNA damages is reviewed in Chapter 9.

In multicellular organisms, DNA damage in germ cells causes the death of those cells and loss of potential progeny. Meiosis is a key stage of the sexual cycle. It is the process by which germ cells are produced. A distinctive feature of meiosis is an intimate pairing of homologous chromosomes, accompanied by the exchange of DNA between them. This exchange is called recombination. We argue that recombination reflects a particularly effective type of DNA repair process—recombinational repair. In Chapter 10 we review substantial evidence that recombinational repair is efficient at removing various types of DNA damage. Evidence is presented in Chapters 11 and 12 that meiosis is an adaptation that specifically promotes production of germ cells that have been freed of DNA damage through recombinational repair.

As noted in Section I above, sexual reproduction has a second basic characteristic in addition to recombination—outcrossing. We argue in Chapter 13 that outcrossing in diploid organisms is selected because it promotes masking, or complementation, of deleterious mutations. Thus, a full explanation of sex must take into account the influence of both DNA damage and mutation; therefore, from now on we will refer to the DNA repair hypothesis of sex as the DNA repair (and complementation) hypothesis of sex. From the evidence discussed in Chapters 10 and 13, plus evidence on the molecular basis of sex in simple organisms, we propose in Chapter 14 an explanation for the origin and early evolution of sex.

Most earlier theories of aging are consistent with the DNA damage hypothesis of aging, with some small changes in emphasis or content. We review these basically consistent theories in Chapter 15 as well as some hypotheses on aging that are not consistent with the DNA damage hypothesis of aging. Furthermore, we review three general theories on sex. One of these is consistent with the DNA repair (and complementation) hypothesis of sex whereas two are not. In Chapter 16 we present an overview of the major ideas in the first 15 chapters. We discuss the implications of these ideas for future efforts to forestall aging and also describe the future research directions suggested by these ideas. In Section III, below, we discuss the historical background of the DNA damage hypothesis of aging, the DNA repair (and complementation) hypothesis of sex, and some of the prominent competing ideas.

III. *Historical Background*

Historical perspectives on aging, sex, and DNA repair have been presented previously. In particular, reviews were presented by Comfort (1979) on aging, Ghiselin (1988) on sex, and Bernstein (1981) on DNA repair. Here, however, we will briefly outline the historical background of the specific concepts that figure prominently in this book. In Section A, below, we describe the historical background of the idea that DNA damage is the basis of aging. We also review the history of the major conflicting idea that aging is a genetically programmed adaptation, selected for the benefit of the species. In Section B, we review the historical background of the two hypotheses that form the basis of our explanation for sex. These are the hypotheses that recombination is an adaptation for repairing DNA damage and that outcrossing in multicellular organisms is maintained by the benefit of masking deleterious mutations. In addition, we explain the main idea that is in conflict with these hypotheses. This is the idea that sex is an adaptation for promoting genetic variation. We also review the historical background of this conflict.

A. The DNA Damage Hypothesis of Aging

The hypothesis that DNA damage directly causes aging was preceded by the concept that somatic mutation is the basis of aging and that DNA damage is only indirectly involved because it can cause mutations. The somatic mutation hypothesis was first proposed by Failla (1958); other early proponents of the idea were Szilard (1959) and Harmon (1962). As we discuss in Chapter 2, Section I, DNA damage and mutation have fundamentally different properties and consequences. In Chapter 15, Section I.H, we review evidence showing that somatic mutation probably is not the primary cause of aging, although it may account for specific kinds of aging problems, as discussed in Chapter 8, Section III.C.

Alexander (1967) was the first to suggest that DNA damage per se, apart from its role in inducing mutation, may be the primary cause of aging. Alexander noted that the term somatic mutation has a very definite meaning and must not be used to refer to every change involving the cell's DNA. He indicated that a number of experiments independently suggest that the accumulation of somatic mutations throughout life does not play a major role in initiating aging in mammals. However, he proposed that damage to DNA does play a part in the death of postmitotic cells. Alexander postulated that DNA damages accumulate during life in postmitotic cells. These interfere with RNA synthesis and, therefore, protein synthesis until they are lethal to the cell. He also emphasized that at the time (1967) no experimental evidence existed to support such a hypothesis.

Since Alexander's original proposal, the evidence bearing on the DNA damage hypothesis of aging was reviewed by several workers. Gensler and Bernstein (1981) evaluated and unified the numerous diverse lines of argument that support this hypothesis. Tice and Setlow (1985) presented an examination of evidence on DNA damage and repair in aging organisms and cells. Evidence bearing on the DNA damage hypothesis of aging has also been reviewed by Hart et al. (1979), Rothstein (1982: pp. 132–173), Eichhorn (1983), Ames et al. (1985), Hanawalt (1987), Gensler et al. (1987), and Rattan (1989).

The DNA damage hypothesis of aging assumes that aging is a nonadaptive consequence of DNA damage. Therefore, this hypothesis is incompatible with all theories that assume that aging exists because it is beneficial. The idea that aging evolved as a beneficial adaptation for the species traces back to Weismann (1889: p. 24). His argument was based on the supposition that even if individuals were somehow able to maintain immortality they would be progressively worn out by injuries from external factors. From this he argued that worn out individuals are not only valueless to the species but are even harmful, because they take the place of those that are sound. He considered that by the operation of natural selection the life of the hypothetically immortal individual would be shortened by the amount that was useless to the species. Weismann further considered that death is not a primary necessity but that it has been secondarily acquired as an adaptation. He believed that life is endowed with a fixed duration, not because it is contrary to its nature to be unlimited but because unlimited life span would be a "luxury without corresponding advantage."

Kirkwood (1984) has pointed out that the view that aging is adaptive is still widely held and is implicit in the popular idea that aging is a programmed process under its own strict genetic control. He noted that if this idea is to withstand close scrutiny it must explain why, other things being equal, an organism that ages is fitter in a neo-Darwinian sense than one that does not. He then argued that on this crucial test adaptive theories have failed. One part of Kirkwood's argument is that for aging to have arisen adaptively the benefit to the *species* must be more effective in selection than the benefit to *individuals* that results from the reproductive advantages of a longer life. Chronologically older individuals who did not age should have more progeny and, therefore, greater fitness. Superiority of selection based on a species advantage rather than on an advantage to individuals occurs only under special circumstances. Therefore, aging as an adaptive mechanism cannot normally be stably maintained. The issue of programmed aging is discussed further in Chapter 15, Section I.J.

Even if aging is not a programmed adaptation, the rate of aging may still be under genetic control. For instance, by the DNA damage hypothesis, genes that encode enzymes that inactivate DNA-damaging agents or repair damage are adaptive and should be selected for. These genes should reduce the rate of aging. In this limited sense, aging may be regarded as genetically programmed. This, however, does not imply that aging itself is a beneficial adaptation.

B. The DNA Repair (and Complementation) Hypothesis of Sex

The hypothesis that the function of sex is to repair damages in germ line DNA was first proposed by Dougherty (1955). He postulated that the evolution of sex was the result of a single phylogenetic sequence. He thought that sex is advantageous because it allows two damaged DNA molecules to "pool their undamaged parts" and reconstitute an intact unit. Dougherty's idea, however, lay dormant for two decades until it was revived independently by Maynard Smith (1975), Bernstein (1977), Martin (1977), and Walker (1978).

Since 1955, when the idea was first proposed, a much better appreciation of the importance of damage and repair (see Chapters 2 and 9) and a firmer picture of early evolutionary events (Chapter 14) have been achieved. Maynard Smith (1975: p. 190) suggested that the enzymes required for genetic recombination between chromosomes may have evolved originally to repair damaged DNA, and that genetic exchange may have arisen very early in the history of life. Bernstein (1977) presented evidence that germ line recombination may be a manifestation of DNA repair processes. Martin (1977) reviewed observations that related aging with DNA damage, meiosis with DNA repair, and rejuvenation with meiosis. C. Bernstein (1979, 1981) and H. Bernstein (1983) presented evidence that the ability of recombinational repair to efficiently overcome different kinds of damage in various organisms supports the DNA repair hypothesis of sex. In addition, proposals for the evolution of sex (Bernstein *et al.*, 1981) and the origin of sex (Bernstein *et al.*, 1984) based on the DNA repair (and complementation) hypothesis were presented. This hypothesis was also developed by Bernstein *et al.* (1985a) from the perspective of the general problem of random noise in transmitting information. The various lines of evidence bearing on the hypothesis were reviewed by Bernstein *et al.* (1987).

More than 60 years before Dougherty's 1955 proposal that the function of sex is the repair germ line DNA, other workers had proposed the similar idea that the function of sex is rejuvenation. In 1892, Weismann commented that a widely held view among his contemporaries was that the advantage of sex is rejuvenation. Weismann himself did not agree with this view. A major observation supporting the rejuvenation concept was that in some protozoans vitality declines during the course of successive asexual divisions by binary fission, but that after sexual reproduction (conjugation) vitality is restored. Maupas, a principal advocate of the rejuvenation idea, contended that conjugation is needed to ensure the continuation of the species, because it imparts to the animal the power "de renoveler et rejeunir les sources de la vie" (Maupas, 1889 [quoted by Weismann, 1892]). Weismann (1892: p. 197) found it difficult to understand how an "almost exhausted vital force" could be raised again to its original state of activity as the result of a union with another equally exhausted force.

Work during the last 40 years on DNA damage and repair has provided a rational basis for explaining the rejuvenation that occurs during sexual reproduction. We present evidence in Chapter 8, Section I.B, that restoration of vitality in paramecia is likely to be based on the repair of DNA damage during conjugation. Medvedev (1981) reviewed the genetic and biochemical mechanisms bearing on the immortality of the germ line. He concluded that rejuvenation of germ cells occurs, in part, by meiotic recombination and repair—unique processes capable of restoring the integrity of DNA and chromosomes that have damages which are irreversible when they occur in somatic cells.

One way in which DNA damage is repaired involves replacement of damaged information in one DNA molecule by undamaged information from another homologous molecule. The two homologous DNA molecules need not be identical in sequence to exchange information but may contain mutational or allelic variations. The process of physical exchange of parts between two homologous DNA molecules is recombination. Recombination generates genetic change because it breaks up associations of linked alleles and randomly generates new allelic associations. As pointed out by Shields (1982: p. 67), the majority of new associations are likely to be either selectively neutral or disadvantageous, because they were generated by random changes in parental DNAs that had survived previous natural selection. Recombination is one source of the inheritable variation upon which natural selection operates, the other is mutation.

There are two contrasting views on the adaptive advantage of recombination. One view argues that recombinational variation is the primary selective force maintaining sex (for recent examinations of this idea, see Bell, 1988; Bernstein *et al.,* 1988; Crow, 1988; Felsenstein, 1988; Maynard Smith, 1988; Shields, 1988; and Williams, 1988). The alternative view is that, because new allelic associations are more often disadvantageous than advantageous to the individual, the process of recombination would be selected against if not for the benefit of recombinational repair. We think that recombination persists as the central feature of sex because of a trade-off between the benefit to the individual of efficiently repairing germ cell DNA damages (which if unrepaired are deleterious and often lethal) and the disadvantage of generating, as a by-product, recombinational variations (which are most often neutral or mildly deleterious to the individual, but rarely lethal).

The difference in the two points of view about the adaptive advantage of recombination ultimately comes down to whether or not one considers DNA damage in germ cells to be a serious problem that needs to be handled by recombinational repair. If it is not serious, one could suppose that the process of recombination is selected for independently of its repair function, perhaps for the benefit of recombinational variation. On the other hand, if DNA damage is a serious problem leading to the death of gametes, then recombination could be selected solely for its role in repair. In Chapters 2 and 3 we review evidence showing that DNA damage is prevalent and is a likely cause of functional decline. In Chapters 11 and 12 we discuss the importance of DNA damage in relation to germ cell formation.

As mentioned in Section II, above, we assume that outcrossing in diploid organisms is selected because it promotes masking, or complementation, of deleterious mutations. This view has a long history. Charles Darwin, in his book "The Effects of Cross and Self Fertilization in the Vegetable Kingdom" (1889), came to clear and definite conclusions about the adaptive advantage of sex. For instance, on page 462 he concluded that the offspring from the union of two distinct individuals, especially if their progenitors have been subjected to very different conditions, have a great advantage in height, weight, constitutional vigor, and fertility over the self-fertilized offspring from either one of the same parents. He considered that this fact was amply sufficient to account for the development of sexual reproduction. The reduced vigor associated with inbreeding is now recognized as being due largely to the expression of deleterious recessive alleles. The progeny of outcrosses are more vigorous because such deleterious recessive alleles are usually masked. Masking occurs because distantly related parents are unlikely to carry the same deleterious recessive alleles.

Weismann proposed another explanation for outcrossing based on the advantage of inheritable variation. Weismann (1892) was motivated by a need to explain the source of inheritable variation on which natural selection must operate. In the absence of knowledge about mutation (a concept that had not yet emerged), he adopted the reasonable view that outcrossing was the source of inheritable variation. He further contended that production of inheritable variation was *the* adaptive function of sex. Thus, although he was otherwise a staunch Darwinian, Weismann advocated a different view on the adaptive advantage of sex than Darwin himself. Weismann (1892: p. 195) held that the deeper significance of every form of amphimixis—whether occurring in conjugation, fertilization, or in any other way—is the "creation of the hereditary individual variability," which is requisite for the operation of the process of selection and arises from the periodic mingling of two different hereditary substances. This explanation assumes that the advantage of sex is at the level of the evolution of the species rather than at the level of the individual organism's ability to produce viable progeny.

Interestingly, Weismann was aware of the problem that variation did not appear to be advantageous to the individual organism. He noted (Weismann, 1892: p. 213) that conjugation must have had some beginning, and although he believed that in its present form it signifies a source of variability, he considered that it originally must have had some other meaning, because two Monera would scarcely coalesce to ensure variability in their descendants.

Weismann's general explanation for the advantage of sex, rather than Darwin's, has prevailed during most of the past century. Recently, Michod (1986) pointed out the fallacy of reasoning that because variation is essential for evolution selection will tend to produce phenotypes that produce variation. Michod further commented that although most modern evolutionists would agree that this logic is faulty it is precisely what many evolutionary biologists, over the past century, have had in the back of their minds, and to this day it is what motivates biologists to accept the variation hypothesis even though there is little evidence for it.

In recent years, much effort has gone into trying to show an advantage of recombinational variation at the level of competing individual organisms rather than competing groups of organisms. There is, however, widespread skepticism that any particular variation model can provide a general explanation for the adaptive benefit of sex (Maynard Smith, 1978; Williams, 1975). Maynard Smith (1978:

p. 10) commented in his authoritative book "The Evolution of Sex" that he feared the reader may find these models "unsubstantial and unsatisfactory"; however, he noted that they are the best we have. The alternative view (Darwin's), that the advantage of outcrossing is the greater vigor of the progeny of outcrosses, has been presented in a modern genetic interpretation by Bernstein *et al.* (1985b) as part of the repair (and complementation) hypothesis of sex (see Chapter 13, Section III.B). Supporters of this view do not deny the necessity of genetic variation for evolution; however, they consider that the variation produced by sex arises as a natural by-product of the DNA exchanges employed in the repair of DNA damage.

References

Alexander, P. (1967). The role of DNA lesions in processes leading to aging in mice. *Symp. Soc. Exp. Biol.* **21,** 29–50.

Ames, B. N., Saul, R. L., Schwiers, E., Adelman, R., and Cathcart, R. (1985). Oxidative DNA damage as related to cancer and aging: Assay of thymine glycol and hydroxymethyluracil in human and rat urine. *In* "Molecular Biology of Aging: Gene Stability and Expression" (R. S. Sohal, L. S. Birnbaum, and R. G. Cutler, eds.), pp. 137–144. Raven Press, New York.

Bell, G. (1988). Uniformity and diversity in the evolution of sex. *In* "The Evolution of Sex: An Examination of Current Ideas" (B. Levin and R. Michod, eds.), pp. 127–138. Sinauer, New York.

Bernstein, C. (1979). Why are babies young? Meioses may prevent aging of the germ line. *Perspect. Biol. Med.* **22,** 539–544.

Bernstein, C. (1981). Deoxyribonucleic acid repair in bacteriophage. *Microbiol. Rev.* **45,** 72–98.

Bernstein, H. (1977). Germ line recombination may be primarily a manifestation of DNA repair processes. *J. Theor. Biol.* **69,** 371–380.

Bernstein, H. (1983). Recombinational repair may be an important function of sexual reproduction. *BioScience* **33,** 326–331.

Bernstein, H. , Byers, G. S., and Michod, R. E. (1981). Evolution of sexual reproduction: Importance of DNA repair, complementation and variation. *Am. Nat.* **117,** 537–549.

Bernstein, H., Byerly, H. C., Hopf, F. A., and Michod, R. E. (1984). Origin of sex. *J. Theor. Biol.* **110,** 323–351.

Bernstein, H., Byerly, H. C., Hopf, F. A., and Michod, R. E. (1985a). The evolutionary role of recombinational repair and sex. *Int. Rev. Cytol.* **96,** 1–28.

Bernstein, H., Byerly, H. C., Hopf, F. A., and Michod, R. E. (1985b). Genetic damage, mutation and the evolution of sex. *Science* **229,** 1277–1281.

Bernstein, H., Hopf, R. A., and Michod, R. E. (1987). The molecular basis of the evolution of sex. *Adv. Genet.* **24,** 323–370.

Bernstein, H., Hopf, F. A., and Michod, R. E. (1988). Is meiotic recombination an adaptation for repairing DNA, producing genetic variation, or both? *In* "The Evolution of Sex: An Examination of Current Ideas" (B. Levin and R. Michod, eds.), pp. 139–160. Sinauer, New York.

Comfort, A. (1979). "The Biology of Senescence." Elsevier, New York.

Crow, J. F. (1988). The importance of recombination. *In* "The Evolution of Sex: An Examination of Current Ideas" (B. Levin and R. Michod, eds.), pp. 56–73. Sinauer, New York.

Darwin, C. (1889). "The Effects of Cross and Self Fertilization in the Vegetable Kingdom." D. Appleton and Co., New York.

Dougherty, E. C. (1955). Comparative evolution and the origin of sexuality. *Syst. Zool.* **4,** 145–190.

Eichhorn, G. L. (1983). Nucleic acid biochemistry and aging. *In* "Review of Biological Research in Aging," Vol. 1 (M. Rothstein, ed.), pp. 295–303. Alan R. Liss, New York.

Failla, G. (1958). The aging process and carcinogenesis. *Ann. N.Y. Acad. Sci.* **71,** 1124–1135.

Felsenstein, J. (1988). Sex and the evolution of recombination. *In* "The Evolution of Sex: An Examination of Current Ideas" (B. Levin and R. Michod, eds.), pp. 74–86. Sinauer, New York.

Gensler, H. L., and Bernstein, H. (1981). DNA damage as the primary cause of aging. *Q. Rev. Biol.* **56,** 279–303.

Gensler, H. L., Hall, J. D., and Bernstein, H. (1987). The DNA damage hypothesis of aging: Importance of oxidative damage. *In* "Review of Biological Research in Aging," Vol. 3 (M. Rothstein, ed.), pp. 451–465. Alan R. Liss, New York.

Ghiselin, M. T. (1988). The evolution of sex: A history of competing points of view. *In* "The Evolution of Sex: An Examination of Current Ideas" (B. Levin and R. Michod, eds.), pp. 7–23. Sinauer, New York.

Hanawalt, P. C. (1987). On the role of DNA damage and repair processes in aging: Evidence for and against. *In* "Modern Biological Theories of Aging" (M. R. Warner, R. N. Butler, R. L Sprott, and E. L. Schneider, eds.), pp. 183–198. Raven Press, New York.

Harmon, D. (1962). Role of free radicals in mutation, cancer, aging and the maintenance of life. *Radiat. Res.* **16,** 753–763.

Hart, R. W., D'Ambrosio, S. M., Ng, K. J., and Modak, S. P. (1979). Longevity, stability and DNA repair. *Mech. Ageing Dev.* **9,** 203–223.

Haynes, R. H. (1988). Biological context of DNA repair. *In* "Mechanisms and Consequences of DNA Damage Processing" (E. C. Friedberg and P. C. Hanawalt, eds.), pp. 577–584. Alan R. Liss, New York.

Kirkwood, R. B. L. (1984). Towards a unified theory of cellular aging. *Monogr. Dev. Biol.* **17,** 9–20.

Martin, R. (1977). A possible genetic mechanism of aging, rejuvenation, and recombination in germinal cells. *ICN-UCLA Symp. Mol. Cell. Biol.* **7,** 355–373.

Maupas, E. (1889). La rejeunissement chez les cilies. *Arch. Zool. Exp. Gen.* **7(2),** 149–517.

Maynard Smith, J. (1975). "The Theory of Evolution," 3rd ed. Penguin Books Ltd., Harmondsworth, England.

Maynard Smith, J. (1978). "The Evolution of Sex." Cambridge University Press, London.

Maynard Smith, J. (1988). The evolution of recombination. *In* "The Evolution of Sex: An Examination of Current Ideas." (B. Levin and R. Michod, eds.), pp. 106–125. Sinauer, New York.

Medvedev, Z. A. (1981). On the immortality of the germ line: Genetic and biochemical mechanisms. A review. *Mech. Ageing Dev.* **17,** 331–359.

Michod, R. E. (1986). On fitness and adaptedness and their role in evolutionary explanation. *J. Hist. Biol.* **19,** 289–302.

Rattan, S. I. S. (1989). DNA damage and repair during cellular aging. *Int. Rev. Cytol.* **116,** 47–88.

Rothstein, M. R. (1982). "Biochemical Approaches to Aging." Academic Press, New York.

Shields, W. M. (1982). "Philopatry, Inbreeding and the Evolution of Sex." State University of New York Press, Albany.

Shields, W. M. (1988). Sex and adaptation. *In* "The Evolution of Sex: An Examination of Current Ideas." (B. Levin and R. Michod, eds.), pp. 253–269. Sinauer, New York.

Sonneborn, T. M. (1978). The origin, evolution, nature and causes of aging. *In* "The Biology of Aging." (J. A. Behnke, C. E. Finch, and G. B. Moment, eds.), Chapter 21, pp. 361–374. Plenum Press, New York.

Szilard, L. (1959). On the nature of the aging process. *Proc. Natl. Acad. Sci. U.S.A.* **45,** 30–45.

Tice, R. R., and Setlow, R. B. (1985). DNA repair and replication in aging organisms and cells. *In* "Handbook of the Biology of Aging." (C. E. Finch and E. L. Schneider, eds.), pp. 173–224. Van Nostrand Reinhold, New York.

Walker, I. (1978). The evolution of sexual reproduction as a repair mechanism. Part I. A model for self-repair and its biological implications. *Acta Biotheor.* **27,** 133–158.

Weismann, A. (1889). "Essays upon Heredity and Kindred Biological Problems," Vol. I. Clarendon Press, Oxford.

Weismann, A. (1892). "Essays upon Heredity and Kindred Biological Problems," Vol. II. Clarendon, Oxford.

Williams, G. C. (1975). "Sex and Evolution." Princeton University Press, Princeton, New Jersey.

Williams, G. C. (1988). Retrospect on sex and kindred topics. *In* "The Evolution of Sex: An Examination of Current Ideas." (B. Levin and R. Michod, eds.), pp. 287–298. Sinauer, New York.

Chapter Two

DNA Damage

In this chapter, we review the nature of DNA damage and, especially, emphasize the distinction between DNA damage and mutation. We then discuss each of the various kinds of DNA damage that appear to be common in nature.

I. DNA Damages as Physically Distinct Alterations in Polynucleotide Chains

DNA damage is very different from mutation. A DNA damage is a DNA alteration that has an abnormal structure. A mutation, on the other hand, is a change in the DNA sequence rather than a change in DNA structure. Biologically, DNA damages and mutations have different consequences. A damage cannot be replicated; therefore, it cannot be inherited. A mutation is a change in the polynucleotide sequence in which standard base pairs are substituted, added, deleted, or rearranged. Even when mutations are large changes, such as extended deletions, the DNA is still an uninterrupted sequence of standard nucleotide pairs. Because of this, mutations can be replicated; thus mutations, unlike damages, can be inherited.

Another way in which DNA damages and mutations differ is with respect to repair. Because a DNA damage is an abnormal alteration in DNA structure, it can be recognized by enzymes. The altered structure can be removed and replaced; thus, DNA damage can be repaired (discussed further in Chapter 9). On the other hand, a mutation has a regular DNA structure. There is no correction mechanism for converting a new base pair, or an added or deleted sequence, back to the original one; thus mutations, unlike damages, cannot be repaired. In summary, damages can be repaired but not replicated, whereas mutations can be replicated but not repaired.

To clarify further the distinction between DNA damage and mutation, we use as an example the transmission of information from a source to a destination. As discussed in Bernstein *et al.* (1985), strong parallels can be drawn between communication in the English language from ground to satellite and transmitting a working genome from parent to offspring. In both cases, disruptions due to noise, i.e., random influences that change the information sequence in unpredictable ways, are a critical difficulty. These disruptions have two possible results: change of one allowed symbol to another allowed symbol (which corresponds to a mutation), or change of an allowed symbol to a disallowed symbol (which corresponds to genetic damage).

Let us assume that any linear sequence of Latin letters is allowable in the sense of being a proper set of characters for transmitting the English language. Figure 1a illustrates a mutation in the transmission of information in the English language. It is not possible to determine, at the point of reception, that the received linear sequence is incorrect, because it is comprised of allowed symbols. This is the essence of mutation. A mutation cannot be repaired or corrected. Figure 1b illustrates the occurrence of DNA damage. It shows that the damage can be recognized at the point of reception, because the # symbol is not part of the allowed set of symbols. The

a) Mutation

Source sequence: **Bubble**

Received sequence: **Babble**

b) Damage

Source sequence: **Bubble**

Received sequence: **B#bble**

Figure 1 Distinction between mutation and damage in the transmission of information consisting of strings of letters.

damaged information can be recovered if a redundant copy of the linear sequence is also transmitted. Redundancy can be used to replace the obviously damaged information.

In DNA, damages are structural irregularities such as breaks, modified bases, depurinations, and cross-links. DNA is a double-stranded helix, with the two strands having complementary information. The strands are chains of nucleotides, and where an adenine (a type of purine) is present in one strand it is always paired with a thymidine (a type of pyrimidine) in the complementary strand. In shorthand, this is written A : T. Where there is a guanine (a purine) in one strand it is always paired with a cytosine (a pyrimidine) in the complementary strand. This is written G : C.

Many DNA damages only affect one strand of DNA. For example, a depurination, which removes a base but does not interrupt the sugar phosphate backbone of a polynucleotide chain, is represented in Figure 2. Another single-strand damage is a single-strand break. This is represented in Figure 3, with some of the unpairing that can accompany it. Still another type of single-strand damage is a modified base. This may be caused by covalent interaction with a reactive molecule that joins to the base to form an adduct. This is illustrated in Figure 4.

In each of the examples shown above of a single-strand damage, redundant information is present in the complementary strand of DNA that could be used as a template to repair the damage. Double-strand damages, however, lack this nearby redundancy. Redundant information would only be available from another DNA double helix. One type of double-strand damage is a cross-link, in which chemical bonds are added across the two strands of the DNA double helix, and each base involved is changed to an abnormal variant of the standard base. This is represented in Figure 5 by the use of nonstandard symbols and a line spanning the two strands. Another kind of double-strand damage would be a double-strand break (not shown).

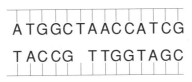

Figure 2   A depuration.

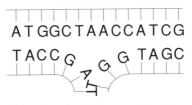

Figure 3   A single-strand break.

II. *Kinds of DNA Damage Prevalent in Nature*

DNA damage can be introduced either by intrinsic processes or by extrinsic chemicals and radiation. We describe next the types of damages that occur in nature and list their frequency of occurrence in mammalian cells in Table I.

A. Spontaneous Depurination and Depyrimidination

Lindahl (1977) proposed that depurination, depyrimidination, and the conversion of cytosine to uracil are frequent kinds of spontaneous (or heat-induced) DNA damages. Apurinic damages can occur under physiological conditions by hydrolytic cleavage of the purines adenine and guanine from the deoxyribose phosphate backbone of DNA. In addition, at neutral pH and physiological solvent conditions, an apurinic site in the DNA chain has an approximate half-life of 100 hr at 37°C before chain breakage occurs. Lindahl (1977) estimated that a mammalian cell at 37°C loses about 10,000 purines and 500 pyrimidines from its DNA during 20 hr by spontaneous hydrolysis. The amount of DNA depurination caused by nonenzymatic (spontaneous) hydrolysis that occurs in a single long-lived, nonreplicating mammalian cell, such as the human nerve cell, during the lifetime of an individual was estimated to be about 10^8 purine bases. This is about 3% of the total number of purines in the cell's DNA (Lindahl and Nyberg, 1972).

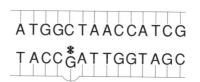

Figure 4   A base modified by an adduct (*).

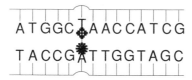

A T G G C A A C C A T C G
T A C C G T T G G T A G C

Figure 5   A cross-link in DNA is a double-strand damage. The bases involved are each changed to an abnormal variant of the standard base.

Tice and Setlow (1985) estimated the rates of appearance of spontaneous DNA damage in a mammalian cell at 37°C from data on DNA in solution obtained by Shapiro (1981). The estimated rates for depurinations and depyrimidinations per cell per hour were 580 and 29, respectively. Converting Lindahl's (1977) and Tice and Setlow's (1985) estimates to a daily rate gives values for depurination of

Table I

Estimated Rates of Occurrence of Endogenous DNA Damages in Mammalian Cells

Damage	Events per Cell per Day	Reference
Depurination	12,000	Lindahl, 1977
	13,920	Tice and Setlow, 1985
Depyrimidination	600	Lindahl, 1977
	696	Tice and Setlow, 1985
Cytosine deamination	192	Tice and Setlow, 1985
Single-strand break	55,200	Tice and Setlow, 1985
O^6-methylguanine	3,120	Tice and Setlow, 1985
Glucose-6-phosphate adduct	2.7	Bucala *et al.*, 1985
Thymine glycol	270	Saul *et al.*, 1987
Thymidine glycol	70	Saul *et al.*, 1987
Hydroxymethyluracil	620	Saul *et al.*, 1987
8-Hydroxydeoxyguanosine	unknown	Richter *et al.*, 1988; Ames, 1989
Unidentified methyl adduct	unknown	Park and Ames, 1988a, 1988b
Double-strand break	8.8	see text, Section H
Interstrand cross-link	8.0	see text, Section H
DNA–protein cross-link	unknown	see text, Section F

12,000/cell/day (Lindahl) and 13,920/cell/day (Tice and Setlow). For depyrimidination the daily rates are 600/cell/day (Lindahl) and 696/cell/day (Tice and Setlow). Thus, DNA is significantly unstable at the temperatures at which mammalian cells normally exist.

B. Cytosine Deaminations

The oxidative deamination of DNA cytosine to uracil occurs spontaneously at an appreciable rate (Lindahl and Nyberg, 1974). Uracil DNA glycosylases, which remove uracil from DNA as the free base and, hence, yield an apyrimidinic site in the DNA, have been isolated from the bacteria *Escherichia coli* and *Bacillus subtilis* (see Friedberg *et al.,* 1978, for review) as well as from mammalian cells and tissues (Lindahl, 1976; Sekiguichi *et al.,* 1976; Anderson and Friedberg, 1980; Friedberg *et al.,* 1978). Tice and Setlow (1985) estimated that the rate of spontaneous deamination of cytosine to uracil is 8/cell/hr based on data of Shapiro (1981). This corresponds to a rate of 192/cell/day.

C. Single-Strand Breaks

Apparently, the most prevalent type of damage in mammalian cells apparently is single-strand breaks. Tice and Setlow (1985) estimated the spontaneous rate to be 2,300/cell/hr, which translates into a rate of 55,200/cell/day. By comparison, the estimated incidence of single-strand breaks due to exogenous background radiation is very small, about 10^{-4}/cell/hr (Setlow, 1987).

D. O^6-Methylguanine

The methylation of guanine to yield O^6-methylguanine occurs as a minor side reaction of the necessary methyl transfer function of S-adenosylmethionine (Barrows and Magee, 1982). Tice and Setlow (1985) estimated the rate of appearance of O^6-methylguanine to be 130/cell/hr, which corresponds to 3,120/cell/day.

E. Glucose and Glucose-6-Phosphate

Bucala *et al.* (1984) presented evidence that glucose, glucose-6-phosphate, and possibly other sugars can react with DNA to produce marked structural and biological alterations. They concluded that because nucleic acids are long-lived molecules in the resting cell, the accumulation of these alterations might contribute to aging. Bucala *et al.* (1985) estimated that on the average the human genome would suffer 2.7 glucose-6-phosphate "hits" per day. They also suggested that additional DNA damage would be anticipated from the reaction of other reducing sugars (e.g., glucose, glyceraldehyde 3-phosphate).

F. Oxidative Damage

The highly reactive superoxide radical (O_2^-) and hydrogen peroxide (H_2O_2) occur as by-products of cellular respiration (Harman, 1981). These compounds are very reactive and can cause oxidative damages in DNA. Saul *et al.* (1987) reviewed evidence that repair of oxidized DNA in mammals leads to the removal of thymine glycol, thymidine glycol, and hydroxymethyluracil. These molecules are then excreted in the urine. The levels of these compounds in human urine indicate the following average rates of removal of these damages in DNA: thymine glycol (270/cell/day), thymidine glycol (70/cell/day), and hydroxymethyluracil (620/cell/day). The sum of these three average rates is 960/cell/day. Saul *et al.* (1987) pointed out that these products are only three of a considerable number of possible oxidative DNA damage products; thus the total number of oxidative DNA hits per cell per day in humans may be considerably more than 1,000. Richter *et al.* (1988) estimated that the oxidized base 8-hydroxydeoxyguanosine is present in the DNA of rat liver at the level of about 1.4×10^5/cell in nuclear DNA and 4.1×10^4/cell in mitochondrial DNA. Thus, about 78% of these oxidative damages were in nuclear DNA. Extrapolating from this, it can be surmised that the majority of oxidative DNA damages are in nuclear DNA. Lesko (1982) has presented evidence that H_2O_2 also produces DNA–protein cross-links. Oxidative DNA damages, we think, have special relevance to aging; therefore, we discuss these types of damage in more detail in Chapter 6.

G. Methyl Adducts

Park and Ames (1988a) identified the methyl adduct 7-methylguanine in the DNA of rat liver. The estimated numbers of adducts per cell in old (24 mo) rats was 21,000 in mitochondrial DNA and 460,000 in nuclear DNA. Subsequently, Park and Ames (1988b) corrected this report to indicate that the major adduct present was not 7-methylguanine, but rather an unidentified adduct with electrochemical properties similar to 7-methylguanine.

H. Double-Strand Damages: Cross-Links and Double-Strand Breaks

A minimum estimate of the rate of occurrence of double-strand damage can be derived from the rates of occurrence of oxidatively modified bases *in vivo*. This is done by extrapolating the relative rates at which different types of DNA damages are produced by H_2O_2 *in vitro* at 37°C to the *in vivo* data. Modified bases, double-strand breaks, and interstrand cross-links are produced *in vitro* by H_2O_2 at a ratio of 120 : 1.1 : 1 (Massie *et al.*, 1972). If the ratio of modified bases to double-strand damages that was measured *in vitro* also applies *in vivo* and the minimum estimated rate of occurrence of oxidatively damaged bases of 960/cell/day is used, then the incidence of double-strand breaks and interstrand cross-links in mammalian cells are 8.8/cell/day and 8.0/cell/day, respectively. To estimate the numbers of double-strand damages in nuclear DNA, we can assume that oxidative damages in general are distributed between nuclear and mitochondrial DNA in the same proportion as the specific oxidative damage 8-hydroxydeoxyguanosine (see above). Because 78% of 8-hydroxydeoxyguanosine residues occur in nuclear DNA, we can calculate that the incidence of double-strand breaks in the nuclear DNA of mammalian cells is 6.9/cell/day and the incidence of interstrand cross-links is 6.2/cell/day.

Bacteria grown at above-optimal temperatures lose viability. Woodcock and Grigg (1972) presented evidence that DNA strand breakage is responsible for thermal inactivation of *E. coli* at 52°C. Double-strand breakage of the DNA constituted a major fraction of the breaks observed during heat treatment. This evidence suggests that spontaneous double-strand damages may be a problem for cell survival in nature, even at normal mammalian cell temperatures.

Double-strand damages can only be repaired by physical recombination with another homologous DNA molecule. Because of this characteristic, double-strand damages may have had a critical role in the evolution of recombination and sex. Therefore, double-strand damages are of special interest, even though they are less frequent than single-strand damages.

I. Exogenous Damages

Some exogenous sources of DNA damage are prevalent. Two natural sources of damage are ultraviolet (UV)-irradiation from the sun and chemicals produced by many plant species as a defense against their predators. There are also man-made sources of DNA-damaging agents.

1. *UV Irradiation* Pyrimidine dimers are a type of DNA damage induced by UV irradiation. Pyrimidine dimers arise when two normal, adjacent pyrimidine bases (C or T) in a strand of DNA become covalently linked. Setlow (1982) and Tice and Setlow (1985) estimated that the rate at which UV irradiation (noon, Texas sun) induces pyrimidine dimers in human skin is 5×10^4/cell/hr. This estimate was based on data on the inactivation of bacteria by the UV in sunlight (Harm, 1969) corrected for transmission of skin. Furthermore, Setlow (1987) estimated that the maximum rate of repair of normal skin cells was 5×10^4/cell/hr, a rate that should be barely able to cope with the rate of introduction of damage to skin in full sunlight.

Any aging effect of UV should be limited to the skin, because UV light cannot penetrate to tissues beneath the skin. However, for microorganisms exposed to sunlight, UV-irradiation is a danger to survival. UV light may have been a particular hazard during early evolution due to the lack of a shielding ozone layer. Sagan (1973) examined the flux of solar UV light penetrating the primitive reducing atmosphere of the earth. He concluded that a mean lethal dose at 2,600 Å would be delivered to unprotected microorganisms of the type existing today in 0.3 sec or less, and that this could have posed a major problem for the early evolution of life.

2. *DNA-Damaging Chemicals in Plants* Ames (1983) has pointed out that plants in nature synthesize toxic chemicals in large amounts. These chemicals are apparently produced as a primary defense against bacterial and fungal predators as well as insect and other animal predators. He further described 16 examples of DNA-

damaging chemicals found in plants that are part of the human diet. No estimate is available for how frequently these plant toxins introduce DNA damage in humans or other plant eaters.

3. *Man-Made DNA-Damaging Agents* Various man-made chemicals have been shown to be mutagenic (e.g., see McCann *et al.*, 1975). Presumably, the mutagenic action occurs through DNA damage that is imperfectly repaired (see Chapter 9, Section II.G.1). We do not have an estimate of the frequency with which they cause DNA damage in mammalian cells in nature.

J. Overall Occurrence of Damage

The estimated rates of occurrence of the different kinds of DNA damage known to occur at significant frequencies in mammalian cells are summarized in Table I. The overall rate of damage, obtained by summing the individual rates listed, is about 72,000/cell/day. The major causes of the DNA damages listed in Table I are spontaneous hydrolysis and interaction with reactive oxygen species. These causes of damage are likely to be ubiquitous in nature.

K. The Rate of Repair and Accumulation of Damage

The net rate of accumulation of a particular type of DNA damage depends on both the rate of its occurrence and the rate of its removal by repair enzymes, as has been pointed out by Hart and Setlow (1974). Approximate maximum rates of repair in some human cells for certain types of damage have been summarized by Tice and Setlow (1985). The maximum rate of repair of single-strand breaks was estimated to be 2×10^5/cell/hr. The maximum rate of repair of O^6-methylguanines in lymphocytes was estimated to be 10^4–10^5/cell/hr based on data of Waldstein *et al.* (1982). These maximum repair rates, when compared with the respective rates of occurrence, suggest that repair activity is adequate to cope with these damages in the cells studied. Thus, the number of single-strand breaks and O^6-methylguanine nucleotides present in these cells may be small on average. However, in Chapter 4, Sections III–IV, we present evidence that in some specific mammalian tissues repair is unable to prevent accumulation of single-strand breaks and certain other types of damages.

References

Ames, B. N. (1983). Dietary carcinogens and anticarcinogens. *Science* **221**, 1256–1264.

Ames, B. N. (1989). Endogenous DNA damage as related to cancer and aging. *Mutat. Res.* **214**, 41–46.

Anderson, C. T. M., and Friedberg, E. C. (1980). The presence of nuclear and mitochondrial uracil–DNA glycosylase in extracts of human KB cells. *Nucleic Acid Res.* **8**, 875–888.

Barrows, L. R., and Magee, P. N. (1982). Nonenzymatic methylation of DNA by *S*-adenosylmethionine *in vitro*. *Carcinogenesis* **3**, 349–351.

Bernstein, H., Byerly, H. C., Hopf, F. A., and Michod, R. E. (1985). The evolutionary role of recombinational repair and sex. *Int. Rev. Cytol.* **96**, 1–28.

Bucala, R., Model, P., and Cerami, A. (1984). Modification of DNA by reducing sugars: A possible mechanism for nucleic acid aging and age-related dysfunction in gene expression. *Proc. Natl. Acad. Sci. USA* **81**, 105–109.

Bucala, R., Model, P., Russel, M., and Cerami, A. (1985). Modification of DNA by glucose 6-phosphate induces DNA rearrangements in an *Escherichia coli* plasmid. *Proc. Natl. Acad. Sci. USA* **82**, 8439–8442.

Friedberg, E. C., Bonura, T., Cone, R., Simmons, R., and Anderson, C. (1978). Base excision repair in DNA. *In* "DNA Repair Mechanisms, ICN-UCLA Symposia on Molecular and Cellular Biology" (P. C. Hanawalt, E. C. Friedberg, and F. C. Fox, eds.), pp. 163–173. Academic Press, New York.

Harm, W. (1969). Biological determination of the germicidal activity of sunlight. *Radiat. Res.* **40**, 63–69.

Harman, D. (1981). The aging process. *Proc. Natl. Acad. Sci. USA* **78**, 7124–7128.

Hart, R. W., and Setlow, R. B. (1974). Correlation between deoxyribonucleic acid excision repair and lifespan in a number of mammalian species. *Proc. Natl. Acad. Sci. USA* **71**, 2169–2173.

Lesko, S. A. (1982). Deoxyribonucleic acid-protein and deoxyribonucleic acid interstrand cross-links induced in isolated chromatin by hydrogen peroxide and ferrous ethylenediaminetetracetate chelates. *Biochemistry* **21**, 5010–5015.

Lindahl, T. (1976). New class of enzymes acting on damaged DNA. *Nature* **259**, 64–66.

Lindahl, T. (1977). DNA repair enzymes acting on spontaneous lesions in DNA. *In* "DNA Repair Processes" (W. W. Nichols and D. G. Murphy, eds.), pp. 225–240. Symposia Specialists, Miami.

Lindahl, T., and Nyberg, B. (1972). Rate of depurination of native DNA. *Biochemistry* **11**, 3610–3618.

Lindahl, T., and Nyberg, B. (1974). Heat induced deamination of cytosine residues in DNA. *Biochemistry* **13**, 3405–3410.

Massie, H. R., Samis, H. V., and Baird, M. B. (1972). The kinetics of degradation of DNA and RNA by H_2O_2. *Biochim. Biophys. Acta* **272**, 539–548.

McCann, J., Choi, E., Yamasaki, E., and Ames, B. N. (1975). Detection of carcinogens as mutagens in the Salmonella/microsome test: Assay of 300 chemicals. *Proc. Natl. Acad. Sci. USA* **72**, 5135–5139.

Park, J.-W., and Ames, B. N. (1988a). 7-Methylguanine adducts in DNA are normally present at high levels and increase on aging: Analysis by HPLC with electrochemical detection. *Proc. Natl. Acad. Sci. USA* **85**, 7467–7470.

Park, J.-W., and Ames, B. N. (1988b). Correction. *Proc. Natl. Acad. Sci. USA* **85**, 9508.

Richter, C., Park, J.-W., and Ames, B. N. (1988). Normal oxidative damage to mitochondrial and nuclear DNA is extensive. *Proc. Natl. Acad. Sci. USA* **85**, 6465–6467.

Sagan, C. (1973). Ultraviolet selection pressure on the earliest organisms. *J. Theor. Biol.* **39**, 195–200.

Saul, R. L., Gee, P., and Ames, B. N. (1987). Free radicals. DNA damage, and aging. *In* "Modern Biological Theories of Aging" (H. R. Warner, R. N. Butler, R. L. Sprott and E. L. Schneider, eds.), pp. 113–129. Raven Press, New York.

Sekiguchi, M., Hayakawa, H., Makino, F., Tanaka, K., and Okada, Y. (1976). A human enzyme that liberates uracil from DNA. *Biochem. Biophys. Res. Commun.* **73**, 293–299.

Setlow, R. B. (1982). DNA repair, aging and cancer. *Natl. Cancer Inst. Monogr.* **60**, 249–255.

Setlow, R. B. (1987). Theory presentation and background summary. *In* "Modern Biological Theories of Aging" (H. R. Warner, R. N. Butler, R. L. Sprott, and E. L. Schneider, eds.), pp. 177–182. Raven Press, New York.

Shapiro, R. (1981). Damage to DNA caused by hydrolysis. *In* "Chromosome Damage and Repair" (E. Seeberg and K. Kleppe, eds.), pp. 3–18. Plenum Press, New York.

Tice, R. R., and Setlow, R. B. (1985). DNA repair and replication in aging organisms and cells. *In* "Handbook of the Biology of Aging" (C. E. Finch and E. L. Schneider, eds.), pp. 173–224. Van Nostrand Reinhold, New York.

Waldstein, E. A., Cao, E.-H., Bender, M. A., and Setlow, R. B. (1982). Abilities of extracts of human lymphocytes to remove O^6-methylguanine from DNA. *Mutat. Res.* **95**, 405–416.

Woodcock, E., and Grigg, G. W. (1972). Repair of thermally induced DNA breakage. *Nature* **237**, 76–79.

Chapter Three

Immediate Consequences of DNA Damage

In this chapter, we review evidence that the presence of DNA damage in mammalian cells interferes with transcription and replication and causes cell death. Cell death and/or progressive impairment of cell function resulting from DNA damage, we argue in Chapters 4–7, is the cause of aging in multicellular organisms.

I. *Transcription Termination*

The simplest way to determine whether or not DNA damage interferes with transcription is to measure the effect of damaging cellular DNA on overall RNA synthesis. The deleterious effect of UV light on cells is thought to be mainly due to DNA damage. Thus, mouse cells were irradiated with UV light and production of RNA was examined. Total RNA decreased in a dose-dependent manner to approximately 35% of the untreated level at a UV dose of 50 J/m² (Hackett and Sauerbier, 1975). In a different study using monkey kidney cells, RNA synthesis was reduced to 30% by the same dose of UV (Nocentini, 1976). These initial studies indicated that DNA damage indeed interferes with RNA transcription. Several studies have also shown that pyrimidine dimers, a particular kind of DNA damage induced by UV light, terminate transcription in eukaryotic cells (Giorno and Sauerbier, 1976; Sauerbier and Hercules, 1978; Hackett *et al.*, 1981; Mayne and Lehman, 1982).

In further work, the permanence of this blockage of RNA transcription was examined. In human skin fibroblasts, UV-irradiation caused an immediate depression in the rate of RNA synthesis (Mayne, 1984). However, RNA synthesis rates recovered, returning to 90% of unirradiated levels, within 90 min in normal cells. This result indicates that the effect of DNA damage on transcription can be temporary. Studies with DNA synthesis inhibitors have suggested that DNA repair processes are involved in the recovery. Thus, this recovery in transcription efficiency is probably directly tied to the removal of DNA damage, which presumably had blocked transcription.

In addition to UV light, DNA-damaging chemicals have also been examined for their ability to block RNA transcription. For instance, calf thymus DNA, modified by a dihydrodiol epoxide derivative of benzo(a)pyrene, has been shown to be less efficiently transcribed *in vitro* by *E. coli* RNA polymerase than unmodified DNA (Leffler *et al.*, 1977). In these experiments, the RNA transcripts decreased progressively in chain length as the extent of DNA modification increased. In a different study, rat hepatic DNA, treated with N-acetoxy-2-acetylaminofluorene, also was much less effective as a template for RNA transcription (Zieve, 1973). A number of chemical agents have been found to bind preferentially to transcriptionally active regions of mammalian chromosomes (Moses *et al.*, 1976; Arrand and Murray, 1982). Yu (1983) has reported that the DNA-damaging chemical aflatoxin B_1 binds preferentially to the transcriptionally active DNA regions of rat liver nucleolar chromatin. Once bound, aflatoxin B_1 has also been shown to inhibit RNA synthesis. This suggests that transcriptionally active regions of DNA may be especially vulnerable to DNA-damaging agents and that this damage may then inhibit further RNA synthesis.

Although the naturally produced DNA damages listed in Table I were not tested (and not all natural damages that tend to remain in DNA have been identified), natural damages, particularly the more bulky ones, probably would inhibit transcription in a manner similar to the induced damages described above.

In conclusion, the results described in this section show that in mammalian cells DNA damage inhibits RNA transcription, but this inhibition can be overcome in time, presumably by DNA repair processes. Transcriptionally active regions of the DNA appear to be especially vulnerable to attack by DNA-damaging agents, perhaps because of their uncoiled configuration.

II. *Impairment of Replication*

In addition to blocking transcription, UV-irradiation also inter-feres with DNA replication. This interference with DNA replication has been shown in *E. coli* (Rupp and Howard-Flanders, 1968), in phage phiX174 (Benbow *et al.*, 1974; Villani *et al.*, 1978), and in phage T7 (Burck *et al.*, 1979). DNA replication in mammalian cells also is inhibited by the products of UV-irradiation. For instance, Cleaver (1969) incubated hamster cells with tritiated deoxythymidine after treatment with various doses of UV light. The incorporation of tritiated deoxythymidine into cellular DNA is a measure of DNA replication. He found a dose-dependent reduction in the incorpora-tion of tritiated deoxythymidine into DNA, a reduction reflecting blocked DNA replication.

As reviewed by Buhl *et al.* (1974), a number of investiga-tors have shown that the DNA synthesized shortly after the UV-irradiation of *E. coli*, human, and other mammalian cell lines has a lower molecular weight than DNA synthesized in nonirradiated cells. The sizes of these low-molecular weight DNA segments in *E. coli*, mouse, and human cells approximated the average distance between UV-induced pyrimidine dimers in the parental DNA. These data support the notion that pyrimidine dimers interrupt normal DNA replication and that synthesis resumes beyond the dimer.

Buhl *et al.* (1974) showed that after UV-irradiation replication can eventually bypass these blocking pyrimidine dimers. By 24 hr after UV-irradiation of mammalian cells, the DNA synthesized in these cells was in normal-sized segments, even though pyrimidine dimers were still present. This evidence indicates that UV-induced thymine dimers can block DNA replication and that this blockage can be overcome eventually by postreplication repair processes. These processes may involve the bypass of pyrimidine dimers by the DNA replication mechanism rather than the removal of the blocking dimers. The mechanism of bypass is discussed further in Chapter 9, Section II.G.2.

Sarasin and Hanawalt (1980) studied replication of UV-irradiated simian virus 40 (SV40) in monkey kidney cells. Their results indicated that replication was blocked at the first pyrimidine dimer encountered in one parental strand. Synthesis, however, was not inhibited along the complementary parental strand at the same position in the DNA. The size of the newly synthesized DNA strands, which at first was equivalent to the inter dimer spacing, increased markedly at later times. This indicates an eventual *trans*-dimer continuity of daughter strands. These experiments provide further evidence that pyrimidine dimers mechanically interfere with DNA synthesis and that this interference is only temporary.

Dahle *et al.* (1980) studied the inhibition and recovery of DNA synthesis in UV-irradiated hamster cells by the technique of DNA fiber autoradiography. Their results also indicated that UV-irradiation produces damages that temporarily block chain growth of DNA that is synthesized during the first 5 hr after exposure. Within 10 hr after exposure, however, the damages were either repaired or modified (or the replicative enzymes altered) so that DNA chain growth was no longer impeded.

Moore and Strauss (1979) have shown that *in vitro* DNA synthesis catalyzed by DNA polymerase I of *E. coli* is terminated one nucleotide before any pyrimidine dimer in the DNA template strand. Furthermore, Moore *et al.* (1981) found that the termination of DNA synthesis when using UV-irradiated templates occurs one nucleotide before the pyrimidine dimer when DNA synthesis was catalyzed by either *E. coli* DNA polymerase III, phage T4 DNA polymerase, or human DNA polymerase alpha. This suggests that pyrimidine dimers tend to block DNA replication similarly over a wide range of species.

Oxidative stress, or treatment of DNA with hydrogen peroxide, leads to the formation of thymine glycol residues (Chapter 2, Section II.F, and Chapter 6, Section I) and urea residues in DNA (Kow and Wallace, 1985). Ide *et al.* (1985) have shown that thymine glycol and urea residues block DNA replication. Sequencing gel analysis of the products synthesized by *E. coli* DNA polymerase I and T4 DNA polymerase revealed that DNA synthesis terminated right at thymine glycol sites in the template strand, and one nucleotide before urea sites. Rouet and Essigman (1985) have also reported evidence that the *E. coli* DNA polymerase I Klenow fragment is blocked at thymine glycol sites in the DNA template.

Apurinic–apyrimidinic sites in DNA are a frequent kind of

spontaneous DNA damage (Chapter 2, Section II.A). Strauss *et al.* (1982) have shown that DNA replication is terminated one base before apurinic or apyrimidinic sites. These experiments were performed *in vitro* with DNA synthesis catalyzed by phage T4 DNA polymerase or *E. coli* polymerase I. Chemicals that methylate DNA produce 0^6-methylguanine damages at a significant rate (Chapter 2, Section II.D). Abbotts *et al.* (1988) reported that DNA synthesis catalyzed by *E. coli* DNA polymerase I was blocked by the addition of a methyl group to the 0^6 position of guanine, addition of a methyl group to the N^3 position of thymine, or addition of an ethyl group to the phosphate moiety of template DNA. These alterations also strongly blocked elongation catalyzed by human DNA polymerase beta. The authors speculated that when these DNA polymerases "stall" at positions opposite methylated bases it may allow more time for methyl removal before the extension of synthesis.

Stalling of the DNA polymerase at damaged sites may help the cell avoid mutations. A misincorporation is the insertion of an incorrect nucleotide in the newly synthesized strand opposite a nucleotide in the template strand. An incorporated nucleotide is incorrect if its pairing violates the base pairing rules: A pairs with T, G pairs with C. Misincorporations are often promoted by damages in a template base. It is important for a cell to defend against misincorporations because these become mutations after the next round of replication, and mutations are often deleterious to the cell. Therefore, stalling of the DNA polymerase at a site of damage may allow more time for removal of the damage and, thus, avoid a potential mutation.

In mammalian cells, DNA synthesis is initiated on individual chromosomes at multiple sites, each called an origin of replication. When double-stranded DNA is replicated, a new single strand is synthesized as a partner to each of the two parental strands. A DNA polymerase catalyzes synthesis of the new strands in such a way that these strands are complementary copies of the parental strands. For the DNA polymerase to act, the two parental DNA strands must first become at least partially unwound at an origin of replication. If an origin occurs at a position in the DNA other than at the ends, DNA synthesis can proceed from the origin in two directions. If synthesis is in only one direction, the region of DNA synthesis is referred to as a unidirectional replication fork; if synthesis is carried out in both directions, the replication is referred to as bidirectional. In mammalian chromosomes, replication usually proceeds bidirectionally from a replication origin. The segment of DNA replicated from a given origin is a unit of replication. These units of replication are referred to as replicons. Each mammalian chromosome consists of multiple tandem replicons. In mammalian cells, DNA damage not only blocks the chain elongation step of DNA replication but also the initiation step occurring at a replication origin. Cleaver *et al.* (1983) concluded that within 1 hr after UV-irradiation *in vivo,* the relative impact of damage on elongation and initiation depends on the average number of damages per replicon; with low numbers (two or less) of damages per replicon, inhibition of initiation predominates; with high numbers of damages per replicon, blockage of DNA chain elongation becomes important. The mechanism by which DNA damages block initiation is uncertain (Friedberg, 1985).

A mechanism by which thymine dimers may block DNA synthesis has been suggested by Villani *et al.* (1978). They observed a large increase in the turnover of nucleotide triphosphates to free monophosphates during *in vitro* synthesis by *E. coli* DNA polymerase I on bacteriophage phiX174 DNA following UV-irradiation. They proposed that this nucleotide turnover is due to "idling" by the DNA polymerase. Idling consists of, first, the incorporation of nucleotides and, second, the subsequent excision of nucleotides opposite UV-induced damages by the 3' to 5' "proofreading exonuclease" activity of the polymerase. The idling of the polymerase prevents replication past pyrimidine dimers and the potentially mutagenic event that would result.

III. *Decreased Cell Survival*

Studies that focus on cell survival rather than on RNA transcription or DNA replication have also been carried out. Cells that are defective in DNA repair presumably retain more damages in their DNA when treated with UV light. Cell survival after UV-irradiation was compared using either cells defective in DNA repair or wild-type cells. Both hamster (Cleaver, 1969) and human (Maher *et al.*, 1976) repair-defective cells were more sensitive to killing by UV light than wild-type cells. Yeast is widely used as a model for studies of eukaryotic cell DNA repair. It has been calculated that the dose of UV required to reduce the survival of repair-proficient wild-type cells to

37% (i.e., the dose that delivers an average of one lethal hit per cell, assuming a Poisson distribution of hits) induces about 27,000 pyrimidine dimers per haploid genome (Cox and Game, 1974). A mutant strain with defects in the three known pathways by which pyrimidine dimers are repaired was much more sensitive to UV. It was calculated that on the average only one or, at most, two unrepaired pyrimidine dimers per haploid genome are lethal to the cell (Cox and Game, 1974). These data suggest that the wild-type yeast strain can efficiently repair on the order of all 27,000 of these dimers; however, if only one or two dimers cannot be repaired then cell death occurs.

Mammalian cells can be infected by SV40 virus DNA leading to the formation of viable progeny viruses. This process is referred to as transfection. Transfection experiments were performed with SV40 DNA, which was artificially modified to contain DNA damage in specific domains of the genome. These experiments allowed an estimate of the relative sensitivity of different genetic functions to inactivation by UV-induced DNA damage (Brown and Cerutti, 1986). The lethal effect of DNA damage depends on the location of the damage within the genome. DNA damages that occurred within a short DNA sequence encoding a transcriptional regulatory region and a gene necessary for the initiation of DNA replication exerted a maximal inactivating effect on the viral DNA. The damages least effective in inactivating viral DNA were those that occurred within a region of the genome encoding functions that are transcribed late in infection. The transcription of these regions presumably occurs after most damages are repaired and after intact new copies of DNA have been synthesized. A sevenfold difference in sensitivity to DNA damage existed between those different regions exhibiting maximal and minimal sensitivity to DNA damage. Brown and Cerutti (1986) concluded from these results that transcriptional promoters and enhancers were most sensitive to DNA damage and that the elongation step of DNA replication (as distinct from the initiation and termination steps) is least sensitive to DNA damage.

Even though elongation was less sensitive to blockage by DNA damage, blockages of elongation can occur at any position in the DNA. In contrast, only a fraction of damages occur within actively transcribed genes or within nontranscribed regions of DNA that govern gene expression or other important events of chromosome metabolism. Thus, the proportion of total lethality attributable to the blockage of DNA replication was found to be about 45%. The proportion of total lethality attributable to interference with the regulation of transcription or blockage of transcription of vital genes was also found to be about 45%. The remainder of the lethal effect was apparently due to blockage of the termination step of DNA replication. This analysis suggests that, for cellular genomes generally, DNA damages that block transcription and replication contribute comparably to the lethal consequence of DNA damage.

References

Abbotts, J., Zon, G., and Wilson, S. H. (1988). Studies of DNA repair DNA polymerases on templates with defined damage. *J. Cell. Biochem.* **12A,** 331.

Arrand, J. E., and Murray, A. M. (1982). Benzpyrene groups bind preferentially to the DNA of active chromatin in human lung cells. *Nucleic Acids Res.* **10,** 1547–1555.

Benbow, R. M., Zuccarelli, A. J., and Sinsheimer, R. L. (1974). A role for single-strand breaks in bacteriophage phiX174 genetic recombination. *J. Mol. Biol.* **88,** 629–651.

Brown, T., and Cerutti, P. (1986). Ultraviolet radiation inactivates SV40 by disrupting at least four genetic functions. *EMBO J.* **5,** 197–203.

Buhl, S. N., Setlow, R. B., and Regan, J. D. (1974). DNA repair in *Potorous tridactylus. Biophys. J.* **14,** 791–803.

Burck, K. B., Scarba, D. G., and Miller, R. C., Jr. (1979). Electron microscopic analysis of partially replicated bacteriophage T7 DNA. *J. Virol.* **32,** 606–613.

Cleaver, J. E. (1969). DNA repair in Chinese hamster cells of different sensitivities to ultraviolet light. *Int. J. Radiat. Biol.* **16,** 277–285.

Cleaver, J. E., Kaufman, W. K., Kapp, L. N., and Park, S. D. (1983). Replicon size and excision repair as factors in the inhibition and recovery of DNA synthesis from ultraviolet damage. *Biochim. Biophys. Acta* **739,** 207–215.

Cox, B. S., and Game, J. C. (1974). Repair systems in *Saccharomyces. Mutat. Res.* **26,** 257–264.

Dahle, D., Griffiths, T. D., and Carpenter, J. G. (1980). Inhibition and recovery of DNA synthesis in UV-irradiated Chinese hamster V-79 cells. *Photochem. Photobiol.* **32,** 157–165.

Friedberg, E. C. (1985). "DNA Repair." W. H. Freeman and Company, New York.

Giorno, R., and Sauerbier, W. (1976). A radiobiological analysis of the transcription units for heterogeneous nuclear RNA in cultured murine cells. *Cell* **9,** 775–783.

Hackett, P. B., and Sauerbier, W. (1975). The transcriptional organization of the ribosomal RNA genes in mouse L cells. *J. Mol. Biol.* **91**, 235–256.

Hackett, P. B., Varmus, H. E., and Bishop, J. M. (1981). Repair of lesions which cause premature termination of transcription in chicken embryo cells irradiated with ultraviolet light. *Virology* **112**, 752–756.

Ide, H., Kow, Y. W., and Wallace, S. (1985). Thymine glycols and urea residues in M13 DNA constitute replicative blocks *in vitro*. *Nucleic Acids Res.* **13**, 8035–8052.

Kow, Y. W., and Wallace, S. S. (1985). Exonuclease III recognizes urea residues in oxidized DNA. *Proc. Natl. Acad. Sci. USA* **82**, 8354–8358.

Leffler, S., Pulkrabak, P., Grunberger, D., and Weinstein, I. B. (1977). Template activity of calf thymus DNA modified by a dihydrodiol epoxide derivative of benzo(a)pyrene. *Biochemistry* **16**, 3133–3136.

Maher, V. M., Curren, R. D., Quellette, L. M., and McCormick, J. J. (1976). Role of DNA repair in the cytotoxic and mutagenic action of physical and chemical carcinogenesis. *In* "In vitro Metabolic Activation in Mutagenic Testing" (F. J. deSerres, J. R. Fouts, J. R. Bend, and R. M. Philpot, eds.), pp. 313–336. North Holland, New York.

Mayne, L. V. (1984). Inhibitors of DNA synthesis (aphidicolin and araC/HU) prevent the recovery of RNA synthesis after UV-irradiation. *Mutat. Res.* **131**, 187–191.

Mayne, L. V., and Lehman, I. R. (1982). Failure of RNA synthesis to recover after UV irradiation: An early defect in cells from individuals with Cockayne's syndrome and xeroderma pigmentosum. *Cancer Res.* **42**, 1473–1478.

Moore, P., and Strauss, B. S. (1979). Sites of inhibition of *in vitro* DNA synthesis in carcinogen and UV treated phiX174 DNA. *Nature* **278**, 664–666.

Moore, P. D., Bose, K. K., Rabkin, S. D., and Strauss, B. S. (1981). Sites of termination of *in vitro* DNA synthesis on ultraviolet- and *N*-acetylaminofluorene-treated phiX174 templates by prokaryotic and eukaryotic DNA polymerases. *Proc. Natl. Acad. Sci. USA* **78**, 110–114.

Moses, M. L., Webster, R. A., Martin, G. D., and Spelsberg, T. C. (1976). Binding of polycyclic aromatic hydrocarbons to transcriptionally active nuclear subfractions of AKR mouse embryo cells. *Cancer Res.* **36**, 2905–2910.

Nocentini, S. (1976). Inhibition and recovery of ribosomal RNA synthesis in ultraviolet irradiated mammalian cells. *Biochim. Biophys. Acta* **454**, 114–128.

Rouet, P., and Essigman, J. M. (1985). Possible role for thymine glycol in the selective inhibition of DNA synthesis on oxidized DNA templates. *Cancer Res.* **45**, 6113–6118.

Rupp, W. D., and Howard-Flanders, P. (1968). Discontinuities in the

DNA synthesized in an excision-defective strain of *Escherichia coli* following ultraviolet irradiation. *J. Mol. Biol.* **31,** 291–304.

Sarasin, A. R., and Hanawalt, P. C. (1980). Replication of ultraviolet-irradiated simian virus 40 in monkey kidney cells. *J. Mol. Biol.* **138,** 299–319.

Sauerbier, W., and Hercules, K. (1978). Gene and transcription unit mapping by radiation effects. *Annu. Rev. Genet.* **12,** 329–363.

Strauss, B., Rabkin, S., Sagher, D., and Moore, P. (1982). The role of DNA polymerase in base substitution mutagenesis on non-instructional templates. *Biochimie* **64,** 829–838.

Villani, G., Boiteux, S., and Radman, M. (1978). Mechanisms of ultraviolet-induced mutagenesis: Extent and fidelity of *in vitro* DNA synthesis on irradiated templates. *Proc. Natl. Acad. Sci. USA* **75,** 3037–3041.

Yu, F.-L. (1983). Preferential binding of aflatoxin B_1 to the transcriptionally active regions of rat liver nucleolar chromatin. *Carcinogenesis* **4,** 889–893.

Zieve, F. J. (1973). Effects of the carcinogen *N*-acetoxy-2-fluorenylacetamide on the template properties of deoxyribonucleic acid. *Mol. Pharmacol.* **9,** 658–699.

Chapter Four

Accumulation of DNA Damage in Somatic Cells

The DNA damage hypothesis of aging predicts that DNA damage accumulates in cells of critical tissues, giving rise to the manifestations of aging. We first consider the classification of mammalian cell types based on rate of division. We then review evidence showing that long-lived, nondividing, differentiated cells (including muscle cells and neurons as well as other cell types) have a low DNA repair capacity. This suggests that these cell types accumulate DNA damage with age. We review evidence that this is the case and argue that these nondividing or slowly dividing cells govern the rate of mammalian aging.

I. Cell Types Based on Rate of Division

Bowman (1985) classified the populations of cells from rat, mouse, and humans into four types according to their reproductive behavior. The first cell type consists of rapidly renewing cells that have a generation time of <30 days. This group includes duodenum and colon epithelial cells, hemopoietic (bone marrow) cells, and lymphopoietic cells (small lymphocytes) of the thymus and spleen. The second cell type consists of slowly renewing cells that have a generation time of >30 days but less than the mean life span of the animal. This group includes the cells of the respiratory tract epithelium and the fibroblasts found in the dermal connective tissue. The third cell type demonstrates some renewal but at such a slow rate that not all of the cells renew during the life span of the animal. This group includes smooth muscle cells and the glial cells found in most areas of the brain. The fourth cell type consists of static or nonrenewing cell populations that are terminally differentiated. This group includes neurons of all types, cardiac muscle cells, skeletal muscle cells, molar odontoblasts, and Sertoli cells of the testis.

II. *Low DNA Repair in Postmitotic Muscle and Neuronal Tissue*

Alexander (1967) was the first to speculate that DNA repair is turned off during mammalian development as cells differentiate to the postmitotic state. This, he thought, leads to an accumulation of DNA damage, a progressive decline in the capacity for RNA synthesis, and consequent aging. Alexander's view has since been largely confirmed in postmitotic muscle and neuronal cells, as indicated by the evidence reviewed in this chapter and the succeeding one.

A. Low Repair in Muscle

Here we discuss the evidence that DNA repair capacity of muscle cells declines as these cells become terminally differentiated. Each of the studies are also summarized in Table II. When chick embryonic pectoral myocytes differentiate to the postmitotic state, UV-induced excision repair is reduced twofold (Stockdale, 1971). Similarly, Hahn *et al.* (1971) demonstrated a deficiency in DNA repair synthesis in differentiated skeletal muscle cells from the rat. When exposed to the monofunctional alkylating agent methyl methanesulfonate, these muscle cells did not perform repair DNA synthesis. This contrasts with undifferentiated fibroblasts isolated from the same animals that did perform repair DNA synthesis following exposure to methyl methanesulfonate.

Lampidis and Schaiberger (1975) showed that primary heart muscle cell cultures derived from newborn rats are able to proliferate and undergo excision repair of UV-induced damages. However, they found that this DNA repair process was absent in postmitotic heart fibers taken from young adult rats.

Neidermuller (1985) studied repair synthesis in several tissues of the rat after treatment with *N*-nitroso-methylurea (NMU). NMU

Table II
DNA Repair in Postmitotic Muscle

Tissue	Animal	Damaging Agent	Comment	Reference	Repair Low in Postmitotic Cells
Pectoral muscle	chick	UV	repair declines twofold upon differentiating to postmitotic state	Stockdale, 1971	yes
Skeletal muscle	rat	methyl methanesulfonate	differentiated skeletal muscle has low repair compared with differentiated fibroblasts	Hahn *et al.*, 1971	yes
Cardiac muscle	rat	UV	ability to repair UV damages lost as cells differentiate to postmitotic state	Lampidis and Schaiberger, 1975	yes
Skeletal muscle	rat	*N*-nitrosomethylurea	repair capacity in muscle 15% of spleen and liver	Neidermuller, 1985	yes
Muscle cells in culture	rat	4-NQO	ability to carry out 4-NQO-induced repair synthesis declined 50% as cells differentiated from myoblasts to myotubes	Chan *et al.*, 1976	yes
		X-rays	no loss of repair during differentiation		no
Muscle cells in culture	rat	UV	ability to carry out repair of UV damages unchanged as cells differentiated from myoblasts to myotubes	Koval and Kaufman, 1981	no

Abbreviations: 4-NQO, 4-nitroquinoline-1-oxide; UV, Ultraviolet light.

methylates DNA bases at a variety of positions, producing, for example, 7-methylguanine and O^4-methylthymine, which then are removed by excision repair. He found that in the skeletal muscle of young adult (9 and 18 mo) rats the level of unscheduled DNA synthesis (a measure of excision repair) is low; i.e., unscheduled DNA synthesis was 15% of that found in spleen and lung cells.

The previous four studies describe work done with cells derived from intact animals. Two other studies were performed with cultured muscle cells. Chan *et al.* (1976) studied levels of repair in cultured muscle cells as they developed from the undifferentiated myoblast form to the differentiated myotube form. In this study, repair synthesis was measured by the incorporation of [^{3}H]thymidine into unreplicated DNA. In addition, the repair of DNA strand breaks was examined by sedimentation in an alkaline sucrose gradient. To induce DNA repair, cells were treated with the DNA-damaging agent 4-nitroquinoline-1-oxide (4-NQO). The level of repair synthesis in response to 4-NQO treatment declined by about 50% after the cells had fused into myotubes. Also, the single-strand breaks induced by 4-NQO were mostly repaired in the myoblasts, but not in the differentiated myotubes. In contrast, when single-strand breaks were induced by X-rays both myoblasts and myotubes were equally able to repair them. This implies that the lack of rejoining of 4-NQO-induced breaks may be a consequence of some peculiarity associated with 4-NQO. Koval and Kaufman (1981) measured unscheduled DNA synthesis in response to UV light several times during the differentiation of cultured rat skeletal muscle cells. The results obtained were similar to those obtained with X-rays in the above study by Chan *et al.* (1976). The amount of DNA repair synthesis did not change as the cells fused and differentiated from postmitotic perfusion myoblasts to multinucleated contracting myotubes in culture. Why cells in culture (Koval and Kaufman, 1981) gave different results with respect to repair of UV damages from cells in intact animals (Stockdale, 1971; Lampidis and Schaiberger, 1975) is unclear.

In summary, these studies indicate that repair levels are lower in postmitotic muscle than in other tissues of adult animals (Hahn *et al.*, 1971; Niedermuller, 1985), and that repair capacity is lower in postmitotic muscle cells than in proliferating muscle cells from embryonic or newborn rats (Stockdale, 1971; Lampidis and Schaiberger, 1975). Myoblasts undergoing differentiation to multinucleated contracting myotubes in culture decline in the capacity to repair 4-NQO damages, but not X-ray or UV-induced damages (Chan *et al.*, 1976; Koval and Kaufman, 1981).

B. Low Repair in Neuronal Tissue

As with terminally differentiated muscle cells, a number of studies indicate that DNA repair is low in terminally differentiated neuronal cells. These studies are also summarized in Table III.

Table III

DNA Repair in Postmitotic Neuronal Tissue

Tissue	Animal	Damaging Agent	Comment	Reference	Repair Low in Postmitotic Cells
Cerebellum	rabbit	UV and NAAF	repair of damage in neurons undetectable	Lieberman and Forbes, 1973	yes
Brain	rat	O^6-ethylguanine	repair of damage in brain low relative to liver	Goth and Rajewsky, 1974	yes
Brain	rat	O^6-methylguanine	repair of damage in brain low relative to liver	Margison and Kleihues, 1975	yes
Brain	rat	O^6-methylguanine	alkyl acceptor repair enzyme low in brain relative to liver and spleen	Woodhead *et al.,* 1985	yes
Neural retina	chick	NAAAF and methylmethanesulfonate	repair capacity declined 10-fold as cells differentiate to the postmitotic state	Karran *et al.,* 1977	yes
Brain	hamster	UV	repair capacity declined threefold upon differentation to post-mitotic state; 9.6% of adult lung, 22% of adult kidney	Gensler, 1981b	yes
Brain	rat	NMU	repair capacity in brain 15% of spleen and lung	Niedermuller, 1985	yes
Cerebellum	human	spontaneous	lack of DNA turnover during lifetime indicates low repair	Slatkin *et al.,* 1985	yes
Neuroblastoma cells	mouse	UV	repair declines as cells differentiate in culture	McCombe *et al.,* 1976	yes
Neuroblastoma cells	human	UV and adducts	repair declines as cells differentiate in culture	Jensen and Linn, 1988	yes

Abbreviations: NAAAF, *N*-acetoxy-2-acetylaminofluorene; NMU, *N*-nitrosomethylurea; UV, ultraviolet light.

Lieberman and Forbes (1973) observed that neurons from the rabbit cerebellum did not exhibit repair synthesis in response to UV irradiation or *N*-acetoxy-2-acetylaminofluorene (NAAAF) treatment. Non-neuronal cells present in the same tissue preparation did undergo repair, however. Goth and Rajewsky (1974) administered a pulse of ethylnitrosourea to rats and then measured retention in DNA of O^6-ethylguanine and other ethylated bases. The O^6-ethylguanine–guanine ratio of brain DNA, although initially only about 0.65 of that for liver DNA, was about 11-fold greater than liver DNA at 114 hr after the ethylnitrosourea treatment. This suggests that these types of DNA damages are repaired by liver cells, but not repaired as efficiently by brain cells. In other experiments, Margison and Kleihues (1975) showed that after 5 wk of weekly injections with N-^{3}[H] methyl-N-nitrosourea, rat brain had >100 times the level of O^6-

methylguanine in its DNA than liver. They interpreted this result as reflecting a low capacity to repair O^6-methylguanine damages in the brain, similar to the interpretation with O^6-ethylguanine. The repair of O^6-alkylguanine residues in DNA can be accomplished by the transfer of the alkyl group to a cysteine residue of an acceptor protein (Chapter 9, Section II.F.2). Woodhead *et al.* (1985) found, in rats, that acceptor protein activity levels were highest in liver and somewhat less in spleen; the levels in brain and kidney were still lower. The level of acceptor protein in brain cells averaged about 43% of the level in liver.

Karran *et al.* (1977), working with chick neural retina, reported a developmental decline in excision repair capacity for both NAAAF-induced and methyl methanesulfonate-induced DNA damages. Replicative DNA synthesis in primary cell cultures declined about 1,000-fold between days 6 and 15 of embryogenesis. Excision repair in response to both chemicals declined 10-fold during this same interval. By embryogenic day 16, these chick neural retinal cells became postmitotic. This study suggests that excision repair capacity declines as this type of neuronal cell differentiates from the mitotic to the postmitotic state.

Gensler (1981a, 1981b) found that unscheduled DNA synthesis in response to UV irradiation (a measure of pyrimidine dimer excision repair) is markedly lower in mature hamster brain than in mature hamster lung or kidney cells. The capacity of brain cells to carry out excision repair declined threefold early in life (between ages 1 and 7 wk). The excision repair level then remained low, at about 9.6% of the level in the adult lung and 22% of that in the adult kidney. In the rat, which is closely related to the hamster, neuronal cells become postmitotic at approximately 2–3 wk after birth (Fish and Winnick, 1969). This suggests that the early threefold decline in repair observed in hamsters may be related to the transition from the mitotic to the postmitotic state.

Niedermuller (1985, also see this chapter, Section II.A) described DNA repair synthesis in several tissues of the rat. In his study, Neidermuller showed that the capacity to repair NMU-induced DNA damages in brain was low. For example, the level for brain in rats aged 9 and 18 mo was about 15% of the repair capacity in spleen and lung.

Mouse neuroblastoma cells, which can be induced to undergo

differentiation in culture, were used by McCombe *et al.* (1976) as a model to investigate the effects of UV on differentiated nerve cells. Differentiated neuroblastoma cells were found to be extremely sensitive to UV compared with proliferating cells from the same clone. Neuroblastoma cells in the differentiated mode were capable of carrying out some excision repair of DNA damage, but they apparently had much less repair capability than proliferating cells.

Jensen and Linn (1988) found that human neuroblastoma cells that had differentiated in culture removed DNA adducts of benzo (a)pyrene, benzo(a)pyrenediolepoxide, and N^7-methylguanine at a considerably lower rate than did undifferentiated mitotic cells. A dramatic decline in the capacity for DNA repair synthesis in response to UV-irradiation was also observed when these cells differentiated.

Another type of study, based on human populations, was performed to assess the DNA repair potential of neuronal tissue after the neuronal cells became postmitotic. In this work, an estimate of the overall level of repair in neuronal DNA was obtained by studying the turnover of DNA in the human brain. Slatkin *et al.* (1985) first noted that human senescence is often associated with impairment of central nervous system function. They argued that if neuronal DNA is metabolically stable, so that very little is turned over during DNA repair synthesis, then neurological senescence might be due to progressive accumulation of abnormalities such as chemical adducts in neuronal DNA. To test this hypothesis, they then measured the stability of human DNA using carbon–isotope ratios.

Stable carbon–isotope ratios, $^{13}C/^{12}C$, are slightly but consistently greater in North Americans than in Europeans. The difference has been attributed to the relative proportions of two photosynthetic pathways in the food chains of these populations. Human migration from Europe to North America provided Slatkin *et al.* (1985) with the basis for a natural stable carbon isotope experiment capable of detecting turnover of long-lived body constituents such as neuronal DNA without the exposure of human subjects to radioactive tracers.

About 90% of cell nuclei in the human cerebellum are in small neurons of the cortical granular layer. If adult human neuronal DNA is metabolically stable, its $^{13}C/^{12}C$ ratio should reflect the $^{13}C/^{12}C$ ratio in the maternal diet during fetal development because almost all neurons are formed during maturation of fetal brain and do not undergo cell division thereafter. Thus, it may be predicted that the $^{13}C/^{12}C$ ratio of whole cerebellar DNA from adult European-born Americans and adult European-born Europeans would reflect the maternal diet; i.e., it would be the same in the two populations if turnover of neuronal DNA in the mature cerebellum does not take place. To test this hypothesis, stable carbon–isotope ratios were determined in postmortem cerebellar specimens of adult American-born Americans, European-born Americans, and European-born Europeans. The $^{13}C/^{12}C$ ratios in cerebellar DNA from European-born Americans were found to be closer to the $^{13}C/^{12}C$ ratios in cerebellar DNA from European-born Europeans than from American-born Americans. The authors concluded that a large proportion of the DNA in small neurons of the cerebellum undergoes no metabolic turnover during the human life span. Other studies, reviewed by Slatkin *et al.* (1985), also have shown that *in vivo* most mouse brain neuronal DNA is metabolically stable.

Altogether, low DNA repair capacity was shown in all 10 studies on neuronal tissue (for summary, see Table III). The results with neuronal tissue, both *in vivo* and in culture, were even clearer than the results on postmitotic muscle tissue (reviewed in this chapter, Section II.A).

C. Repair Is Not Completely Shut Off in Differentiated Neuronal Tissue

Although much evidence indicates that the DNA repair capacity of adult neurons is less than that of other tissues, some repair appears to occur. For instance, Korr and Schultz (1989) have presented evidence for low levels of DNA repair in various types of cells of mouse brain *in vivo*. DNA polymerase-beta, an enzyme thought to be involved in DNA repair, has been isolated from rat cortex neurons (Waser *et al.*, 1979). This was found to be the major DNA polymerase occurring in neuronal nuclei of adult rats, because the replicative DNA polymerase-alpha is absent from postmitotic neurons. Karran *et al.* (1977) showed that as chick retinal tissue differentiates to the postmitotic state (between ages 6 and 14 days) the activity of DNA polymerase-alpha declines 20-fold, whereas polymerase-beta declines only 3-fold. This suggests that in the postmitotic brain, repair is less completely shut off than replication. Jensen and Linn (1988) showed that when human neuroblastoma cells differentiate, in culture, to a postmitotic state the levels of the repair enzymes DNA polymerase-beta and uracil DNA glycosylase remain unchanged. Furthermore, apurinic/apyrimidinic endonuclease activity, which is also involved in DNA repair (Chapter 9, Section II.B), increases about threefold upon differentation.

Even though the level of DNA repair for some types of damage in neurons is low, this repair might be sufficient to cope with DNA damages if they occur at a low rate in these cells. Thus, to relate DNA damage to aging, a net accumulation of damages needs to be demonstrated. Evidence for such accumulation is reviewed in this chapter, Section III.B.

D. Low DNA Repair May Reflect Competition for Energy Resources

In this section, we discuss an evolutionary rationale for low DNA repair in brain cells and muscle cells. Kirkwood (1977) and Kirkwood and Holliday (1979), in considering the evolution of aging, suggested that mortality may be due to an energy-saving strategy of reduced error regulation in somatic cells. According to Kirkwood (1977), accuracy in the germ line is vital for genome survival, but a high level of accuracy in somatic cells may be a luxury the genome does better to forego. Although this idea was proposed in support of the error catastrophe theory of aging (see Chapter 15, Section I.I), it serves equally well as an explanation for the depressed DNA repair found in both muscle cells and brain cells.

Hahn *et al.* (1971) argued in a similar way. They pointed out that although differentiated fused muscle cells can synthesize RNA *de novo,* such synthesis is not required for muscle contraction; therefore, they suggested that the integrity of the genome in these fused cells may not be required. In such cells, the ability to repair DNA could be regarded as a discardable luxury. Likewise, neurons may enter the postmitotic state early in development because of selective pressure to divert energy resources from replication and cell division to characteristic neuronal functions such as transmission of impulses. At least some of the enzymes involved in DNA replication are probably also required in DNA repair, suggesting that repair would decline concurrently with the turnoff of replication.

In general, the reduction or turnoff of repair in postmitotic cells may be part of a comprehensive redirection of cellular resources associated with terminal differentiation. In accord with the ideas of Kirkwood and Holliday (1979) and of Hahn *et al.* (1971), the advantages of removing DNA damages may be insufficient to select for a higher repair capacity in such cells.

III. *Accumulation of DNA Damage in Adult Muscle and Neuronal Tissue*

A. Muscle

We discuss here five studies of accumulation of DNA damage in muscle with age. These studies are also summarized in Table IV. Price *et al.* (1971) examined the template activity of DNA in sections of ethanol-fixed heart muscle fibers, as well as brain and liver tissues, from young (3–4 mo) and old (30–35 mo) mice. Single-strand breaks in the DNA of these tissues can act as initiation points for DNA synthesis. Acid denaturation was used on the DNA to enhance the template activity of breaks. Calf thymus DNA polymerase was then used to catalyze DNA synthesis on the mouse template DNA. Calf thymus DNA polymerase is known to require partially denatured template DNA with a 3'-hydroxyl group facing a single strand template region, which is at least 20 nucleotides long. Newly synthesized DNA using the mouse template may then be labeled by incorporation of tritium-labeled deoxyribonucleotide triphosphates and measured by autoradiography. The amount of DNA synthesis that occurred with this assay indicated that single-strand breaks accumulate in the DNA of mouse heart muscle fibers. Thus, significantly more single-strand breaks were found in the DNA from heart muscle fibers of old mice than in the equivalent DNA from young mice. The results using brain and liver template DNA were similar and will be discussed later in this chapter.

Chetsanga *et al.* (1976) studied DNA from myocardial cells of mice. They found that the DNA from myocardial cells of young mice sedimented in a narrow peak in an alkaline CsCl density gradient at

Table IV
Accumulation of DNA Damage in Adult Muscle with Age

Tissue	Animal	DNA Damage	Comment	Reference	Accumulation Shown
Heart Muscle	mouse	DNA strand breaks	increased in old (30–35 mo) compared with young (3–4 mo) mice; measured template activity for DNA polymerase	Price *et al.*, 1971	yes
Heart muscle	mouse	single-strand regions	S$_1$ endonuclease-sensitive sites increase as mice age from 6 to 30 mo	Chetsanga *et al.*, 1976	yes
Heart muscle	mouse	single-strand regions	S$_1$ endonuclease-sensitive sites do *not* increase from age 5 to 22 mo	Mori and Goto, 1982	no
Muscle	human	single-strand regions	S$_1$ endonuclease-sensitive sites accumulate with age	Zahn *et al.*, 1987	yes
Heart muscle	mouse	methylated guanine	ninefold increase occurred between age 2 and 39 mo	Gaubatz, 1986	yes

the expected density of 1.70 g/cm^3. In contrast, the DNA from the heart cells of senescent (20, 25, and 30 mo) mice became broadly distributed in alkaline CsCl gradients. This mode of sedimentation indicates that mouse DNA becomes partially fragmented as the mouse ages. When the native DNAs of myocardial cells from 6-, 20-, and 30-mo-old mice were treated with single-strand-specific S$_1$ endonuclease (Chetsanga *et al.*, 1976), a progressive increase in sensitivity to digestion by the enzyme was found with age of the mice. This indicates that mice accumulate single-strand regions in their heart DNA as they age. Mori and Goto (1982), however, could not repeat this S$_1$ endonuclease result. The basis for this discrepancy is unclear.

Zahn *et al.* (1987) studied age-correlated DNA damage in human muscle tissue. DNA damage was determined as DNA alterations that give rise to complete molecular breaks during the treatment of purified DNA solutions with single-strand-specific nucleases. The DNA was derived from samples of human muscle of individuals, most of whom were undergoing surgical treatment. Three different deoxyribonucleases were used: S_1 endonuclease, nuclease Bal 31, and pea endonuclease. Each nuclease gave similar results. The lengths of the double-stranded pieces obtained after nuclease treatment were determined from electron micrographs. The average molecular weight of the double-stranded DNA between two single-strand breaks was determined by measuring 20–200 molecules for each individual. The 470 individuals contributing to the data were 1–91 yr old. It was found that the average molecular weight of the double-stranded DNA between two single-strand breaks significantly decreases with age. This implies that single-strand breaks accumulate with age in human muscle tissue.

Gaubatz (1986) pointed out that if genetic damages accumulate faster than they can be repaired, the cell carrying these DNA damages will eventually become defective in maintaining homeostasis. He noted that this situation should be particularly serious for cells that do not divide after they have differentiated to their terminal forms (e.g., heart muscle cells). To test these ideas, he used a ^{32}P-postlabeling technique to measure the level of modified nucleotides in mouse myocardial DNA as a function of age. The postlabeling analysis indicated that particular modified nucleotides increased in number about ninefold in heart DNA from mice between ages 2 mo and 39 mo. The damage in this case was found to be an altered nucleotide, presumably a methylated form of guanosine monophosphate. This result suggests that the level of the altered nucleotide is greatly elevated in senescent mouse heart tissue compared with young mouse heart tissue.

In conclusion, the studies presented in this section, with one exception (Mori and Goto, 1982), indicate that mammals accumulate DNA damages in their muscle cell DNA as they age.

B. Neuronal Tissue

Here we discuss 11 studies of accumulation of DNA damage in neuronal tissue with age. They are also summarized in Table V.

All of these studies examine whether or not single-strand breaks accumulate in brain DNA. A number of different methods of detection were used. The study by Price *et al.* (1971), mentioned in the previous section, showed that single-strand breaks, which can act as initiation points for DNA synthesis, accumulate with age in mouse brain, heart, and liver tissues. More single-strand breaks were found in brain cells than in liver or heart cells. Modak and Price (1971) used an experimental approach similar to that of Price *et al.* (1971). They found that single-strand break points, again which act as initiation points for DNA synthesis, increase both with X-irradiation and when individuals age.

Wheeler and Lett (1974) used alkaline sucrose gradient sedimentation, which denatures double-stranded DNA, to obtain estimates of the sizes of the single strands present between single-strand breaks in the DNA. They studied the DNA from the cerebella of beagle dogs. Comparing dogs aged from 0.13 to 10 yr, they found that the size of DNA single-strands declined. A decline in the size of single-stranded DNA indicates an increased frequency of breaks per strand. Ono *et al.* (1976) also used alkaline sucrose gradient centrifugation to obtain estimates of the sizes of single strands within duplex DNA. They examined DNA from four tissues of mouse, using animals of various ages. They found no differences between young (1–2 mo) and older (22 mo) mice with respect to splenic, thymic, or cerebellar DNA; i.e., these authors found no evidence for an increase in single-strand breaks in brain with mouse age. However, liver DNA from older mice was significantly smaller than that from younger mice. Again, the decrease in single-stranded DNA size is an indication of an increase in the frequency of single-strand breaks as these animals age. Ono *et al.* (1976) suggested that the difference between their alkaline sucrose gradient results and the mouse DNA template measurements of Price *et al.* (1971) (see above) might be due to the entirely different methods employed as well as the difference in the age of the old mice used (22 mo for Ono *et al.;* 35 mo for Price *et al.*).

Chetsanga *et al.* (1977) showed that DNA from the brain of young (6 mo) mice sedimented monodispersely in an alkaline sucrose gradient, the peak fraction corresponding to an average molecular weight of 120×10^6 daltons. The DNA from 30-mo-old mice sedimented polydispersely, banding in four peaks corresponding to aver-

Table V

Accumulation of DNA Damage in Neuronal Tissue with Age

Tissue	Animal	DNA Damage	Comment	Reference	Accumulation Shown
Brain	mouse	single-strand breaks	accumulated in all tissues, but more in brain than in heart and liver; compared young (3–4 mo) with old (30–35 mo) mice	Price *et al.*, 1971	yes
Brain	mouse	single-strand breaks	accumulated in brain from age 3 to 35 mo	Modak and Price, 1971	yes
Brain	beagle dogs	single-strand breaks	increased from age 0.13 to 10 yr	Wheeler and Lett, 1974	yes
Brain	mouse	single-strand breaks	no change up to age 22 mo	Ono *et al.*, 1976	no
Brain	mouse	single-strand breaks	sevenfold increase in nuclease-sensitive sites occurred between ages 6–15 mo and 30 mo	Chetsanga *et al.*, 1977	yes
Cerebral	rat	single-strand breaks	nuclease-sensitive sites increased eightfold between ages 2 mo and 2 yr	Murthy *et al.*, 1976	yes
Brain	mouse	single-strand regions	nuclease-sensitive regions increased from 2.0% of DNA to 2.9% with age	Mori and Goto, 1982	yes
Brain	mouse	single-strand regions	using an immunofluorescent technique, found an increase in single-strand regions from age 4 wk to 1 yr	Nakanishi *et al.*, 1979	yes
Brain	mouse	single-strand breaks	no increase from age 1 to 20 mo	Su *et al.*, 1984	no
Brain	rat	single-strand breaks	no increase from age 0 to 36 mo	Mullaart *et al.*, 1990	no
Photoreceptor cells of retina	rabbit	single-strand breaks	old rabbits showed changes in DNA equivalent to young rabbits treated with several 100-rads X-rays	Bergtold and Lett, 1985	yes

age molecular weights of 140×10^6, 70×10^6, 15×10^6 and 3×10^6 daltons. Again, the smaller the single-stranded DNA pieces, which double-stranded DNA falls into upon denaturation, the greater the frequency of single-strand breaks. In addition, they found that about 2% of the DNA from the brain of 6–15-mo-old mice was sensitive to digestion by S_1 endonuclease, which specifically cleaves single-stranded DNA. The percentage of DNA found to be S_1 endonuclease-sensitive increased to 10% by 20 mo and to 14% by 30 mo. Again, these results imply that single-strand regions accumulate in the DNA of brain cells with age.

Murthy *et al.* (1976) determined the percentage of single-strand regions in cerebral cortex DNA of young and old rats by an assay that measures sensitivity to S_1 endonuclease. The percentage of single-strand regions in purified DNA was 4.1% at 2 mo and 31.2% at 2 yr. When chromatin, rather than purified DNA, was examined, the percentages were 0% single-strand regions at 2 mo and 14.2% at 2 yr. The differences between purified DNA and chromatin DNA may result from proteins bound to DNA in chromatin, which can restrict hydrolysis of DNA by S_1 endonuclease.

Mori and Goto (1982) used the single-strand-specific S_1 endonuclease assay of Murthy *et al.* (1976) to examine mouse DNA. They isolated DNA from brain, liver, kidney, heart, and spleen tissues, and examined these types of DNA from mice of different ages. Brain DNA from mice aged 15–30 mo contained 2.9% single-strand regions, which was significantly higher than that from younger animals, which contained 2.0% single-strand regions. DNA from liver, kidney, heart, and spleen did not show significant age-associated changes.

Nakanishi *et al.* (1979) used touch smears of brain cells and liver cells of young (4 wk) and senescent (1 yr) mice. The smears were stained with antibody to cytidine nucleoside by an indirect immunofluorescence technique. The antibody reacts only with denatured or single-stranded DNA. Cellular DNA binding by the antibody was only seen in the brain and liver tissues of aged animals, indicating the presence of single-stranded DNA damages. No such DNA damages were detected in the epithelial cells of the gastrointestinal tract at any age. The authors suggested that this type of DNA alteration may accumulate in slowly renewing and nonreplenishing tissues as a function of age. Also, they suggested that tissues made up of cells that are continually replaced will not accumulate such DNA alterations.

Su *et al.* (1984), using alkaline sucrose sedimentation to measure single-strand breaks, found that DNA damage did not measurably accumulate in mouse brain tissue with age. A similar lack of accumulation of single-strand breaks in rat brain was found by Mullaart *et al.* (1990) using the technique of alkaline elution. Bergtold and Lett (1985) used the technique of zonal ultracentrifugation of DNA through reoriented alkaline sucrose gradients. They considered this technique to be the most sensitive method yet devised for measuring damage to mammalian chromosomal DNA. They observed an age-related decline in median sedimentation coefficient of rabbit retinal photoreceptor cell DNA, indicating that DNA breaks accumulate as animals age. The oldest rabbits showed changes in the photoreceptor cells of their retina equivalent to the effects of several hundred rads of X-photons in young rabbits.

In conclusion, we have reviewed eight studies on neuronal tissue showing accumulation of DNA damage with age and three studies in which no increase in DNA damage was observed. Thus, the majority of evidence indicates that DNA damages accumulate in neuronal tissue as mammals age.

IV. Accumulation of DNA Damage in Liver

In this section, we review evidence that mammalian liver retains a substantial repair capacity with age but, nevertheless, accumulates DNA damage.

A. DNA Repair in Liver

DNA repair capacity with respect to O^6-ethylguanine and O^6-methylguanine DNA damages is substantially higher in liver than in brain (Goth and Rajewsky, 1974; Margison and Kleihues, 1975; Woodhead *et al.*, 1985; see this chapter, Section II.B., for further discussion). Indeed, liver appears to have a generally high repair capacity (Niedermuller, 1985). Ono and Okada (1978) obtained results indicating that hepatic cells of old (14 and 22 mo) mice were able to rejoin radiation-induced DNA strand breaks as quickly as those of young (2 mo) mice; however, only radiation-induced DNA breaks, and not age-associated DNA strand breaks, could be rejoined.

B. Accumulation of DNA Damage in Liver

The liver is composed of several cell types including hepatocytes (parenchymal cells), endothelial cells, and specialized macrophages (Kupffer cells). Hepatocytes, the major liver cell type, are able to divide, but their turnover time (defined as the time necessary to replace the number of cells present in the entire cell population) is long: 400–450 days in the rat (Bowman, 1985). DNA damages might be expected to accumulate in such slowly dividing cells by analogy to muscle and neuronal cells, which do not divide at all and which accumulate DNA damage.

A total of 14 studies have been carried out on the accumulation of DNA damage in liver cells. These studies are summarized in Table VI. Eight of these studies reported an accumulation of single-strand breaks in liver DNA with age. Three of these studies reported no accumulation of single-strand breaks in liver DNA. Three additional studies reported an accumulation of altered bases.

The three studies that found no accumulation of DNA damages (Dean and Cutler, 1978; Finch, 1979; Mori and Goto, 1982) were specifically designed to either support or refute the previous work of Chetsanga et al. (1975), who found an increase with age in S_1 endonuclease-sensitive regions of DNA in liver. This endonuclease is specific for single-strand regions of DNA, and their result suggested that single-strand regions accumulate with age. Because three studies refute the finding of Chetsanga et al. (1975), S_1 endonuclease-sensitive single-strand regions probably do not increase with age in liver.

No evidence refutes the remaining studies (Price et al., 1971; Massie et al., 1972; Ono et al., 1976; Nakanishi et al., 1979; Park and Ames, 1988a, 1988b; Sharma and Yamamoto, 1980; Su et al., 1984; Lawson and Stohs, 1985; Mullaart et al., 1988, 1990; Randerath et al., 1989; Gupta et al., 1990), which showed increases in DNA damage as measured by techniques not involving S_1 endonuclease digestion. We review three of the more recent of these studies below. A fourth study, by Lawson and Stohs (1985), is reviewed in Chapter 6, Section II.

Table VI
Accumulation of DNA Damage in Liver with Age

Tissue	Animal	DNA Damage	Comment	Reference	Accumulation Shown
Kupffer cells of liver	mouse	single-strand breaks	increased in old (30–35 mo) compared with young (3–4 mo) mice; measured template activity for DNA polymerase	Price *et al.*, 1971	yes
Liver	rat	single- and double-strand breaks	increased as rats aged from 39 to 400 days; used alkaline sedimentation velocity	Massie *et al.*, 1972	yes
Liver	mouse	single-strand regions	S_1 endonuclease-sensitive regions increased from age 1 to 30 mo	Chetsanga *et al.*, 1975	yes
Liver	mouse	single-strand breaks	increased from age 2 to 14 mo; used alkaline sucrose	Ono *et al.*, 1976	yes
Liver	mouse	single-strand regions	increased from age 4 wk to 1 yr; used immunofluorescent technique	Nakanishi *et al.*, 1979	yes
Liver	mouse and rat	oxidized base(s)	increase in modified base from age 3 to 28 wk in mouse and from age 3 to 31 wk in rat	Sharma and Yamamoto, 1980	yes
Liver	mouse	single-strand regions	S_1 endonuclease-sensitive regions did *not* increase from age 4.5 to 31 mo	Dean and Cutler, 1978	no
Liver	mouse	single-strand regions	S_1 endonuclease-sensitive regions did *not* increase from age 10 to 30 mo	Finch, 1979	no
Liver	mouse	single-strand regions	S_1 endonuclease-sensitive regions did *not* increase from age 5 to 22 mo	Mori and Goto, 1982	no

(continued)

Table VI (*Continued*)

Tissue	Animal	DNA Damage	Comment	Reference	Accumulation Shown
Liver	mouse	alkali-labile sites	increase (60–80%) as mice aged from 1 to 20 mo; used alkaline sucrose gradient sedimentation	Su *et al.*, 1984	yes
Liver	mouse	single-strand breaks	increase from age 6 to 18 mo; used alkaline elution	Lawson and Stohs, 1985	yes
Liver parenchymal cells	rat	alkali-labile sites	damage increased 80–90% as rats aged 6 to 36 mo	Mullaart *et al.*, 1988, 1990	yes
Liver	rat	methyl adduct of base	increase 2.5-fold as rats age from 6 to 24 mo	Park and Ames, 1988a, 1988b	yes
Liver	rat and hamster	DNA adducts (I-compounds)	increase from age 1 to 10 mo	Randerath *et al.*, 1989; Gupta *et al.*, 1990	yes

Mullaart *et al.* (1988) used a highly sensitive alkaline elution assay to quantitate alkali-labile sites found in freshly isolated DNA of postmitotic and mitotic liver cells taken from young and old rats. This assay detects single-strand breaks or damages that are converted to breaks under alkaline conditions. Both depurination sites and several forms of oxygen-induced base damages are alkali-sensitive. Thus, such bases will be detected in the alkaline elution assay. Mullaart *et al.* (1988) found that the DNA taken from postmitotic parenchymal cells from old rats contained more alkali-labile sites than DNA from parenchymal cells taken from young rats. Such an age-associated difference was not observed in the mitotically active nonparenchymal liver cells taken from young and old rats. Thus, long-lived postmitotic cells exhibit more pronounced symptoms of cellular senescence than actively dividing cells, even when those cells are isolated from the same organ.

Recent studies by Park and Ames (1988a; 1988b) concluded that a methylated base occurs at a high level in rat liver nuclear and mitochondrial DNA. Young (6 mo) rats contained this methylated base at a level of one residue per 31,000 bases in mitochondrial DNA and one residue per 105,000 bases in nuclear DNA. These residue levels increased about 2.5-fold in old (24 mo) rats. In mitochondrial DNA, the average numbers of modified bases per cell was estimated to be 10,000 for young rats and 21,000 (2.1-fold higher) for old rats. In nuclear DNA, the average numbers of modified bases per cell were 174,000 for young rats and 460,000 (2.6-fold higher) for old rats.

The study by Su *et al.* (1984) not only showed that DNA damage accumulates with age in mouse liver but also that it accumulates faster in a mouse species with a short life span than in a second mouse species with a longer life span. The method used to measure DNA damage was alkaline sucrose sedimentation. This method measures single-strand breaks plus alkali-labile bonds. The two species of mouse, *Mus musculus* and *Peromyscus leucopus,* are from different families. Although they are closely related taxonomically, they differ significantly in maximum achievable life span (2.5 times greater in *P. leucopus*). The accumulation of DNA damage in liver and kidney cells with age was greater in the shorter-lived species (*M. musculus*).

C. Increased Incidence of Chromosomal Aberrations in Liver

In addition to the studies on DNA damage in liver cells described in the previous section, several studies show that the incidence of chromosomal aberrations (e.g., bridges between chromosomes, fragments of chromosomes) in animal liver cells increases with age. These studies are listed in Table VII and are described below. The methods used involved partial hepatectomy or injection with CCl_4 prior to sacrifice of the animal to stimulate liver regeneration. Generally, though not always, aberrations increased more rapidly with age in short-lived than in long-lived animals (Brooks *et al.,* 1973). Life span in the four species studied increased in the following order: mouse, Chinese hamster, guinea pig, and dog. Generally, more chromosomal aberrations were evident in the shorter-lived species at a given age (i.e., mouse was highest and the dog lowest,

Table VII
Incidence of Chromosomal Aberrations in the Liver of Different Animal Species with Age

Animal	Change in Incidence of Chromosomal Aberrations with Age	Reference	Accumulation Shown
Mouse	increase from age 4.5 to 12 mo	Stevenson and Curtis, 1961	yes
Mouse	increase from age 50 to 700 days in two strains; faster increase in short-lived strain	Crowley and Curtis, 1963	yes
Dog	slow rate of increase up to age 8 yr	Curtis *et al.*, 1966	yes
Guinea pig	increase up to age 1,400 days	Curtis and Miller, 1971	yes
Hamster	linear increase from age 78 to 1,100 days	Brooks *et al.*, 1973	yes

although the relative order of Chinese hamster and guinea pig was reversed).

In addition to these experiments, the accumulation of DNA damage in two inbred mouse strains having different average life spans was studied by Crowley and Curtis (1963). This study revealed that chromosomal aberrations in the liver of the long-lived strain increased rather slowly with age, but those for the short-lived strain increased more rapidly. The two mouse strains used were different from those Su *et al.* (1984) used to study the accumulation of DNA damage in mouse liver with age (see previous section). However, the results of the two studies are consistent because both indicated that damage accumulates faster in the mouse strain with shorter life span.

Curtis (1963) studied the effect of whole-body X-irradiation on life-shortening in the mouse. He found that the amount of life-shortening and the frequency of chromosome aberrations in the liver were related to dose rate, dose pattern, and type of exposure in the same way. This suggested a causal relationship between chromosome aberrations and aging. Curtis (1963) regarded chromosome aberrations to be an index of mutation and, therefore, that his studies supported the somatic mutation theory of aging. However, as outlined below, chromosomal aberrations arise from DNA damage and may be regarded as an index of DNA damage as well as mutation.

Chromosomal aberrations may arise from double-strand breaks in DNA. Studies by Bryant (1983, 1985) indicate that X-irradiated mammalian cells undergo a mode of death in which double-strand breaks in the DNA cause chromosomal aberrations. These chromosomal aberrations are lethal as a result of loss of genetic material in the form of chromosomal fragments or as a result of chromosome bridge formation. Bryant (1985) commented that in mammalian cells the DNA damage responsible for both cell death and chromosome aberrations in irradiated cells may be the double-strand break. Natarajan *et al.* (1986) also summarized evidence that chromosome aberrations arise from double-strand breaks. In Chapter 2, Section II.H, we reviewed evidence that double-strand damages induced by oxidative reactions or heat may be a problem for cell survival in nature. Thus, the increase in chromosome aberrations with age noted above may have originated with endogenous double-strand breaks caused by oxidative metabolism or heat.

D. Polyploidy

Most adult mammalian liver parenchymal cells are polyploid. In rat (Alfert and Geschwind, 1958) about 90% and in mouse (Inamdar, 1958) about 85% of such cells are polyploid. It has been shown that increases in ploidy in plant and animal species reduces sensitivity to ionizing radiation (Tul'tseva and Astaurov, 1958; Sakharov *et al.*, 1960). This suggests that redundant genome copies protect against DNA damage. Gahan (1977) has proposed, for polyploid liver cells, that if one genome copy becomes damaged the cell would not suffer irreversibly because additional genome copies could cover the loss. Gahan further suggested that polyploidy in mammalian liver parenchymal cells would reduce the rate of onset of deleterious changes, thus allowing the cells to achieve a long life span.

In contrast to the idea that polyploidy may be an adaptation to protect against the effects of DNA damage, Nakanishi and Fujita (1977) proposed that polyploidy in liver cells may be a direct consequence of DNA damage. To test this idea, they administered chemicals that cause DNA cross-links to newborn rats whose liver cells were actively proliferating in a diploid state. The two chemicals tested were actinomycin D and mitomycin C. They found, with both chemicals, that the number of polyploid liver cells increased compared with an untreated control, thus supporting their hypothesis. Although it is unclear which of these two general explanations (Gahan, 1977, or Nakanishi and Fujita, 1977) is correct, both explanations include the concept that natural DNA damage is a problem for liver cells.

V. Accumulation of DNA Damage in Other Tissues

A. Kidney

Su *et al.* (1984) found that DNA damage accumulated with age in the kidney of the short-lived mouse species *M. musculus* but not in the kidney of the long-lived species *P. leucopus*. Randerath *et al.* (1989) reported an increase in the number of DNA adducts (I-compounds) present in rat and hamster kidneys as these animals age from about 1 to 10 mo.

B. Lymphocytes

Doggett *et al.* (1981), in a review of the cellular and molecular aspects of immune system aging, pointed out that lymphocytes exist in a noncycling state for years and are consequently susceptible to the types of accumulation of DNA damage found with other nondividing cells. At the same time, nondividing lymphocytes are unable to weed out damaged cells by competition, as occurs with cells that rapidly turn over. Nevertheless, lymphocytes may at any moment throughout the life span of the animal be called upon to differentiate, proliferate, and secrete products at a maximal rate in a life-or-death race with an invading infectious agent.

There have been a number of reports of an age-associated increase in spontaneous chromosomal aberrations in lymphocytes (see Williams and Dearfield, 1982, for references). Two recent studies are described here. Fenech and Morley (1985) have shown that as humans age from newborn to 82 yr spontaneous chromosome aberrations in cultured peripheral blood lymphocytes increase by approximately fourfold. The aberrations measured were acentric chromosomal fragments (termed micronuclei) that had not been incorporated into the main nuclei at cell division. Because the incidence of micronuclei is regarded as a measure of DNA damage, these results suggest that DNA damage increases with age in peripheral blood lymphocytes. Hedner *et al.* (1982) also found that the frequency of spontaneous chromosome aberrations in peripheral blood lymphocytes increases significantly with age in humans. The types of aberrations examined included chromatid and isochromatid gaps, chromatid breaks, acentric fragments, chromatid interchanges, ring chromosomes, dicentric chromosomes, and pericentric inversions. The increase in these aberrations with age may reflect an accumulation of DNA damage.

VI. *Conclusions*

Our conclusions, based on the preponderance of evidence reviewed in this chapter, are (1) postmitotic muscle and neuronal cells have a low capacity to repair DNA, and these cells accumulate DNA damage with age; (2) slowly dividing liver cells also accumulate DNA damage with age, and the rate of accumulation is greater when an animal's life span is shorter; and (3) lymphocytes and kidney cells may also accumulate DNA damage. Because DNA damages block gene transcription and are deleterious to cells (Chapter 3), we would expect DNA damage to cause a progressive decline in the function of muscle, brain, liver, kidney, and lymphocytes with age. In the next chapter, we review evidence for such functional declines.

References

Alexander, P. (1967). The role of DNA lesions in processes leading to aging in mice. *Symp. Soc. Exp. Biol.* **21,** 29–50.

Alfert, M., and Geschwind, I. I. (1958). The development of polysomaty in rat liver. *Exp. Cell Res.* **15,** 230–270.

Bergtold, D. S., and Lett, J. T. (1985). Alterations in chromosomal DNA and aging: An overview. *In* "Molecular Biology of Aging: Gene Stability and Gene Expression" (R. S. Sohol, L. S. Birnbaum, and R. G. Cutler, eds.), pp. 23–26. Raven Press, New York.

Bowman, P. D. (1985). Aging and the cell cycle *in vivo* and *in vitro. In* "Handbook of Cell Biology of Aging" (V. J. Cristofalo, ed.), pp. 117–136. CRC Press, Boca Raton, Florida.

Brooks, A. L., Mead, D. K., and Peters, R. F. (1973). Effect of aging on the frequency of metaphase chromosome aberrations in the liver of the Chinese hamster. *J. Gerontol.* **28**, 452–454.

Bryant, P. E. (1983). 9-Beta-D-arabino-furanosyladenine increases the frequency of X-ray induced chromosome abnormalities in mammalian cells. *Int. J. Radiat. Biol.* **46**, 57–65.

Bryant, P. E. (1985). Enzymatic restriction of mammalian cell DNA: Evidence for double-strand breaks as potentially lethal lesions. *Int. J. Radiat. Biol.* **48**, 55–60.

Chan, A. C., Ng, S. K. C., and Walker, I. G. (1976). Reduced DNA repair during differentiation of a myogene cell line. *J. Cell Biol.* **70**, 685–691.

Chetsanga, C. J., Boyd, V., Peterson, L., and Rushlow, K. (1975). Single-stranded regions in DNA of old mice. *Nature* **253**, 130–131.

Chetsanga, C. J., Tuttle, M., and Jacobini, A. (1976). Changes in structural integrity of heart DNA from aging mice. *Life Sci.* **18**, 1405–1412.

Chetsanga, C. J., Tuttle, M., Jacobini, A., and Johnson, C. (1977). Age-associated structural alterations in senescent mouse brain DNA. *Biochim. Biophys. Acta* **474**, 180–187.

Crowley, C., and Curtis, H. J. (1963). The development of somatic mutations in mice with age. *Proc. Natl. Acad. Sci. USA* **49**, 626–628.

Curtis, H. J. (1963). Biological mechanisms underlying the aging process. *Science* **141**, 686–694.

Curtis, H. J., and Miller, K. (1971). Chromosome aberrations in liver cells of guinea pigs. *J. Gerontol.* **26**, 292–293.

Curtis, H. J., Leith, J., and Tilley, J. (1966). Chromosome aberrations in liver cells of dogs of different ages. *J. Gerontol.* **21**, 268–270.

Dean, R. G., and Cutler, R. G. (1978). Absence of significant age-dependent increase of single stranded DNA extracted from mouse liver nuclei. *Exp. Gerontol.* **13**, 287–292.

Doggett, D. L., Chang, M.-P., Makinodan, T., and Strehler, B. L. (1981). Cellular and molecular aspects of immune system aging. *Mol. Cell. Biochem.* **37**, 137–156.

Fenech, M., and Morley, A. A. (1985). The effect of donor age on spontaneous and induced micronuclei. *Mutat. Res.* **148**, 99–105.

Finch, C. E. (1979). Susceptibility of mouse liver DNA to digestion by S_1 nuclease: Absence of age-related change. *Age* **2**, 45–46.

Fish, I., and Winnick, M. (1969). Cellular growth in various regions of developing rat brain. *Pediatr. Res.* **3**, 407–412.

Gahan, P. B. (1977). Increased levels of euploidy as a strategy against rapid aging in diploid mammalian systems: An hypothesis. *Exp. Gerontol.* **12**, 133–136.

Gaubatz, J. W. (1986). DNA damage during aging of mouse myocardium. *J. Mol. Cell. Cardiol.* **18,** 1317–1320.

Gensler, H. L. (1981a). The effect of hamster age on UV-induced unscheduled DNA synthesis in freshly isolated lung and kidney cells. *Exp. Gerontol.* **16,** 59–68.

Gensler, H. L. (1981b). Low level of UV-induced unscheduled DNA synthesis in postmitotic brain cells of hamsters: Possible relevance to aging. *Exp. Gerontol.* **16,** 199–207.

Goth, R., and Rajewsky, M. (1974). Persistence of O^6-ethylguanine in rat brain DNA: Correlation with nervous system-specific carcinogenesis by ethylnitrosourea. *Proc. Natl. Acad. Sci. USA* **71,** 639–643.

Gupta, K. P., vanGolen, K. L., Randerath, E., and Randerath, K. (1990). Age-dependent covalent DNA alterations (I-compounds) in rat liver mitochondrial DNA. *Mutat. Res.* **237,** 17–27.

Hahn, G. M., King, D., and Yang, S.-J. (1971). Quantitative changes in unscheduled DNA synthesis in rat muscle cells after differentiation. *Nature (London) New Biol.* **230,** 242–244.

Hedner, K. B., Hogstedt, B., Kolnig, A., Mark-Vendel, E., Strombeck, B., and Mitelman, F. (1982). Sister-chromatid exchanges and structural chromosome aberrations in relation to age and sex. *Hum. Genet.* **62,** 305–309.

Inamdar, N. B. (1958). Development of polyploidy in mouse liver. *J. Morph.* **103,** 65–90.

Jensen, L., and Linn, S. (1988). A reduced rate of bulky DNA adduct removal is coincident with differentiation of human neuroblastoma cells induced by nerve growth factor. *Mol. Cell. Biol.* **8,** 3964–3968.

Karran, P., Moscona, A., and Strauss, B. (1977). Developmental decline in DNA repair in neural retinal cells of chick embryos. *J. Cell. Biol.* **74,** 274–286.

Kirkwood, T. B. (1977). Evolution of aging. *Nature* **270,** 301–304.

Kirkwood, T. B. L., and Holliday, R. (1979). The evolution of aging and longevity. *Proc. R. Soc. Lond. B* **205,** 531–546.

Korr, H., and Schultz, B. (1989). Unscheduled DNA synthesis in various types of cells of the mouse brain in vivo. *Exp. Brain Res.* **74,** 573–578.

Koval, T. M., and Kaufman, S. J. (1981). Maintenance of DNA repair capacity in differentiating rat muscle cells *in vitro. Photochem. Photobiol.* **33,** 403–405.

Lampidis, T. J., and Schaiberger, G. E. (1975). Age-related loss of DNA repair synthesis in isolated rat myocardial cells. *Exp. Cell Res.* **96,** 412–416.

Lawson, T., and Stohs, S. (1985). Changes in endogenous DNA damage in aging mice in response to butylated hydroxyanisole and oltipraz. *Mech. Ageing Dev.* **30,** 179–185.

Lieberman, M., and Forbes, P. D. (1973). Demonstration of DNA repair in normal and neoplastic tissues after treatment with proximate chemical carcinogens and ultraviolet radiation. *Nature (London), New Biol.* **241**, 199–201.

Margison, G. P., and Kleihues, P. (1975). Chemical carcinogenesis in the nervous system: Preferential accumulation of O^6-methylguanine in rat brain DNA during repetitive administration of N-methyl-N-nitrosourea. *Biochem. J.* **148**, 521–525.

Massie, H. R., Baird, M. B., Nicolosi, R. J., and Samis, H. V. (1972). Changes in the structure of rat liver DNA in relation to age. *Arch. Biochem. Biophys.* **153**, 736–741.

McCombe, P., Lavin, M., and Kidson, C. (1976). Control of DNA repair linked to neuroblastoma differentiation. *Int. J. Radiat. Biol.* **29**, 523–531.

Modak, S. P., and Price, G. B. (1971). Exogenous DNA polymerase-catalyzed incorporation of deoxyribonucleotide monophosphates in nuclei of fixed mouse-brain cells. *Exp. Cell Res.* **65**, 289–296.

Mori, N., and Goto, S. (1982). Estimation of the single stranded region in the nuclear DNA of mouse tissues during aging with special reference to the brain. *Arch. Gerontol. Geriatr.* **1**, 143–150.

Mullaart, E., Boerrigter, M. E. T. I., Brouwer, A., and Berends, F. (1988). Age-dependent accumulation of alkali-labile sites in DNA of post-mitotic but not in that of mitotic rat liver cells. *Mech. Ageing Dev.* **45**, 41–49.

Mullaart, E., Boerrigter, M. E. T. I., Boer, G. J., and Vijg, J. (1990). Spontaneous DNA breaks in the rat brain during development and aging. *Mutat. Res.* **237**, 9–15.

Murthy, M. R., Bharucha, A. D., Jacob, J., and Roux-Murthy, P. K. (1976). Molecular biological models in geriatric neurobiology. *In* "Neuro-psychopharmacology" (P. Deniker, C. Radowco-Thomas, and A. Villeneuve, eds.), pp. 1615–1622. Pergamon Press, Oxford.

Nakanishi, K., and Fujita, S. (1977). Molecular mechanism of polyploidization and binucleate formation of the hepatocyte. *Cell. Struct. Funct.* **2**, 261–265.

Nakanishi, K., Shima, A., Fukuda, M., and Fujita, S. (1979). Age associated increase of single-stranded regions in the DNA of mouse brain and liver cells. *Mech. Ageing Dev.* **10**, 273–281.

Natarajan, A. T., Darroudi, F., Mullenders, L. H. F., and Meijers, M. (1986). The nature and repair of DNA lesions that lead to chromosomal aberrations induced by ionizing radiations. *Mutat. Res.* **160**, 231–236.

Niedermuller, H. (1985). DNA repair during aging. *In* "Molecular Biology of Aging: Gene Stability and Gene Expression" (R. S. Sohal, L. S. Birnbaum, and R. G. Cutler, eds.), pp. 173–193. Raven Press, New York.

Ono, T., and Okada, S. (1978). Does the capacity to rejoin radiation induced DNA breaks decline in senescent mice? *Int. J. Radiat. Biol.* **33**, 403–407.

Ono, T., Okada, S., and Sugahara, T. (1976). Comparative studies of DNA size in various tissues of mice during the aging process. *Exp. Gerontol.* **11**, 127–132.

Park, J.-W., and Ames, B. N. (1988a). 7-Methylguanine adducts in DNA are normally present at high levels and increase on aging: Analysis by HPLC with electrochemical detection. *Proc. Natl. Acad. Sci. USA* **85**, 7467–7470.

Park, J.-W., and Ames, B. N. (1988b). Correction. *Proc. Natl. Acad. Sci. USA* **85**, 9508.

Price, G. B., Modak, S. P., and Makinodan, T. (1971). Age-associated changes in the DNA of mouse tissue. *Science* **171**, 917–920.

Randerath, K., Liehr, J. G., Gladek, A., and Randerath, E. (1989). Age-dependent covalent DNA alterations (I-compounds), in rodent tissues: Species, tissue and sex specificities. *Mutat. Res.* **219**, 121–133.

Sakharov, V. V., Mansurova, V. V., Platonova, R. N., and Shcherbakov, V. K. (1960). Detection of physiological protection against ionizing radiations in the autotetraploids of seed buckwheat. *Biophysics (Biofizika)* **5**, 632–641.

Slatkin, D. N., Friedman, L., Irsa, A. P., and Micca, P. L. (1985). The stability of DNA in human cerebellar neurons. *Science* **228**, 1002–1004.

Sharma, R. C., and Yamamoto, O. (1980). Base modification in adult animal liver DNA and similarity to radiation-induced base modification. *Biochem. Biophys. Res. Commun.* **96**, 662–671.

Stevenson, K. G., and Curtis, H. J. (1961). Chromosome aberrations in irradiated and nitrogen mustard treated mice. *Radiat. Res.* **15**, 774–784.

Stockdale, F. E. (1971). DNA synthesis in differentiating skeletal muscle cells: Initiation by ultraviolet light. *Science* **171**, 1145–1147.

Su, C. M., Brash, D. E., Turturro, A., and Hart, R. W. (1984). Longevity-dependent organ-specific accumulation of DNA damage in two closely related murine species. *Mech. Ageing Dev.* **27**, 239–247.

Tul'tseva, N. M., and Astaurov, B. L. (1958). Increased radioresistance of *Bombyx Mori L.* polyploids and the general theory of the biological action of ionizing radiation. *Biophysics (Biofizika)* **3**, 183–197.

Waser, J., Hubscher, U., Kuenzle, C. C., and Spadari, S. (1979). DNA polymerase beta from brain neurons in a repair enzyme. *Eur. J. Biochem.* **97**, 361–368.

Wheeler, K. T., and Lett, J. T. (1974). On the possibility that DNA repair is related to age in non-dividing cells. *Proc. Natl. Acad. Sci. USA* **71**, 1862–1865.

Williams, J. R., and Dearfield, K. L. (1982). DNA damage and repair in aging mammals. *In* "Handbook Series in Aging, Section D: Biological Sciences, Vol. I, Biochemistry" (J. R. Florin and R. C. Adelman, eds.), pp. 25–48. CRC Press, Boca Raton, Florida.

Woodhead, A. D., Merry, B. J., Cao, E., Holehan, A. M., Grist, E., and Carlson (1985). Levels of O^6-methylguanine acceptor protein in tissues of rats and their relationship to carcinogenicity and aging. *J. Natl. Cancer Inst.* **75,** 1141–1145.

Zahn, R. K., Reinmuller, J., Beyer, R., and Pondeljak, V. (1987). Age-correlated DNA damage in human muscle tissue. *Mech. Ageing Dev.* **41,** 73–114.

Chapter Five

The DNA Damage Hypothesis
of Aging Applied to Mammals

In this chapter, we briefly discuss some theoretical considerations in support of the hypothesis that aging in mammals is caused by DNA damage. We then summarize numerous experimental studies in mammals that are consistent with the expectations of this hypothesis. By the DNA damage hypothesis of aging, an accumulation of DNA damage is expected to lead to reduced RNA transcription. This, in turn, should cause a reduction in protein synthesis. If synthesis of essential proteins is inhibited, then cells should lose function and/or viability. Reduced function and/or viability of cells in any tissue should result in loss of the higher-order function of that tissue. We review experimental evidence for each of these points (reduced RNA transcription, protein synthesis, cell viability, tissue function) in the nondividing cells of the brain and muscle, and in infrequently dividing cells of the liver, kidney, and lymphocytes. This evidence is consistent with a cause and effect relationship between DNA damage and the progressive impairment of function that defines aging. We also discuss strategies employed in some tissues to compensate for the effects of DNA damage.

I. Theoretical Considerations in Support of DNA Damage as the Cause of Aging in Mammals

Aging in mammals can be discussed from three points of view. At the molecular level, we can ask what types of changes in biomolecules would be expected to be progressive, irreversible, and in

comformity with observed phenomena of aging. From an evolutionary or population genetics point of view, we can ask about the
balance of selective forces favoring investment in damage repair
versus investment in reproductive ability. From a physiological perspective, we can ask which cell types should show aging if DNA
damage is the basic cause of aging.

A. Molecular Argument for DNA Damage as the Basis of Aging

The concept that it is damage to DNA, rather than damage to
other biomolecules or structures, that causes aging is based on the
following considerations. Because DNA contains the master code,
loss of its ability to be transcribed for vital genes would be lethal to
the cell. By contrast, other macromolecules (e.g., RNA, proteins) or
supermolecular structures (e.g., membranes, ribosomes) should, in
principle, be replaceable by new synthesis if the DNA remains functional. DNA is particularly vulnerable because it is ordinarily present
in only two copies per diploid cell (compared with the multiple
redundant copies of other macromolecules), and each of the many
nucleotide pairs making up each gene is a potential target of damage.

B. Evolutionary Considerations in Aging

Kirkwood (1984) approached the question of the absolute efficiency of repair from an evolutionary perspective. He concluded by a
mathematical modeling approach that the strategy of maximum fitness is one that balances maintenance of the soma against reproductive effort in such a way that the investment of resources in maintenance and repair of the soma is always less than what is required for
indefinite survival. On this line of reasoning, it should not be cost-
effective to evolve repair processes in the somatic line that approach
100% efficiency. Thus, some structures, such as DNA, would be
expected to accumulate damages.

C. Expectations from Cell Physiology

If cellular aging is the major determinant of organismal aging,
which types of cells in a multicellular organism are responsible?
Consider the expected effects of unrepaired DNA damages in three

types of cells: (1) long-lived postmitotic cells, (2) rapidly dividing cells, and (3) infrequently dividing cells. In differentiated postmitotic cells, if DNA damages are not repaired with 100% efficiency, they will accumulate with time. Such cells should then experience progressive impairments of function; i.e., they should age. Apparent examples of such types of cells are neurons and muscle myocytes.

In a rapidly dividing population of cells, if the rate of occurrence of unrepaired lethal damage is low compared with cell generation time (<0.5 unrepaired lethal DNA damages/cell/generation), then the damaged cells can be replaced by the duplication of undamaged ones. Populations of such cells, in principle, may not experience the progressive impairments of function that define aging. Apparent examples of such cells are duodenum and colon epithelial cells, hemopoietic cells of the bone marrow, and lymphopoietic cells of the thymus and spleen.

We would expect slowly dividing cells to have characteristics falling in between the two extreme types. There should be a spectrum of types, with those cells having the longest generation times being most like postmitotic cells. As one example, liver hepatocytes have a long generation time, and they accumulate DNA damage (see Chapter 4, Section IV.B).

II. Experimental Evidence in Support of the Hypothesis that DNA Damage is the Cause of Aging in Mammals

A. Reduced Transcription and/or Protein Synthesis

In Chapter 3, we reviewed evidence that DNA damage can interfere with transcription and cause cell death. In Chapter 4, we presented evidence that DNA damages accumulate with age in nondividing and in infrequently dividing cells such as those in brain, muscle and, liver. We expect that the DNA damages in these cells should cause reduced transcription with age. If gene transcription is reduced, this should lead to reduced protein synthesis with age. Most of the data available from postmitotic and infrequently dividing cells fit these expectations.

Observations of reduced transcription and reduced protein synthesis have either been made with regard to overall synthesis or with regard to specific products. The studies in which overall decreases in transcription were reported are discussed below and summarized in Table VIII. The studies in which decreases in the transcription and protein synthesis of particular products were reported are also discussed below and summarized in Table IX. The studies described below are divided, for discussion purposes, by different tissue types.

   1. *Reduced Transcription and/or Protein Synthesis in Brain*
Brain cell DNA is actively transcribed. Brown and Church (1971) isolated unique (nonrepetitive) DNA base sequences by hydroxylap-

Table VIII
Change in General Transcription in Animal Tissues with Age

Animal	Tissue	Change in Transcription	Reference
Rat	liver nuclei	transcription declines 90% from age 1 to 30 mo	Devi *et al.*, 1966
Mouse	liver and muscle nuclei	transcription declines at least 50% from age 8–9 to 25–30 mo	Britton *et al.*, 1972
Mouse	brain and liver	decrease in transcription of unique DNA sequences	Cutler, 1975
Rat	liver nuclei	transcription decreases 1.6- to 2.7-fold from age 6 to 31 mo	Castle *et al.*, 1978
Rat	liver nuclei	transcription decreases 75% from age 6 to 24 mo	Bolla and Denckla, 1979
Mouse	brain cortex, epithelial cells of choroid plexus, and liver hepatocytes	decline in transcription from age 3 to 18–24 mo; 31% decline for brain	Fog and Pakkenberg, 1981
Rat	skeletal muscle	decline in heavy polyribosomes between ages 2 and 24 mo	Pluskal *et al.*, 1984
Rat	brain cortex	transcription of mRNA decreases 30–40% between ages 1.5 and 26 mo	Zs-Nagy and Semsei, 1984
Human	brain	mRNA from older individuals less efficient in stimulating protein synthesis	Sajdel-Sulkowska and Marotta, 1985
Rat	liver nuclei and hepatocytes	transcription declines 25–42% in nuclei and 41–49% in hepatocytes from age 12 to 22–26 mo	Park and Buetow, 1990

Table IX

Changes in Transcription of Specific Genes in Animal Tissues with Age

Animal	Tissue	Change in Transcription	Reference
Rabbit	liver	cytochrome P450 and its specific mRNA decline from age 0 to 830 days	Dilella *et al.*, 1982
Male rat	liver	gradual decline and ultimate loss of α-globulin and its mRNA from age 20 to 800 days	Roy *et al.*, 1983
Rat	liver	increase in albumin mRNA synthesis correlated with increase in albumin synthesis with age	Horbach *et al.*, 1984
Rat	liver	decline in mRNA for α-globulin (98%), aldolase (30%), and cytochrome P450 (50%); 30% increase in albumin mRNA as rats age from 6 to 29 mo; changes in proteins parallel changes in mRNA	Richardson *et al.*, 1985b
Rat	spleen lymphocytes	induction of interleukin-2 and its mRNA declines as rats age from 6 to 24 mo	Richardson *et al.*, 1985b
Mouse	submandibular glands	epidermal growth factor protein and mRNA decline 50 and 75% from age 12 to 26–28 mo	Gresik *et al.*, 1986
Rat	liver	tyrosine transferase and its mRNA declines about 50% from age 10 to 25 mo; tryptophan oxygenase and its mRNA do not decline	Wellinger and Guigoz, 1986
Rat	liver	decline in mRNA for catalase (30%), superoxide dismutase (30%), and metallothionein (50%) as rats age from 6 to 37 mo	Semsei and Richardson, 1986
Rat	lymphocytes	induction of interleukin-2 mRNA decreases 85% between 5 and 29 mo and parallels a decline in interleukin-2 production	Wu *et al.*, 1986

atite chromatography from tritium-labeled mouse DNA. Total RNA purified from various tissues of adult mice was hybridized to the unique DNA, and the DNA–RNA hybrids analyzed on hydroxylapatite. Ten percent of the unique DNA was complementary to brain RNA while less than 2% hybridized to RNA from liver, kidney, and spleen. This indicates that the spectrum of RNA molecules present in adult brain includes transcripts from five times as many unique DNA sequences as in adult liver, kidney, or spleen. Similar results were obtained by Hahn and Laird (1971). Therefore, brain cell function should be especially vulnerable to DNA damages that interfere with transcription.

Cutler (1975) found that after sexual maturity in the mouse, transcription of unique DNA sequences in brain and liver decreased. Sajdel-Sulkowska and Marotta (1985) examined human postmortem brains for intact functional mRNA. They found a tendency for mRNA from older individuals to stimulate protein synthesis less efficiently than younger brain mRNA. Zs-Nagy and Semsei (1984) measured the *in vivo* incorporation of [³H]uridine into total RNA and poly(A)⁺ RNA (mRNA) of the brain cortex of female CYF rats. The incorporation of radioactivity into total RNA and poly(A)⁺RNA decreased 30–40% between 1.5 and 26 mo of age. Fog and Pakkenberg (1981) studied the uptake of the nucleic acid precursor [³H]uridine in different mouse tissues with age. Newborn, young (3 mo) and old (18–24 mo) mice were given [³H]uridine orally. Uptake was studied autoradiographically in nerve cells of the fifth layer of the parietal cortex, the epithelial cells of the choroid plexus, the hepatocytes of the liver, and the epithelium of the small intestine. Significant decreases were found in the old mice in the first three tissues. The decline in [³H]uridine uptake in nerve cells of the cortex was 31% in old versus young mice. This decline indicates reduced RNA synthesis with age.

Lower transcription levels should result in decreased brain protein synthesis. Dwyer *et al.* (1980) found that brain protein synthesis significantly declined in forebrain, cerebellum, and brain stem when young mature (3 mo) rats were compared with old (22.5 mo) rats. The incorporation of (³H)L-lysine into forebrain protein was reduced 11% in 10.5-mo-old rats compared with 3-mo-old rats. A further reduction of 9% occurred between 16.5 and 22.5 mo. These results suggested that a significant and progressive decline in the rate of brain protein synthesis occurs during aging. Fando *et al.* (1980) studied brain protein synthesis *in vivo*, in brain slices, and in cell-free systems in rats aged 1, 16, and 24 mo. They observed a highly significant reduction in amino acid incorporation with advancing age in all three types of assay.

Ingvar *et al.* (1985) studied protein synthesis in the whole rat brain as well as in 39 regions of the brain by means of a quantitative autoradiographic method. A decline of protein synthesis of 17% was found in whole rat brain as the animals aged from 6 to 23 mo. In addition, a significant decrease in protein synthesis was observed with increasing age in 17 of the 39 regions studied. The measured protein synthesis could reflect immediate functional demands of cerebral structures as well as the longer-term processes necessary for structural maintenance within the nervous system. The regional declines in protein synthesis may reflect a decrease in the number of cells of the region and/or an atrophy of existing cells. Either neurons or glial cells could have been affected. Ekstrom *et al.* (1980) found a 56% decline in protein synthesis of rat brain as they aged from 6 to 32 mo. Suzuki *et al.* (1964) found that the amino acid incorporation activities of microsomal preparations among five human subjects were more active in the younger (33 and 34 yr) subjects than in older (54, 62, and 70 yr) ones.

There have been a number of reports of reduced capacity to synthesize specific products in the brain of aging mammals. Meier-Ruge *et al.* (1976) found a decline in acetylcholinesterase activity in aging rat brain. Perry *et al.* (1977) have reported a decline in choline acetyltransferase (which catalyzes synthesis of acetylcholine) in the hippocampus of the aging brain. Glucose oxidation in the brain of rats declines about 40% as they age from 3 mo to 2 yr (Patel, 1977).

2. Reduced Transcription and/or Protein Synthesis in Muscle Pluskal *et al.* (1984) examined rat skeletal muscle. They found a decline in the content of heavy polyribosomes in muscle cells as the rats aged from 2 to 24 mo. They also found a 40% reduction in protein synthesis as the animals aged. The decline in polyribosome content may represent blockages of transcription by DNA damage. Sonntag *et al.* (1985) showed that protein synthesis in rat diaphragm muscle declined 26% as rats aged from 3 to 21 mo. Crie *et al.* (1981) showed that protein synthesis in rat heart declined 46% as the rats aged from 12 to 24 mo. Furthermore, Crie *et al.* (1981) listed nine studies performed between 1965 and 1980 in which skeletal muscle protein synthesis and protein degradation declined with age as well as three studies in which no decline was found. Makrides (1983) listed 10 studies on protein synthesis in rat or mouse heart muscle: In seven of these protein synthesis decline with age, in one it increased, and in two no change was reported.

Specific enzymes in muscle have also been shown to decline with age. For example, Bass *et al.* (1975) reported a decrease in the glycolytic activity of the fast extensor digitorum longus, in the form of reduced triosephosphate dehydrogenase, lactate dehydrogenase, and glycerol-3-phosphate dehydrogenase activity with age. The slow soleus showed a decrease in TPDH and a decrease in the aerobic enzymes malate dehydrogenase and citrate synthase.

3. Reduced Transcription and/or Protein Synthesis in Liver

About 60% of the cells of the liver are hepatocytes. As mentioned previously (Chapter 4, Section IV.B), liver hepatocytes can divide, but their turnover time is long (400–450 days in the rat) (Bowman, 1985). The DNA synthetic index (the percentage of cells that are labeled after a single injection of [3H]thymidine) in rat liver declines with age (Post and Hoffman, 1964). After birth there is an exponential decline in the number of rat liver cells engaged in duplication, along with a lengthening of the replication time. After 24 wk of age in the rat, when full growth has occurred, there is little cell division, even though the organ retains this capacity when challenged by injury.

Evidence for the accumulation of DNA damage in liver was summarized in Chapter 4, Section IV.B. An increase in DNA damage should cause transcription to decline. Most experimental evidence available is consistent with this expectation (see also Tables VIII and IX for summary).

Several studies have shown that RNA synthesis by liver nuclei isolated from various strains of senescent (24–31 mo) rats or mice (25–30 mo) is decreased compared, respectively, with nuclei isolated from young or adult rats or mice (Bolla and Denckla, 1979; Britton *et al.*, 1972; Castle *et al.*, 1978; Devi *et al.*, 1966). Park and Buetow (1990a) showed that RNA synthesis in nuclei isolated from senescent (22–26 mo) female Wistar rat liver decreased compared with nuclei from adult (12 mo) liver. The rate of mRNA and rRNA synthesis decreased to 25–42% in the senescent nuclei. A similar result was obtained when isolated intact rat hepatocytes were used instead of hepatocyte nuclei (Park and Buetow, 1990b). In this case, RNA synthesis declined to 41–49% in senescent (25 mo) compared with adult (12 mo) female Wistar rats.

Semsei and Richardson (1986) measured the levels of mRNA species in rat liver coding for proteins involved in the detoxification of free radicals (e.g., catalase, superoxide dismutase [SOD], metallo-thionein [MT]) RNA was isolated from liver tissue of 6–37-mo-old rats and the levels of catalase, SOD, and MT mRNA were determined by dot-blot hybridization using complementary DNA probes to these genes. Catalase and SOD mRNA levels decreased 30% by age 37 mo, while MT mRNA decreased 50% with age.

Roy *et al.* (1983) found that α-globulin and its mRNA declined gradually and were ultimately lost in the male rat liver during aging. Similarly, Dilella *et al.* (1982) found that cytochrome P450 and its specific mRNA declined in aging rabbit liver. However, Horbach *et al.* (1984) found an increase, rather than a decrease, in rat liver albumin mRNA, which correlated with an increase in synthesis of albumin. The explanation for this increase is unknown. Richardson *et al.* (1985b) showed that as rats aged from 6 to 29 mo the liver mRNAs for α-globulin, aldolase, and cytochrome P450 declined by 98%, 30%, and 50%, respectively; however, albumin mRNA increased about 30%. These experiments also indicated that the age-related changes in the levels of the specific mRNA species were similar to changes in the proteins coded for by these mRNAs. For example, the levels of albumin mRNA and the synthesis of albumin by hepatocytes increase after 24 mo of age, the age-related decrease in α-globulin mRNA was similar to the decrease in the content of α-globulin protein in the liver, and the decrease in cytochrome P450 appeared to be paralleled by a decrease in cytochrome P450 mRNA.

Wellinger and Guigoz (1986) studied the expression of the genes coding for tyrosine aminotransferase and tryptophan oxygenase with age in rat liver. These enzymes are induced in response to cold stress. The induction of tyrosine aminotransferase activity in young (10 mo) rats was about twice that observed with old (25 mo) rats. This difference between the two age groups was also observed when the steady-state level of tyrosine aminotransferase mRNA was measured by hybridization with a specific DNA probe. In contrast to the results with tyrosine aminotransferase, induced tryptophan oxygenase enzyme and mRNA levels did not show an age-dependent change. In another study, Birchenall-Sparks *et al.* (1985) found that over-all protein synthesis declined 55% as rats aged from 2.5 to 19 mo. (Some further references to other studies showing changes in protein synthesis with age in liver are given in this chapter, Section II.A.5.)

4. Reduced Transcription and/or Protein Synthesis in Lymphocytes, Parotid and Submandibular Glands, and Kidney As noted in Chapter 4, Section V.B, lymphocytes may exist *in vivo* in a nondividing state for years until stimulated to differentiate and proliferate. Interleukin 2 (IL-2) plays a very important role in lymphocyte proliferation and, thus, in the immune response. Richardson *et al.* (1985b) measured the induction of IL-2 expression by spleen lymphocytes stimulated with the mitogen concanavalin A. The induction of IL-2 and IL-2 mRNA by concanavalin A was approximately 80% lower in lymphocytes from 29-mo-old rats than for lymphocytes from 6-mo-old rats. Similar results were obtained by Wu *et al.* (1986). In humans, Tollefsbol and Cohen (1985) showed that protein synthesis in lymphocytes declines 50% from age 28 to 78 yr.

Gresik *et al.* (1986) compared the levels of epidermal growth factor (EGF) protein and EGF mRNA in the submandibular glands of mature (12 mo) and senescent (26–28 mo) male mice. The submandibular glands of the older mice contained approximately 50% less EGF and 75% less EGF mRNA than those of 12-mo-old male mice. With advancing age, there was a 20% reduction in the absolute volume of the granular convoluted tubule compartment, which is the exclusive site of EGF and EGF mRNA in the gland.

Kim and Arisumi (1985) measured age-related changes in the rate of synthesis of total proteins and the specific secretory protein amylase in the parotid glands of young (2 mo) and old (24 mo) rats. The difference in the rate of incorporation of radioactive leucine into acid insoluble proteins of the gland indicated that the rate of total protein synthesis declined by about 30% with age. The rate of synthesis of amylase was found to decline by a similar amount.

In Chapter 4, Section V.A, we described evidence for accumulation of DNA damage in mouse kidney with age. This would be expected to result in decreased gene expression. Ricketts *et al.* (1985) found that the rate of total protein synthesis by suspensions of rat kidney cells declined 60% between 4 and 31 mo of age. The decline in protein synthesis was accompanied by an increase in urinary protein excretion, which was interpreted as a loss of kidney function.

5. Other Reviews Indicating That Reduction of Transcription and/or Protein Synthesis with Age Makrides (1983) reviewed numerous studies on changes in protein synthesis with age in specific tissues. The data on muscle and brain were most consistent in indicating a decline in protein synthesis with age. Richardson *et al.* (1985a) summarized 14 studies performed between 1981 and 1984 on the effect of age on protein synthesis. Richardson and Semsei (1987) summarized a further 20 studies, including some on plants and invertebrates, performed between 1980 and 1986. Some of these studies are described above. Their data indicate that age-related changes in the expression of a specific protein can be correlated with changes in the level of the mRNA coding for the protein. Thus, they thought that the action of aging on gene expression occurs primarily at the level of transcription. In aggregate, the two summaries of Richardson and coworkers indicated that 11 studies found a decline in protein synthesis in muscle, 2 found a decline in brain, and 9 found a decline in liver. Other tissues or cell types showing a decline were lymphocytes, parotid and submaxillary glands, and kidney. Based on their overview of the data, Richardson and Semsei (1987) suggested that an age-related decline in protein synthesis is a universal phenomenon.

B. Reduction of Cell Viability and Loss of Tissue Function in Cells That Accumulate DNA Damage

1. *Loss of Cell Viability and Tissue Function in Brain*   Neurons may die when transcription of an essential gene is prevented. Sabel and Stein (1981) examined the number and size of neurons in subcortical structures in groups of rats that were 90 and 880 days of age. Their analysis revealed a brain shrinkage of 10% as well as neuronal loss in six of the seven regions examined. In these six regions, the percentage of neurons remaining in the oldest rats compared with the youngest rats varied from 32 to 53%. Significant neuronal shrinkage was also observed in three regions. This atrophy could be responsible for alterations in the animals' emotional and physiological functions.

Brizzee (1984) summarized 11 different studies in humans and 3 in rats that showed loss of neurons in various portions of the brain with age. He noted that as the number of brain cells decreases with age, signal strength may be reduced and the noise level might increase as the number of nerve cells and synapses available for progressive refinement of the signals received is reduced.

Johnson (1985) compiled an extensive tabulation of studies

showing nerve cell and fiber loss with aging. The studies were arranged in groups by type of physical change, part of the nervous system examined, and species. In 17 groups of studies, a loss of neurons was observed, and in 6 groups of studies no loss was observed. Most of the studies involved humans, but other mammalian species were included as well. Buetow (1985) also compiled an extensive summary of studies on the numbers of fibers versus age in the nervous system. He concluded that, for humans, neuronal loss occurs usually by age 60 to 70 yr in the spinal nerves and in some but not all cranial nerves.

Evidence for extensive loss of dendritic systems have been reported by Scheibel *et al.* (1975). Progressive loss of dendrite elements would be expected eventually to diminish the total range and number of cortical output patterns. Intracortical modulatory loops would be lost in increasing measure. Such losses might not initially be obvious due to the enormous redundancy of the systems involved. Eventually, the total amount of deficient dendrite elements would begin to manifest itself in signs of impoverished output, such as decreasing motor strength, lack of dexterity and agility, failing cognitive performance, and problems in association, retrieval, recall, etc.

Glick and Bondareff (1979) compared the numbers of synapses in the cerebellar cortex of adult (12 mo of age) and senescent (25 mo) male rats. The total number of axodendritic synapses was found to be 24% lower in the senescent rats as compared with adults. Synaptic loss was postulated to have a role in the age-related decline in brain function.

Johnson (1985) compiled evidence indicating that the following physical changes are characteristic of the aging brain.

1. Shrinkage of the brain and loss of total brain weight.

2. Nerve cell loss and disorganization of cortical and nuclear cellular patterns.

3. Shrinkage and "sclerosis" of the individual nerve cell, including shrinkage of dendritic arborizations.

4. Irregularities of axons such as torpedo like swellings, spheroidal swellings, and changes in the myelin sheaths.

5. Accumulation of intraneuronal (cytoplasmic) lipofuscin (yellowish to brown granular material composed partly of oxidized lipids).

6. Presence of neurofibrillary tangles or changes.

7. Presence of neuritic (senile) plaques. (The senile or neuritic plaque is a conglomerate lesion with a central core of an abnormal, extracellular, fibrillar protein known as amyloid.)

8. Changes in intracellular organelles and cytoplasm of neurons, including Golgi apparatus, mitochondria, nucleus, and endoplasmic reticulum.

9. Changes in neuroglia and the neuronal microenvironment.

10. Changes in brain blood vessels, these changes being more specifically correlated with aging as opposed to ordinary arteriosclerosis.

11. Changes in afferent and efferent nerve terminals outside the central nervous system (e.g., a decline in the numbers of taste buds, olfactory cells, and nerve fibers).

Among the many infirmities of old age in humans, the decline of brain function may have the most significant consequences. Cognitive dysfunction impairs the ability to reason and remember. Brain biochemistry is at the center of efforts to understand and ameliorate the difficulties of advanced age.

The brain regulates many organ, tissue, and cell functions by means of hormones and nerves. Neurohormones are secreted by the hypothalamus of the brain. For instance, the adrenal cortical hormone stimulates the adrenal cortex to synthesize and secrete cortisol and 17-ketosteroids; thyrotropin stimulates the release of iodothyronines by the thyroid; follicle-stimulating hormone stimulates growth of the ovarian follicle; and luteinizing hormone is synergistic with FSH in promoting the maturation of follicles and secretion of estrogen (Hymer, 1979). Each of these target endocrine glands in turn secretes its respective hormone(s), which are active in additional tissues and in negative feedback loops with the hypothalamus, pituitary gland, or both. By means of such interactions, changes in the aging brain could trigger sequential events throughout the body.

Homeostatic regulation is the capacity to sustain a constant internal equilibrium in spite of environmental fluctuations. The maintenance of homeostatic regulation is one means of measuring the activity of nerve tissue feedback systems. Homeostatic control systems may remain adequate under basal or resting conditions, but with age they may become inadequate when confronted with an environmental challenge. Examples are thermoregulation (Krag and Kountz, 1950; Pickering, 1936), regulation of acid-base equilibrium in the blood (Adler *et al.*, 1968), and regulation of blood glucose levels (Silverstone *et al.*, 1957). Some gerontologists view the progressive decline in homeostatic capacity of the body, a decline that reflects impairment of interaction between brain and other tissues, as central to the aging process (Timiras, 1972; Shock, 1977). This decline in homeostatic capacity leads to death when the body is unable to recover from an insurmountable change in the internal equilibrium. This view is consistent with the presence of multiple age-associated disabilities in aging humans (Burek, 1978). As pointed out by Hazzard and Bierman (1978: p. 237), it may be difficult to assign a specific cause of death in the elderly because most aged patients carry multiple diagnoses in their medical charts, and additional problems lurk just beneath the clinical horizon at any time. Therefore, death of an aged person may seem almost random, regardless of its proximate cause. Furthermore, Timiras (1972: p. 543) commented that we are largely ignorant of the actual cause of death in normal old age, although failure of homeostatic mechanism appears to be the most probable candidate.

2. *Loss of Cell Viability and Tissue Function in Muscle* Although loss of brain function may be most central to aging, the generalized atrophy of many tissues during senescence is most evident in skeletal muscle because it is the largest single tissue of the body of the mammal. Skeletal muscles consist of long, thin muscle fibers, each of which is a single unusually large cell formed by the fusion of many separate cells. When muscle fibers of aged animals are examined with the electron microscope, ultrastructural disorganization is evident. These ultrastructural changes have been reported

in a number of studies, which have been summarized by Goldspink and Alnaqeeb (1985). Structural changes in both the muscle fiber and the neuromuscular junction correspond well with the reduced function of the locomotor system. In addition to structural changes in the muscle fibers, a loss of muscle fibers during aging has been reported for different muscles including the rat soleus (Gutmann *et al.*, 1968) and the rat extensor digitorum longus muscle (Alnaqeeb and Goldspink, 1980). Lexell *et al.* (1983) studied the effects of aging in humans on the total number and size of muscle fibers. The two groups of humans studied at autopsy had been physically healthy males. The first group of six individuals had a mean age of 72 yr and the second group a mean age of 30 yr. The size of the muscles of the older individuals was 18% smaller and the total number of fibers was 25% lower than those of the young individuals. In general then, the various changes at the cellular level may account for the most apparent physiological changes associated with aging in muscle: a decline in strength and a slowing down of the contractile process (Goldspink and Alnaqeeb, 1985; Florini, 1987).

3. *Loss of Cell Viability and Tissue Function in Liver* Tauchi and Sato (1978) reviewed evidence that in the aging liver the number of hepatic cells decreases. In humans, the number of hepatic cells begins to decrease in the sixth decade, decreases significantly in the eighth decade, and markedly thereafter. Dice and Goff (1986) summarized the age-related changes in human liver morphology and function: decreased weight, decreased blood flow, delayed regeneration, decreased responsiveness to certain hormones, altered metabolism of certain drugs, increased proliferation of bile ducts, and increased fibrosis. These authors considered that the liver may be somewhat protected from the effects of aging because of its ability to regenerate. Nevertheless, they concluded that the molecular explanations for age-related alterations in hepatocytes probably apply to other cell types that have little or no ability to regenerate (e.g., muscle, brain).

4. *Loss of Function in the Immune System* Doggett *et al.* (1981) have reviewed the aging of the mammalian immune system with emphasis on the cellular and molecular levels. Immune system functional decline is a prominent feature of aging. There is an age-

related decline of the regulatory control of the system. According to Doggett *et al.* (1981), this loosening of regulatory controls results in a decrease in the suppression of autoreactivity and monoclonal gammopathy as well as a decrease in the activation of the differentiation necessary for a primary response. For further discussion, see Chapter 15, Section I.F.

C. Overview

In Chapter 3, we presented evidence for a causal connection between DNA damage, interference with transcription and replication, and cell death. In Chapter 4, Sections II and III, and in this chapter we reviewed evidence that in the brain the level of DNA repair is low, endogenous DNA damages accumulate with age, mRNA synthesis declines, and general protein synthesis as well as the synthesis of critical specific proteins is reduced. Cell loss occurs, tissue function declines, and functional impairments occur, which are directly related to the central processes of aging. For the brain, a direct cause and effect relationship between the accumulation of DNA damage and decline in function is a reasonable explanation for the evidence presented above.

We have similarly reviewed evidence in Chapter 4, Sections II and III, and in this chapter indicating that in postmitotic muscle cells the level of DNA repair is low, DNA damage accumulates with age, transcription decreases, general protein synthesis as well as synthesis of some specific proteins declines, cellular structure deteriorates, and cells die. These declines parallel a reduction in the strength of muscles and speed of contraction. Again, a direct cause and effect relationship from DNA damage to muscular performance has not yet been proven but does provide a reasonable explanation of the evidence.

In addition, we have reviewed the evidence from studies on liver that indicate that DNA damage accumulates with age (Chapter 4, Section IV). As reviewed in this chapter, this DNA damage may interfere with transcription, leading to a decline in the expression of most hepatic genes, cell loss, and degenerative changes in liver morphology and function.

Evidence for accumulation of DNA damage in lymphocytes is reviewed in Chapter 4, Section V.B, and in this chapter we reviewed evidence for a decline in gene expression and function in the immune system, which may reflect this damage accumulation.

III. Mechanisms of Compensating for DNA Damage

A. Cellular Redundancy in Brain May Compensate for DNA Damage

One means of coping with unrepaired DNA damages may be cellular redundancy (Gensler and Bernstein, 1981). More cells occur in the brain than are necessary for normal functioning. For instance, Smith and Sugar (1975) reported the results of a comprehensive neuropsychological examination of a patient in his twenties who had the left hemisphere of his brain removed for seizures as a 5.5-yr-old boy. This patient had above-normal adult language and intellectual capabilities. Glassman (1987) noted that the phenomenon of behavioral resistance to massive brain damage indicates that there is redundancy in neural tissue. He hypothesized that a larger brain provides long-term reliability or insurance against occasional small errors that could prove fatal in a challenging environment. A theoretical investigation by Glassman, based on this hypothesis, led to the conclusion that the human brain may be at least twice as large as it would have to be for short-term survival.

Hoffman (1983) reviewed evidence that the maximum potential life span of a mammal is proportional to the product of its relative brain size (encephalization) and the reciprocal of its metabolic rate per unit weight. (The inverse relationship to metabolic rate is discussed in Chapter 6, Section III.A.) The correlation of longevity with relative brain size may reflect the selective advantage of a reserve supply of neurons to compensate for the loss of neuron function resulting from the accumulation of DNA damage. Only after the reserve supply is depleted would the remaining number of functional cells start to become inadequate under resting conditions. However, a partial depletion of the reserve capacity could become evident under stress. The age-associated loss of the ability to maintain physiological homeostasis under stress is consistent with the idea of a loss of reserve capacity (Timiras, 1972) due to a gradual decline in the total number of functional neurons with age.

B. Rapidly Dividing Cell Populations

In contrast to postmitotic muscle and neuronal cell populations, which decline with age, some rapidly dividing cell populations do not appear to decline. According to the extensive data summarized by Buetow (1985), bone marrow and hemopoietic cells do not decline in numbers with age in the mouse and guinea pig. Mori *et al.* (1986) studied hemopoietic stem cells in young and old people ranging in age from 28 to 95 yr. They found no difference in the concentration of granulocyte–macrophage progenitor cells, although they did find a significant decline in the concentration of erythrocyte progenitor cells in elderly people. The turnover time for replacing the hemopoietic bone marrow cells of the mouse is only 1–2 days (Bowman, 1985). Harrison (1979) has reviewed evidence that in mice erythropoietic stem cells have a very large capacity for self-renewal. No significant differences were found when comparing erythrocyte production by marrow stem cell lines from old and young adults. This suggested that little or none of erythropoietic stem cell proliferative capacity is exhausted in the mouse by a lifetime of normal functioning.

Cell renewal in the epithelial lining of the mammalian small intestine is restricted to the crypts where production of new cells is continuous (Leblond and Stevens, 1948). Miquel *et al.* (1979) noted that most rapidly dividing cells of the crypts are free of the age pigment lipofuscin and have normal fine structure even in old mice. However, as mice age generation time increases and the number of cycling cells in the crypts decreases (Lesher and Sacher, 1968). The factor(s) responsible for this decline are not known.

Earlier in this chapter (Section I.C), we proposed that rapidly dividing cell populations may cope with DNA damage by replacing lethally damaged cells through the replication of undamaged cells. The evidence reviewed in this section suggests that at least some proliferating cell populations can cope in this way. The observation that some dividing cell populations eventually decline (e.g., intestinal crypt cells) may mean that, over the long run, the rate of cell replacement by duplication in these populations is insufficient to keep up with the rate of DNA damage.

IV. Summary of the Stages of Aging Leading to Death in Mammals

From the perspective of the DNA damage hypothesis and the evidence reviewed so far, the aging process in mammals may proceed by the following steps:

1. Endogenous DNA damage occurs at a specific rate.

2. Those cells whose rate of repair is lower than the incident rate of DNA damage accumulate damages.

3. These damages inhibit transcription and replication.

4. Postmitotic and slowly dividing cell populations accumulate DNA damages, resulting in decreased genetic expression and, eventually, cell death. Dividing cell populations may cope with cell loss due to damage by replication of undamaged cells.

5. Decreases in the viable cell populations of tissues eventually impair the abilities of these tissues to function properly.

6. Reductions in specific cellular proteins may have repercussions at the tissue level as well as at the cellular level. When a gene product essential to a cell is lost, cell death ensues. However, some gene products perform composite tissue functions rather than reactions necessary for the individual cell (e.g., formation of neurotransmitters, hormones, growth factors, extracellular enzymes). When such a product is diminished, the reduction may only become apparent when the entire tissue lacks enough functionally capable cells to produce the necessary amount of product. Thus, an initial latent period, during which the functional capacity of the tissue is adequate, is succeeded by a progressive loss of tissue function. The rate of this loss depends in part on cellular redundancy.

7. Tissue dysfunction will be first evident under conditions of stress. Decreased homeostatic capacity and progressive immunodeficiency with age are examples of debilities that are first evident under stress.

8. Inability to control the internal environment, particularly under stress, may be an underlying cause of death in most aged individuals. Furthermore, with age, the individual becomes more fragile and more vulnerable to injury. The principal precipitating injurious factor in death is often trauma and resultant infection, which leads to cardiovascular collapse (King, 1988).

In general, we have argued that the basic factors that determine the maximum life span of a population of individuals are the rate of occurrence of endogenous DNA damage, the rate of DNA repair, the degree of cellular redundancy, and the extent to which the individuals are exposed to physiological stress.

References

Adler, S., Lindeman, R. D., Yiengst, M. J., Beard, E., and Shock, N. W. (1968). Effect of acute loading on urinary acid excretion by the aging human kidney. *J. Lab. Clin. Med.* **72**, 278–289.

Alnaqeeb, M. A., and Goldspink, G. (1980). Interrelation of muscle fiber types, diameter and number in aging white rats. *J. Physiol. Proc.* **310**, 56P.

Bass, A., Gutmann, E., and Hanzlikova, V. (1975). Biochemical and histochemical changes in energy supply–enzyme pattern of muscle of the rat during old age. *Gerontologia* **21**, 31–45.

Birchenall-Sparks, M. C., Roberts, M. S., Staecker, J., Hardwick, J. P., and Richardson, A. (1985). Effects of dietary restriction on liver protein synthesis in rats. *J. Nutr.* **115**, 944–950.

Bolla, R. I., and Denckla, W. D. (1979). Effect of hypophysectomy on liver nuclear ribonucleic acid synthesis in aging rats. *Biochem. J.* **184**, 669–674.

Bowman, P. D. (1985). Aging and the cell cycle *in vivo* and *in vitro*. *In* "Handbook of Cell Biology of Aging" (V. J. Cristofolo, ed.), pp. 117–136. CRC Press, Boca Raton, Florida.

Britton, V. J., Sherman, F. G., and Florini, J. R. (1972). Effect of age on RNA synthesis by nuclei and soluble RNA polymerases from liver and muscle of C57BL/6J mice. *J. Gerontol.* **27**, 188–192.

Brizzee, K. R. (1984). Pathophysiology of aging. *In* "Risk Factors for Senility" (H. Rothchild, ed.), pp. 44–62. Oxford University Press, New York.

Brown, I. R., and Church, R. B. (1971). RNA transcription from nonrepetitive DNA on the mouse. *Biochem. Biophys. Res. Commun.* **42**, 850–856.

Buetow, D. E. (1985). Cell numbers vs. age in mammalian tissues and organs. In "Handbook of Cell Biology and Aging" (V. J. Cristofolo, R. C. Adelman, and G. S. Roth, eds.), pp. 1–115. CRC Press, Boca Raton, Florida.

Burek, J. P. (1978). In "Pathology of Aging Rats," pp. 29–167. CRC Press, West Palm Beach, Florida.

Castle, T., Katz, A., and Richardson, A. (1978). Comparison of RNA synthesis by liver nuclei from rats of various ages. Mech. Ageing Dev. 8, 383–395.

Crie, J. S., Millward, D. J., Bates, P. C., Griffin, E., and Wildenthal, K. (1981). Age-related alterations in cardiac protein turnover. J. Mol. Cell. Cardiol. 13, 589–598.

Cutler, R. G. (1975). Transcription of unique and reiterated DNA sequences in mouse liver and brain tissues as a function of age. Exp. Gerontol. 10, 37–60.

Devi, A., Lindsey, P., Raina, P. L., and Sarkar, N. K. (1966). Effect of age on some aspects of the synthesis of ribonucleic acid. Nature 212, 474–475.

Dice, F. J., and Goff, S. A. (1986). Aging and the liver. In "The Liver, Biology and Pathobiology" (I. M. Arias, W. B. Jakoby, H. Pepper, D. Schachter, and D. A. Schafritz, eds.), Chapter 71, pp. 1245–1258. Raven Press, New York.

Dilella, A. G., Chiang, J. Y. L., and Steggles, A. W. (1982). The quantitation of liver cytochrome P450-LM$_2$ mRNA in rabbits of different ages and after phenobarbital treatment. Mech. Ageing Dev. 19, 113–125.

Doggett, D. L., Chang, M.-P., Makinodan, T., and Strehler, B. L. (1981). Cellular and molecular aspects of immune system aging. Mol. Cell. Biochem. 37, 137–156.

Dwyer, B. E., Fando, J. L., and Wasterlain, C. G. (1980). Rat brain protein synthesis declines during postdevelopmental aging. J. Neurochem. 35, 746–749.

Ekstrom, R., Liu, D. S. H., and Richardson, A. (1980). Changes in brain protein synthesis during the lifespan of male Fischer rats. Gerontology 26, 121–128.

Fando, J. L., Salinas, M., and Wasterlain, C. G. (1980). Age-dependent changes in brain protein synthesis in the rat. Neurochem. Res. 5, 373–383.

Florini, J. R. (1987). Effect of aging on skeletal muscle composition and function. In "Review of Biological Research in Aging," Vol. 3 (M. Rothstein, ed.), pp. 337–358. Alan R. Liss, New York.

Fog, R., and Pakkenberg, H. (1981). Age-related changes in ³H uridine uptake in the mouse. J. Gerontol. 36, 680–681.

Gensler, H. L., and Bernstein, H. (1981). DNA damage as the primary cause of aging. Q. Rev. Biol. 56, 279–303.

Glassman, R. B. (1987). An hypothesis about redundancy and reliability in the brain of higher species: Analogies with genes, internal organs, and engineering systems. *Neurosci. Biobehav. Rev.* **11**, 275–285.

Glick, R., and Bondareff, W. (1979). Loss of synapses in the cerebellar cortex of the senescent rat. *J. Gerontol.* **34**, 818–822.

Goldspink, G., and Alnaqeeb, M. A. (1985). Aging of skeletal muscle. *In* "Handbook of Cell Biology and Aging" (V. J. Cristofolo, R. C. Adelman, and G. S. Roth, eds.), pp. 179–194. CRC Press, Boca Raton, Florida.

Gresik, E. W., Wenk-Salamone, K., Onetti-Muda, A., Gubits, R. M., and Shaw, P. A. (1986). Effect of advanced age on the induction by androgen or thyroid hormone of epidermal growth factor and epidermal growth factor mRNA in the submandibular glands of C57BL/6 male mice. *Mech. Ageing Dev.* **34**, 175–189.

Gutmann, E., Hanzlikova, V., and Jackoubek, B. (1968). Changes in the neuromuscular system during old age. *Exp. Gerontol.* **3**, 141–146.

Hahn, W. E., and Laird, C. D. (1971). Transcription of nonrepeated DNA in mouse brain. *Science* **173**, 158–161.

Harrison, D. E. (1979). Proliferative capacity of erythropoietic stem cell lines and aging: An overview. *Mech. Ageing Dev.* **9**, 409–426.

Hazzard, W. R., and Bierman, E. L. (1978). Old age. *In* "The Biological Ages of Man from Conception through Old Age" (D. W. Smith, E. L. Bierman and N. M. Robinson, eds.), pp. 229–239. W. B. Saunders, Philadelphia.

Hoffman, M. A. (1983). Energy metabolism, brain size and longevity in mammals. *Q. Rev. Biol.* **58**, 495–512.

Horbach, G. J., M. J. Princen, H. M. G., Van Der Kroef, M., Van Bezooijen, C. F. A., and Yap, S. H. (1984). Changes in the sequence content of albumin mRNA and its translational activity in the rat liver with age. *Biochem. Biophys. Acta.* **783**, 60–66.

Hymer, W. C. (1979). Introduction to endocrine control systems (Chap. 1) and Control of the anterior lobe of the pituitary gland (Chap. 2). *In* "Endocrine Control Systems" (H. E. Morgan, ed.) [Section 7 of "Best and Taylor's Physiological Basis of Medical Practice" (J. R. Brobeck, ed.)], pp. (7-1)–(7-23). William and Wilkins, Baltimore.

Ingvar, M. C., Maeder, P., Sokoloff, L., and Smith, C. B. (1985). Effects of aging on local rates of cerebral protein synthesis. Sprague-Dawley Rats. *Brain* **108**, 155–170.

Johnson, R. J. (1985). Anatomy of the aging cell. *In* "Handbook of Cell Biology and Aging" (V. J. Cristofolo, R. C. Adelman, and G. S. Roth, eds.), pp. 149–178. CRC Press, Boca Raton, Florida.

Kim, S. K., and Arisumi, P. P. (1985). The synthesis of amylase in parotid glands of young and old rats. *Mech. Ageing Dev.* **31**, 251–266.

King, D. W. (1988). Pathology and ageing. *In* "Human Ageing Research: Concepts and Techniques" (B. Kent and R. N. Butler, eds.), pp. 325–340. Raven Press, New York.

Kirkwood, T. B. L. (1984). Towards a unified theory of cellular aging. *Monogr. Dev. Biol.* **17**, 9–20.

Krag, C. L., and Kountz, W. B. (1950). Stability of body function in the aged. I. Effect of exposure of the body to cold. *J. Gerontol.* **5**, 227–235.

Leblond, C. P., and Stevens, C. E. (1948). The constant renewal of the intestinal epithelium in the albino rat. *Anat. Rec.* **100**, 357–377.

Lesher, S., and Sacher, G. A. (1968). Effects of age on cell proliferation in mouse duodenal crypts. *Exp. Gerontol.* **3**, 211–217.

Lexell, J., Henriksson-Larsen, K., Winblad, B., and Sjostrom, M. (1983). Distribution of different fiber types in human skeletal muscles: Effects of aging studied in whole muscle cross sections. *Muscle Nerve* **6**, 588–595.

Makrides, S. C. (1983). Protein synthesis and degradation during aging and senescence. *Biol. Rev.* **58**, 343–422.

Meier-Ruge, W., Reichlmeier, K., and Iwangoff, P. (1976). Enzymatic and enzyme histochemical changes of the aging animal brain and consequences for experimental pharmacology on aging. *In* "Neurobiology of Aging" (R. D. Terry and S. Gershon, eds.), pp. 379–387. Raven Press, New York.

Miquel, J., Economos, A. C., Bensch, K. G., Atlan, H., and Johnson, J. E., Jr. (1979). Review of cell aging in *Drosophila* and mouse. *Age* **2**, 78–88.

Mori, M., Tanaka, A., and Sato, N. (1986). Hematopoietic stem cells in elderly people. *Mech. Ageing Dev.* **37**, 41–47.

Park, G. H., and Buetow, D. E. (1990a). RNA synthesis by nuclei and chromatin isolated from adult and senescent Wistar rat liver. *Gerontology* **36**, 61–75.

Park, G. H., and Buetow, D. E. (1990b). RNA synthesis by hepatocytes isolated from adult and senescent Wistar rat liver. *Gerontology* **36**, 76–83.

Patel, M. S. (1977). Age-dependent changes in oxidative metabolism in rat brain. *J. Gerontol.* **32**, 643–646.

Perry, E. K., Perry, R. H., Gibson, P. H., Blessed, G., and Tomlinson, B. E. (1977). A cholinergic connection between normal aging and senile dementia in the human hippocampus. *Neurosci. Lett.* **6**, 85–89.

Pickering, G. W. (1936). The peripheral resistance in persistent hypertension. *Clin. Sci.* **2**, 209–235.

Pluskal, M. G., Moreyra, M., Burini, R. C., and Young, V. R. (1984). Protein synthesis studies in skeletal muscle of aging rats. I. Alterations in nitrogen composition and protein synthesis using a crude polyribosome and pH5 enzyme system. *J. Gerontol.* **39**, 385–391.

Post, J., and Hoffman, J. (1964). Changes in the replication times and patterns of the liver cell during the life of the rat. *Exp. Cell Res.* **36,** 111–123.

Richardson, A., and Semsei, I. (1987). Effect of aging on translation and transcription. *In* "Review of Biological Research in Aging," Vol. 3 (M. Rothstein, ed.), pp. 467–483. Alan R. Liss, New York.

Richardson, A., Roberts, M. S., and Rutherford, M. S. (1985a). Aging and gene expression. *In* "Review of Biological Research in Aging," Vol. 2 (M. Rothstein, ed.), pp. 395–419. Alan R. Liss, New York.

Richardson, A., Rutherford, M. S., Birchenall-Sparks, M. C., Roberts, M. S., Wu, W. T., and Cheung, H. T. (1985b). Levels of specific messenger RNA species as a function of age. *In* "Molecular Biology of Aging: Gene Stability and Gene Expression," Vol. 29 (R. S. Sohal, L. S. Birnbaum, and R. G. Cutler, eds.), pp 229–242. Raven Press, New York.

Ricketts, W. G., Birchenall-Sparks, M. C., Hardwick, J. P., and Richardson, A. (1985). Effect of age and dietary restriction on protein synthesis by isolated kidney cells. *J. Cell Physiol.* **125,** 492–498.

Roy, A. K., Nath, T. S., Motwani, N. M., and Chatterjee, B. (1983). Age-dependent regulation of the polymorphic forms of alpha2u-$_{globin}$. *J. Biol. Chem.* **258,** 10123–10127.

Sabel, B., and Stein, D. G. (1981). Extensive loss of subcortical neurons in the aging rat brain. *Exp. Neurol.* **73,** 507–516.

Sajdel-Sulkowska, E. M., and Marotta, C. A. (1985). Functional messenger RNA from the postmortem human brain: Comparison of aged normal with Alzheimer's disease. *In* "Molecular Biology of Aging: Gene Stability and Gene Expression" (R. S. Sohal, L. S. Birnbaum, and R. G. Cutler, eds.), pp. 243–256. Raven Press, New York.

Scheibel, M. E., Lindsay, R. D., Tomiyasu, U., and Scheibel, A. B. (1975). Progressive dendritic changes in aging human cortex. *Exp. Neurol.* **47,** 392–403.

Semsei, I., and Richardson, A. (1986). Effect of age on the expression of genes involved in free radical protection. *Fed. Proc.* **45,** 217.

Shock, N. W. (1977). Systems integration. *In* "Handbook of the Biology of Aging" (C. E. Finch and L. Hayflick, eds.), pp. 639–665. Van Nostrand Reinhold, New York.

Silverstone, F. A., Brandfonbrener, M., Shock, N. W., and Yiengst, M. J. (1957). Age differences in the intravenous glucose tolerance test and the response to insulin. *J. Clin. Invest.* **36,** 504–514.

Smith, A., and Sugar, O. (1975). Development of above normal language and intelligence, 21 years after left hemispherectomy. *Neurology* **25,** 813–818.

Sonntag, W. E., Hylka, V. M., and Meites, J. (1985). Growth hormone restores protein synthesis in skeletal muscle of old rats. *J. Gerontol.* **40,** 689–694.

Suzuki, K., Korey, S. R., and Terry, R. D. (1964). Studies on protein synthesis in brain microsomal system. *J. Neurochem.* **11,** 403–412.

Tauchi, H., and Sato, T. (1978). Hepatic cells of the aged. *In* "Liver and aging" (K. Kitani, ed.), pp. 3–19. Elseview/North Holland Biomedical Press, New York.

Timiras, P. S. (1972). Decline in homeostatic regulation. *In* "Developmental Physiology and Aging" (P. S. Timiras, ed.), pp. 542–563. Macmillian, New York.

Tollefsbol, T. O., and Cohen, H. J. (1985). Decreased protein synthesis of transforming lymphocytes from aged humans. *Mech. Ageing Dev.* **30,** 53–62.

Wellinger, R., and Guigoz, Y. (1986). The effect of age on the induction of tyrosine aminotransferase and tryptophan oxygenase genes by physiological stress. *Mech. Ageing Dev.* **34,** 203–217.

Wu, W., Pahlavani, M., Cheung, H. T., and Richardson, A. (1986). The effect of aging on the expression of interleukin 2 messenger ribonucleic acid. *Cell. Immunol.* **100,** 224–231.

Zs-Nagy, I., and Semsei, I. (1984). Centrophenoxine increases the rates of total and mRNA synthesis in the brain cortex of old rats: An explanation of its action in terms of the membrane hypothesis of aging. *Exp. Gerontol.* **19,** 171–178.

Chapter Six

Oxidative DNA Damage as a Potential Major Cause of Aging

In this chapter, we review evidence indicating that oxidative DNA damage is a major cause of aging in mammals (see also Gensler *et al.*, 1987). This evidence includes the occurrence of high levels of oxidative DNA damage in mammals and its possible accumulation with age; a correlation between the incidence of oxidative damage and the life span of animals, and increased oxidative damage and accelerated aging in individuals with Down's syndrome.

I. *The Nature and Cause of Oxidative DNA Damage*

Oxidative damage is a consequence of normal respiratory metabolism and is caused by free radicals produced from molecular oxygen. As shown in Figure 6, the univalent pathway of oxygen reduction generates the superoxide radical (O_2^-), hydrogen peroxide (H_2O_2), the hydroxyl radical ($\cdot OH$), and water (Chance *et al.*, 1979; Halliwell and Gutteridge, 1985a; Freeman, 1984; Floyd *et al.*, 1984).

As reviewed by Imlay and Linn (1988), oxygen radicals can react with DNA at either the sugar-phosphate backbone or at a base. Some results of the reaction of DNA with oxygen radicals are shown in Figures 7 and 8.

The reaction of an oxygen radical with a backbone sugar leads to sugar fragmentation, loss of a base with part of the sugar residue remaining attached, and a strand break (Fig. 7). The attack on a base

$$O_2 + e^- \longrightarrow O_2^-$$
$$O_2^- + e^- + 2H^+ \longrightarrow H_2O_2$$
$$H_2O_2 + e^- + H^+ \longrightarrow H_2O + \cdot OH$$
$$\cdot OH + e^- + H^+ \longrightarrow H_2O$$

Figure 6   Pathway of production of high-energy oxygen species. Reduction of O_2 occurs in four steps. O_2, oxygen; e^-, electron; H^+, proton; H_2O, water; H_2O_2, hydrogen peroxide; O_2^-, superoxide radical, $\cdot OH$, hydroxyl radical.

such as thymine or adenine results in a damaged base. Five damaged base products are shown in Figure 8. Four of them are products of an attack on thymine and one of an attack on adenine.

Mutants of *E. coli* that are defective in repairing single-strand breaks have been identified, and these mutants are very sensitive to killing by H_2O_2 (Imlay and Linn, 1986; Carlsson and Carpenter, 1980; Cunningham *et al.*, 1986). This sensitivity indicates that unrepaired oxidative damages in DNA are lethal to cells. Enzymes for removing oxidatively damaged bases from DNA are common. For example,

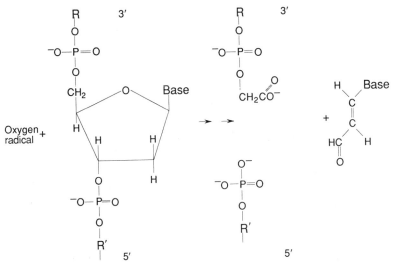

Figure 7   Result of oxygen radical attack on sugar-phosphate backbone of DNA.

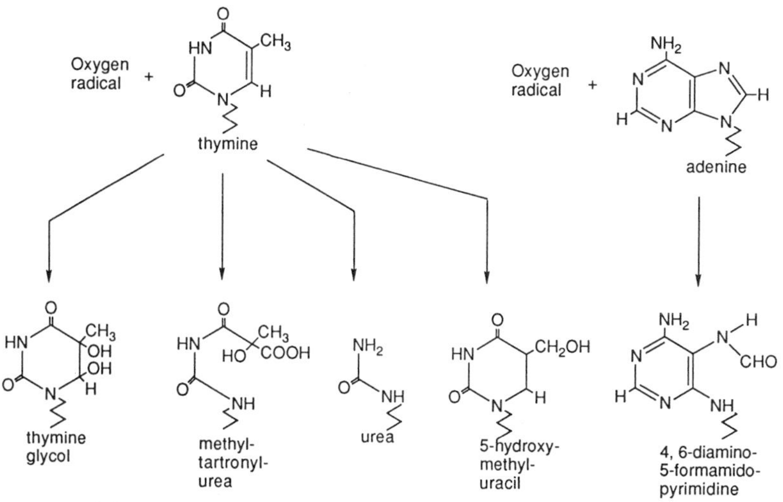

Figure 8   Result of oxygen radical attack on DNA bases.

glycosylases that remove thymine glycol in calf, human, *E. coli,* and *Drosophila* (Breimer, 1983, 1986; Breimer and Lindahl, 1984); hydroxymethyluracil in mouse (Hollstein *et al.,* 1984); and formamidopyrimidine in rodents (Margison and Pegg, 1981) and *E. coli* (Chetsanga and Lindahl, 1979) have been identified. The occurrence of these enzymes in such widely disparate organisms suggests that oxidative damages are an important problem.

Over 90% of the damages induced in DNA by H_2O_2 are altered bases. H_2O_2 also causes single-strand breaks (2–4% of all H_2O_2 damages) and double-strand breaks (0.8–0.9% of all H_2O_2 damages) (Massie *et al.,* 1972). X-irradiation produces a spectrum of DNA damages similar to that produced by oxidative damage. This is a consequence of the absorption of the radiation by the water in cells, which produces reactive intermediates that damage DNA. Three major identifiable types of DNA damage produced by X-irradiation were found to inactivate a small circular double-stranded plasmid. The efficiency of inactivation of the plasmid by each type of damage was as follows: damage to bases, 54 ± 23%; single-strand breaks, 10 ± 15%; and double-strand breaks, 100% (van der Schans *et al.,* 1973). These efficiencies indicate that the relative biological potency of double-strand breaks is greater than that of base damage, which in turn is greater than that of single-strand breaks. H_2O_2 also induces chromosomal aberrations (Schoneich, 1967) and sister-chromatid exchanges (Sperit et al., 1982), presumably as a direct consequence of oxidative DNA damage.

Although damages to DNA by oxygen radicals are well documented, it is not clear which oxidant causes the damage. Figure 6 shows that univalent reductions of oxygen produce O_2^-, H_2O_2, and ·OH. As pointed out by Imlay and Linn (1988), each of these is exempt from spin restriction and is kinetically and thermodynamically proficient at monovalent electron exchanges. At first glance, because O_2^- and H_2O_2 have mild oxidative capabilities and ·OH is a very powerful oxidant (Halliwell and Gutteridge, 1985b), it might be assumed that ·OH would be cytotoxically the most important. On the other hand, as described by Imlay and Linn (1988), the ·OH is produced from H_2O_2 by a Fenton reaction which depends on a transition metal (such as iron or copper). This Fenton reaction can be inhibited by ethanol. About half of H_2O_2-induced DNA nicking can be inhibited by ethanol in vitro using oxidative damage assays. Therefore, Imlay and Linn conclude that only half of oxidative DNA damage is due to diffusable ·OH, whereas the remaining nicks are due to a different oxidant.

The cellular enzymes shown in Figure 9 limit the release of active oxygen species. However, the entrapment is not complete (Chance et al., 1979), which allows some oxygen species to escape the cells' defenses.

Furthermore, the cell membrane is apparently not an obstacle to oxidative damaging agents, because release of active oxygen species from leukocytes can induce DNA strand scissions in adjacent cells (Dutton and Bowden, 1985). Thus, active oxygen species can pass from cell to cell.

Overall, in human cells a large number of oxidative DNA damages occur per cell per day due to oxidative cellular metabolism. Cathcart et al. (1984) and Ames et al. (1985) presented evidence that thymine glycol, deoxythymidine glycol, and hydroxymethyluracil residues are excreted in the urine of humans and rats as a result of their removal during the repair of oxidatively damaged DNA. This allows an estimate of their rate of occurrence. As summarized in Chapter 2, Section II.F, the average human cell may experience more than 1,000 of these oxidative damages per day.

$$2O_2^- + 2H^+ \xrightarrow{\text{superoxide dismutase}} H_2O_2 + O_2$$

$$2H_2O_2 \xrightarrow{\text{catalase}} 2H_2O + O_2$$

$$H_2O_2 + 2GSH \xrightarrow{\text{glutathione peroxidase}} 2H_2O + GSSG$$

$$GSSG + NADPH + H^+ \xrightarrow{\text{glutathione reductase}} NADP^+ + 2GSH$$

Figure 9 Enzymatic defenses against active oxygen species. Reactive species are metabolized by several enzymes, including those shown here. GSH, reduced glutathione; GSSG, oxidized glutathione.

Another oxidized base, 8-hydroxydeoxyguanosine, was shown to occur in rat liver (Richter *et al.*, 1988). As mentioned in Chapter 2, Section II.F, the estimated average number of 8-hydroxydeoxy-guanosine damages per rat liver cell was about 1.4×10^5 in nuclear DNA and about 4.1×10^4 in mitochondrial DNA. The level of 8-hydroxydeoxyguanosine in human urine is of the same order as thymidine glycol (Ames, 1989).

II. *Accumulation of Oxidative DNA Damages with Age*

An oxidatively altered base (which has not yet been structurally identified) has been shown to accumulate in mouse and rat liver DNA with age (Sharma and Yamamoto, 1980). This indicates that at least some oxidatively damaged bases accumulate in DNA over time. On the other hand, hydroxymethyluracil, an oxidation product of thymine, was not found in DNA samples from brain, liver, or small intestinal mucosa cells taken from mice of different ages (Kirsh *et al.*, 1986). This suggests that removal of hydroxymethyluracil by a repair glycosylase may be highly efficient in at least some mammalian cells.

As noted above (this chapter, Section I), about 2–4% of oxidative damages (induced by hydrogen peroxide) are single-strand breaks (Massie *et al.*, 1972). In Chapter 4, Sections III and IV, we reviewed a number of studies showing that single-strand breaks accumulate in muscle, nerve, and liver cells. The observed accumulation of single-strand breaks in postmitotic or slowly dividing cells is expected if oxidative damages are prevalent. However, single-strand breaks may also be produced spontaneously (Chapter 2, Section II.C) and may not require oxidative processes to form.

Work by Lawson and Stohs (1985) suggests that oxidative reactions produce a significant fraction of the single-strand breaks in mouse liver DNA. Starting from age 6 mo, single-strand breaks increase in rat liver DNA, reaching a maximum at age 18 mo. The effects of the antioxidants butylated hydroxyanisole and oltipraz on the prevention of single-strand breaks in rat liver DNA was also measured by these authors. They found that limited administration of either butylated hydroxyanisole or oltipraz to mice significantly reduced the levels of hepatic single-strand DNA breaks.

III. *Inverse Correlation between Incidence of Oxidative DNA Damage and Life Span*

A. Oxidative DNA Damage in Somatic Cells

One important factor correlated with longevity appears to be basal metabolic rate (Harman, 1981; Sohal, 1981; Sullivan, 1982; Tolmasoff *et al.*, 1980). Increasing mammalian metabolic rates correlate with decreasing life spans. Rats, which have a higher specific metabolic rate and shorter life span than humans, were found to excrete 10–15 times more oxidatively modified bases per kilogram of body weight per unit time than humans (Ames *et al.*, 1985). More recently, Saul *et al.* (1987) and Adelman *et al.* (1988) extended this correlation to mouse and monkey (*Macaca fascicularis*). When the average urinary output of thymine glycol (in nanomoles per kilogram body weight per day) was plotted against specific metabolic rate (oxygen consumption in milliliters per gram body weight per hour) for humans, monkey, rat, and mouse, a linear relationship was observed. Humans, had both the lowest metabolic rate and the lowest rate of thymine glycol output. These rates were exceeded in monkey, rat, and mouse in that order. A similar relationship was also found when output of thymidine glycol, rather than thymine glycol, was plotted against metabolic rate. Because metabolic rate correlates inversely with longevity, these data also indicate that urinary output of thymine glycol and thymidine glycol correlates inversely with longevity. Because urinary output of these damaged bases is considered to be a reflection of the incidence of oxidative DNA damage (Adelman *et al.*, 1988), it can be concluded that the incidence of oxidative DNA damage correlates inversely with longevity. As pointed out by Adelman *et al.* (1988), these results are consistent with the hypothesis that DNA is a critical target in the aging process and that larger mammals owe their longer life spans to their lower metabolic rates and correspondingly low rates of oxidative damage formation.

In another line of inquiry, it was found that the feeding of antioxidants to rodents increases their life span under some, though not all, conditions (e.g., Harman, 1961, 1968; Harman and Eddy, 1979; Comfort *et al.,* 1971; Kohn, 1971). These observed increases can be explained by the hypothesis that oxidative free radicals are important in aging. However, the authors making these observations, and others, have expressed several specific reservations about this explanation. Perhaps the most important criticism arises from the observation of Kohn (1971) that when maintenance conditions of mice were optimal, antioxidant intake did not extend life span. This suggests that antioxidants might not affect the intrinsic aging process, but instead act to mitigate some harmful environmental or nutritional factor.

Stohs *et al.* (1986) postulated that antioxidant substances, such as those found in the diet, might protect cellular macromolecules from free radicals and oxidants. Thus, they studied the effects of feeding 18-mo-old mice diets containing the antioxidants butylated hydroxyanisole, oltipraz, anethole dithione, or freeze-dried cabbage and measured cellular levels of DNA damage and lipid peroxidation (malondialdehyde content). They also measured reduced glutathione levels in hepatic cells and glutathione metabolizing enzymes after feeding these diets. Reduced glutathione is believed to play a major role in protecting cellular macromolecules from free radicals and oxidants. After 2 wk, all mice on these diets showed significant decreases in hepatic DNA damage (single-strand breaks) compared with controls not fed antioxidant supplements. The number of single-strand breaks, as a percentage of the control, varied from 27 to 62%, depending on the particular antioxidant in the diet. Also, lipid peroxidation significantly decreased. In addition, Stohs *et al.* (1986) found a significant increase in reduced glutathione content and significant increases in glutathione reductase and glutathione S-transferase activities in liver cells. The authors suggested that dietary antioxidants may act directly to prevent damage to cellular DNA and lipids or may act indirectly by enhancing the formation of reduced glutathione or glutathione-metabolizing enzymes, which subsequently provide cellular protection. Thus, the results of Stohs *et al.* (1986) suggest that diets high in antioxidants are effective in retarding oxidative DNA damage. This may account for the increased life span, reported by some workers, of rodents fed with these agents.

B. Oxidative DNA Damage in the Brain

Several arguments support the idea that oxidative processes cause DNA damage in the brain. *A priori,* more oxidative DNA damages might be expected to accumulate with age in the brain than in other tissues, because little or no turnover of neuronal DNA occurs during the human life span (Slatkin *et al.,* 1985) and oxygen utilization in the brain is high. The human brain represents only 2% of the total body weight, but its rate of oxygen utilization accounts for 20% of the total resting consumption (Iverson, 1979). This enormous expenditure of energy is perhaps due to the maintenance of ionic gradients across the neuronal membrane on which the conduction of impulses in the billions of brain neurons depends.

Cutler (1984, 1985) found an inverse correlation between longevity in different mammalian species and the peroxide-producing potential of the brain. He also found a direct correlation between longevity and the levels of brain carotenoid antioxidants. He concluded that his results support the hypothesis that aging may be caused in part by oxygen radicals initiating peroxidation reactions in the brain. He also speculated that peroxidation defense processes are involved in governing the longevity of mammalian species.

Zs-Nagy and Semsei (1984) showed that as rats age from 13 to 26 mo, the rate of mRNA synthesis in the brain cortex decreases considerably. However, intraperitoneal injection with centrophenoxinol (CPH), a scavenger of hydroxyl free radicals, reversed this tendency. CPH treatment significantly increased the mRNA synthesis rates of old rats, almost to the level seen in 13-mo-old rats. In addition, CPH-treated mice had about a 30% longer median life span than untreated ones (Hochschild, 1973). Although Zs-Nagy and Semsei (1984) explained these effects of CPH in terms of another hypothesis of aging, they may also be understood in terms of a DNA damage hypothesis. That is, CPH might act by preventing the attack of DNA by hydroxyl radicals.

IV. Premature Aging in Down's Syndrome Correlated with Increased Oxidative Damage

A prediction of the hypothesis that oxidative DNA damage causes aging is that metabolic disorders, which cause excessive production of reactive oxygen species and a consequent high level of DNA damages, should lead to accelerated aging. This prediction is apparently fulfilled in patients with Down's syndrome, a disease that exhibits both increased production of oxidative radicals and symptoms of premature aging.

Martin (1978) tabulated 162 genetic syndromes in humans with features of premature aging. (The 21 specific features of aging considered are listed in Table XI.) The disease that correlated best with these criteria, having 15 of the 21 features of premature aging, was Down's syndrome. Wright and Whalley (1984) have also reviewed evidence indicating that the neuropathology of Down's syndrome represents an acceleration of normal aging.

Substantial evidence points to an imbalance in oxidative metabolism and a high level of peroxidation in persons with Down's syndrome. The genetic basis of this disease is trisomy of the distal part of the long arm of chromosome 21 (Sinet, 1982). This chromosomal position is the location of the gene for CuZn superoxide dismutase (CuZn SOD), an enzyme that catalyzes the formation of H_2O_2 (Fig. 9) (Sinet, 1982). In individuals with Down's syndrome, CuZn SOD is found at levels approximately 50% higher than normal. This is true in erythrocytes (Sinet et al., 1975; Kedziora et al., 1986), fibroblasts (Feaster et al., 1977), B and T lymphocytes (Baeteman et al., 1983), polymorphonuclear leukocytes (Bjorksten et al., 1984), and fetal brain (Balazs and Brookshank, 1985). The rate at which transcripts of chromosome 21 genes appear increases in relation to numbers of chromosome 21 that are present (Kurnit, 1979). This suggests that the observed increase in CuZn SOD expression in Down's syndrome individuals results from the trisomy (Sinet, 1982).

Recently, Delabar et al. (1987) and Huret et al. (1987) have located the region of chromosome 21 responsible for Down's syndrome. The critical segment, <2,000–3,000 kilobases long, is located at the interface of regions 21q21 and 21q22.1. Two unrelated individuals with Down's syndrome were found to have a duplication of this small critical segment in one of their two copies of chromosome 21.

Delabar *et al.* (1987) identified three genes within this segment, the CuZn SOD gene, oncogene *ets*-2, and the cystathione-beta-synthase gene. These individuals were also found to have elevated CuZn SOD in their erythrocytes. This evidence suggests that Down's syndrome, including the accelerated aging phenotype, may be due in part to overproduction of CuZn SOD.

As pointed out by Balazs and Brookshank (1985), although CuZn SOD is involved in essential mechanisms of protection against oxygen radicals, it can also cause unwanted effects. The enzyme catalyzes the formation of H_2O_2 from O_2^- (superoxide anion radical). The harm may arise from the activation of H_2O_2 (iron-catalyzed Haber-Weiss reaction), with the generation of the highly reactive $\cdot OH$ or with the generation of $^1O_2$ (singlet oxygen) (Chance *et al.*, 1979). The increased CuZn SOD activity is the likely cause of the elevated level of blood lipid peroxides found in Down's syndrome individuals (Kedziora *et al.*, 1986). This increased oxidation occurs despite induction in erythrocytes of glutathione peroxidase (Fig. 9), an enzyme that reduces oxidative damage by metabolizing H_2O_2 and organic hydroperoxides (Sinet *et al.*, 1975; Sinet, 1982).

Elroy-Stein *et al.* (1986) succeeded in transfecting the human CuZn SOD gene into HeLa and mouse cells. The transfected cells expressed up to sixfold the normal level of SOD. These overproducing cells allowed the study of the cellular response to elevated levels of CuZn SOD in a defined background, distinct from cellular effects caused by gene dosage of other chromosome 21-encoded genes. These cells were more resistant than the parental cells to the lethal effect of paraquat (used as an *in vivo* generator of O_2^-). This shows that the deleterious effect of a large excess over the normal level of superoxide can be prevented by an excess of CuZn SOD activity. The authors point out, however, that under normal conditions (no paraquat) the higher level of CuZn SOD in transfected cells interferes with cellular metabolism due to higher steady-state concentrations of H_2O_2 and other active forms of oxygen ($\cdot OH$ and $^1O_2$). These higher levels presumably explain the observed increase in lipid peroxidation in these cells. The authors suggested that excess CuZn SOD activity may account for some of the clinical symptoms associated with Down's syndrome.

Recently, Groner *et al.* (1988) reported transferring the cloned human gene for CuZn SOD into mouse embryos, resulting in transgenic mice overexpressing the human CuZn SOD gene. They found that the mice synthesize the human enzyme in an active form capable of forming human–mouse enzyme heterodimers. Elevated human CuZn SOD activity was detected in nearly all tissues of the animals expressing the gene, but the levels of activity varied from tissue to tissue. Expression in the liver was quite low, whereas high enzymatic activity was detected in the brain.

The brain of individuals with Down's syndrome appears to be particularly vulnerable to oxidative DNA damage. In other tissues, when hydrogen peroxide production is high (e.g., due to high CuZn SOD), the levels of enzymes which metabolize this compound (e.g., glutathione peroxidase and catalase) are induced. However, in cerebral cortex cells of individuals with Down's syndrome, the activity of CuZn SOD is high, whereas glutathione peroxidase is at an uninduced level (Balazs and Brookshank, 1985). A compensatory response of catalase to increased H_2O_2 production in the cytosolic compartment in the brain cells seems unlikely because catalase activity is both very low in brain cells compared with other tissues (Sinet et al., 1980) and does not seem to correlate with SOD activity (Mavelli et al., 1982). Furthermore, catalase is largely confined in peroxisomes and thus is not readily available to other cellular sites of H_2O_2 production.

In our discussion above, we hypothesized that premature aging in Down's syndrome is caused by excess production of CuZn SOD, leading to increased DNA damage. However, some experiments on X-ray-induced chromosome aberrations in lymphocytes indicate an additional possible source of increased DNA damage as well. The frequency of chromosomal aberrations produced by X-rays is greater in lymphocytes cultured from individuals with Down's syndrome than from normal diploid donors (Sasaki et al., 1970; Countryman et al., 1977; Preston, 1981). Athanasiou et al. (1980) found that lymphocytes from individuals with Down's syndrome were less efficient at repairing single-strand DNA breaks induced by X-rays than lymphocytes from normal individuals. They proposed that the increased frequency of chromosomal aberrations in individuals with Down's syndrome may result from their reduced capacity to repair single-strand breaks. Furthermore, Countryman et al. (1977) proposed that the increased risk for leukemia in individuals with Down's syndrome (Miller, 1966) could reflect a defect in DNA repair. Therefore, it is not clear if the increased DNA damage and increased aging in Down's syndrome individuals is due mainly to higher effective levels of DNA-damaging free radicals, to a reduced DNA repair capacity, or to some combination of both.

To review, the genetic information responsible for Down's syndrome has been located to a small segment of chromosome 21, which includes the CuZn SOD gene. Cells of individuals with Down's syndrome have a high level of CuZn SOD, an increased production of H_2O_2, and an elevated concentration of the very damaging hydroxyl radical $\cdot OH$. The brain of Down's syndrome individuals may be a particularly vulnerable tissue because of its high level of oxidative metabolism, its lack of cell turnover, and the low levels of the H_2O_2-scavenging enzymes glutathione peroxidase and catalase found in the brains of these individuals. Thus, excess production of H_2O_2 and $\cdot OH$ may lead to increased production of DNA damages in the brain, and this may account, at least partly, for the accelerated aging phenotype associated with Down's syndrome.

V. Conclusions

In this chapter, we have reviewed evidence indicating that oxidative DNA damages may be a major cause of aging. Oxidative damages are frequent in nature. At least one type of DNA damage that can be caused by oxidation, the single-strand break, accumulates with age in mammalian brain, muscle, and liver. In liver, this accumulation can be retarded by administration of antioxidants. An increased incidence of oxidatively altered DNA bases correlates with shorter life span, as seen in comparisons of humans, monkey, rat, and mouse. Antioxidants, which decrease the incidence of single-strand breaks in rodents, reportedly increase rodent longevity under some conditions.

Production of oxidative DNA-damaging agents is likely to be a particular problem in the brain because of the high rate of oxygen utilization in this organ. Longevity in mammals correlates inversely with the peroxide-producing potential of the brain. Treatment of rodents with an antioxidant increases the rate of mRNA synthesis in the brain cortex of old rats. This might reflect a lower incidence of DNA damage following antioxidant treatment. In individuals with Down's syndrome, the brain may be especially vulnerable to oxidative DNA damage because an extra copy of the CuZn SOD gene is present in these individuals and there is a consequent high level of intracellular CuZn SOD. This leads to excess production of H_2O_2 and the potent ·OH radical. Excessive levels of ·OH radicals in individuals with Down's syndrome may account, at least in part, for the accelerated aging phenotype of Down's syndrome.

References

Adelman, R., Saul, R. L., and Ames, B. N. (1988). Oxidative damage to DNA: Relation to species metabolic rate and lifespan. *Proc. Natl. Acad. Sci. USA* **85**, 2706–2708.

Ames, B. N. (1989). Endogenous DNA damage is related to cancer and aging. *Mutat. Res.* **214**, 41–46.

Ames, B. N., Saul, R. L., Schwiers, E., Adelman, R., and Cathcart, R. (1985). Oxidative DNA damage as related to cancer and aging: Assay of thymine glycol, thymidine glycol and hydroxymethyluracil in human and rat urine. *In* "Molecular Biology of Aging: Gene Stability and Gene Expression" (R. S. Sohal, L. S. Birnbaum, and R. G. Cutler, eds.), pp. 137–144. Raven Press, New York.

Athanasiou, K., Sideris, E. G., and Bartsocas, C. (1980). Decreased repair of X-ray induced DNA single-strand breaks in lymphocytes in Down syndrome. *Pediatr. Res.* **14**, 336–338.

Baeteman, M. A., Baret, A., Cortiere, A., Rebuffel, P., and Mattei, J. F. (1983). Immunoreactive Cu-SOD and Mn-SOD in lymphocyte subpopulations from normal and trisomy 21 subjects according to age. *Life Sci.* **32**, 895–902.

Balazs, R., and Brookshank, B. W. (1985). Neurochemical approaches to the pathogenesis of Down syndrome. *J. Ment. Defic. Res.* **29**, 1–14.

Bjorksten, B., Marklund, S. L., and Hagglof, B. (1984). Enzymes of leukocyte oxidative metabolism in Down's syndrome. *Acta Pediatr. Scand.* **73**, 97–101.

Breimer, L. H. (1983). Urea-DNA glycosylase in mammalian cells. *Biochemistry* **2**, 4192–4203.

Breimer, L. H. (1986). A DNA glycosylase for oxidized thymine residues in *Drosophila melanogaster*. *Biochem. Biophys. Res. Commun.* **134**, 201–204.

Breimer, L. H., and Lindahl, T. (1984). DNA glycosylase activities for thymidine residues damaged by ring saturation, fragmentation, or ring contraction are functions of endonuclease II in *Escherichia coli*. *J. Biol. Chem.* **259**, 5543–5548.

Carlsson, J., and Carpenter, V. S. (1980). The $recA^+$ gene product is more important than catalase and superoxide dismutase in protecting *Escherichia coli* against hydrogen peroxide toxicity. *J. Bacteriol.* **142**, 319–321.

Cathcart, R., Schwiers, E., Saul, R. L., and Ames, B. N. (1984). Thymine glycol and thymidine glycol in human and rat urine: A possible assay for oxidative DNA damage. *Proc. Natl. Acad. Sci. USA* **81**, 5633–5637.

Chance, B., Sies, H., and Boveris, A. (1979). Hydroperoxide metabolism in mammalian organs. *Physiol. Rev.* **59**, 527–605.

Chetsanga, C., and Lindahl, T. (1979). Release of 7-methylguanine residues whose imidazole rings have been opened from damaged DNA by a DNA glycosylase from *E. coli*. *Nucleic Acids Res.* **6**, 3673–3683.

Comfort, A., Youhotsky-Gore, I., and Pathmanathan, K. (1971). Effect of ethyoxyquin on the longevity of C3H mice. *Nature* **229**, 254–255.

Countryman, P. I., Heddle, J. A., and Crawford, E. (1977). The repair of X-ray-induced chromosomal damage in trisomy 21 and normal lymphocytes. *Cancer Res.* **37**, 52–58.

Cunningham, R. P., Saporito, S. M., Spitzer, S. F., and Weiss, B. (1986). Endonuclease IV (nfo) mutant of *Escherichia coli*. *J. Bacteriol.* **168**, 1120–1127.

Cutler, R. G. (1984). Carotenoids and retinol: The possible importance in determining longevity of primate species. *Proc. Natl. Acad. Sci. USA* **81**, 7627–7631.

Cutler, R. G. (1985). Peroxide-producing potential of tissues: Inverse correlation with longevity of mammalian species. *Proc. Natl. Acad. Sci. USA* **82**, 4798–4802.

Delabar, J. M., Sinet, P. M., Chadefaux, B., Nichole, A., Gegonne, A., Stehelin, D., Fridlansky, F., Creau-Goldberg, N., Turleau, C., and deGrouchy, J. (1987). Submicroscopic duplication of chromosome 21 and trisomy 21 phenotype (Down syndrome). *Hum. Genet.* **76**, 225–229.

Dutton, D. R., and Bowden, G. T. (1985). Indirect induction of a clastogenic effect in epidermal cells by a tumor promoter. *Carcinogenesis* **6**, 1279–1284.

Elroy-Stein, O., Bernstein, Y., and Groner, Y. (1986). Overproduction of human Cu/Zn superoxide dismutase in transfected cells: Extenuation of paraquat-mediated cytotoxicity and enhancement of lipid peroxidation. *EMBO J.* **5**, 615–622.

Faester, W. W., Kwok, L. W., and Epstein, C. J. (1977). Dosage effects of superoxide dismutase-1 in nucleated cells aneuploid for chromosome 21. *Am. J. Hum. Genet.* **29**, 563-570.

Floyd, R. A., Zaleska, M. M., and Harmon, H. J. (1984). Possible involvement of iron and oxygen free radical in aspects of aging in brain. *In* "Free Radicals in Molecular Biology, Aging and Disease" (D. Armstrong,

R. S. Sohal, R. G. Cutler, and T. F. Slater, eds.), pp. 143–161. Raven Press, New York.

Freeman, B. A. (1984). Biological sites and mechanisms of free radical production. In "Free Radicals in Molecular Biology, Aging and Disease" (D. Armstrong, R. S. Sohal, R. G. Cutler, and T. F. Slater, eds.), pp. 43–52. Raven Press, New York.

Gensler, H. L., Hall, J. D., and Bernstein, H. (1987). The DNA damage hypothesis of aging: Importance of oxidative damage. In "Review of Biological Research in Aging," Vol. 3 (M. Rothstein, ed.), pp. 451–465. Alan R. Liss, New York.

Groner, Y., Elroy-Stein, O., Avraham, K. B., Rotman, G., Bernstein, Y., Dafni, N., and Schickler, M. (1988). Overexpression of human CuZn SOD gene in transfected cells and transgenic mice: Implication for Down syndrome pathology. J. Cell Biol. Suppl. 12A, 36.

Halliwell, B., and Gutteridge, J. M. (1985a). Oxygen radicals and the nervous system. Trends Neurosci. 8, 22–26.

Halliwell, B., and Gutteridge, J. M. C. (1985b). "Free Radicals in Biology and Medicine." Clarendon Press, Oxford.

Harman, D. (1961). Prolongation of the normal lifespan and inhibition of spontaneous cancer by antioxidants. J. Gerontol. 16, 247–254.

Harman, D. (1968). Free radical theory of aging. Effect of free radical radiation inhibitors on the mortality of LAF mice. J. Gerontol. 23, 476–482.

Harman, D. (1981). The aging process. Proc. Natl. Acad. Sci. USA 78, 7124–7128.

Harman, D., and Eddy, D. E. (1979). Free radical theory of aging: Beneficial effect of adding antioxidants to the maternal mouse diet on lifespan of offspring: Possible explanation of the sex difference in longevity. Age 2, 109–122.

Hochschild, R. (1973). The aging process. Proc. Natl. Acad. Sci. USA 78, 7124–7128.

Hollstein, M. C., Brooks, P., Linn, S., and Ames, B. N. (1984). Hydroxymethyluracil DNA glycosylase in mammalian cells. Proc. Natl. Acad. Sci. USA 81, 4003–4007.

Huret, J. L., Delabar, J. M., Marlhens, F., Aurias, A., Nicole, A., Berthier, M., Tanzer, J., and Sinet, P. M. (1987). Down syndrome with duplication of a region of chromosome 21 containing the CuZn superoxide gene without detectable karyotypic abnormality. Hum. Genet. 75, 251–257.

Imlay, J. A., and Linn, S. (1986). Bimodal pattern of killing of DNA-repair defective or anoxically grown Escherichia coli by hydrogen peroxide. J. Bacteriol. 166, 519–527.

Imlay, J. A., and Linn, S. (1988). DNA damage and oxygen radical toxicity. Science 240, 1302–1309.

Iverson, L. L. (1979). The chemistry of the brain. *Sci. Am.* **241 (Sept.),** 134–149.

Kedziora, J., Bartosz, G., Gromadzinska, J., Sklodwska, M., Wesowicz, W., and Scianowski, J. (1986). Lipid peroxides in blood plasma and enzymatic antioxidative defense of erythrocytes in Down syndrome. *Clin. Chim. Acta* **154,** 191–194.

Kirsh, M. E., Cutler, R. G., and Hartman, P. E. (1986). Absence of deoxyuridine and 5-hydroxymethyldeoxyuridine in the DNA from three tissues of mice of various ages. *Mech. Ageing Dev.* **35,** 71–77.

Kohn, R. R. (1971). Effect of antioxidants on lifespan of C57BL mice. *J. Gerontol.* **26,** 378–380.

Kurnit, D. M. (1979). Down syndrome: Gene dosage at the transcriptional level in skin fibroblasts. *Proc. Natl. Acad. Sci. USA* **76,** 2372–2375.

Lawson, T., and Stohs, S. (1985). Changes in endogenous DNA damage in aging mice in response to butylated hydroxyanisole and oltipraz. *Mech. Ageing Dev.* **30,** 179–185.

Margison, G. P., and Pegg, A. E. (1981). Enzymatic release of 7-methylguanine from methylated DNA by rodent liver extracts. *Proc. Natl. Acad. Sci. USA* **78,** 861–865.

Martin, G. M. (1978). Genetic syndromes in man with potential relevance to the pathobiology of aging. *Birth Defects: Original Article Series* **14,** 5–39.

Massie, H. R., Samis, H. V., and Baird, M. B. (1972). The kinetics of degradation of DNA and RNA by H_2O_2. *Biochim. Biophys. Acta* **272,** 539–548.

Mavelli, I., Rigo, A., Federico, R., Ciriolo, M. R., and Rotilio, G. (1982). Superoxide dismutase, glutathione peroxidase and catalase in developing rat brain. *Biochem. J.* **204,** 535–540.

Miller, R. (1966). Relation between cancer and congenital defects in man. *New Eng. J. Med.* **275,** 87–93.

Preston, R. J. (1981). X-ray-induced chromosome aberrations in Down lymphocytes and explanation of their increased sensitivity. *Environ. Mutagen.* **3,** 85–89.

Richter, C., Park, J.-W., and Ames, B. N. (1988). Normal oxidative damage to mitochondrial and nuclear DNA is extensive. *Proc. Natl. Acad. Sci. USA* **85,** 6465–6467.

Sasaki, M. S., Tonomura, A., and Matsubara, S. (1970). Chromosome constitution and its bearing on the chromosomal radiosensitivity in man. *Mutat. Res.* **10,** 617–633.

Saul, R. L., Gee, P., and Ames, B. N. (1987). Free radicals, DNA damage, and aging. *In* "Modern Biological Theories of Aging" (H. R. Warner, R. N. Butler, R. L. Sprott, and E. L. Schneider, eds.), pp. 113–129. Raven Press, New York.

Schoneich, J. (1967). The induction of chromosomal aberrations by hydrogen peroxide in strains of ascites tumors in mice. *Mutat. Res.* **4**, 384–388.

Sharma, R. C., and Yamamoto, O. (1980). Base modification in adult liver DNA and similarity to radiation-induced base modification. *Biochem. Biophys. Res. Commun.* **96**, 662–671.

Sinet, P. M. (1982). Metabolism of oxygen derivatives in Down syndrome. *Ann. N.Y. Acad. Sci.* **396**, 83–94.

Sinet, P. M., Michelson, A. M., Bazin, A., Lejeune, J., and Jerome, H. (1975). Increase in glutathione peroxidase activity in erythrocytes from trisomy 21 subjects. *Biochem. Biophys. Res. Commun.* **67**, 910–915.

Sinet, P. M., Heikkila, R. E., and Cohen, G. (1980). Hydrogen peroxide production by rat brain *in vivo. J. Neurochem.* **34**, 1421–1428.

Slatkin, D. N., Friedman, L., Irsa, A. P., and Micca, P. L. (1985). The stability of DNA in human cerebellar neurons. *Science* **228**, 1002–1004.

Sohal, R. S. (1981). Metabolic rate, aging, and lipofuscin accumulation. *In* "Age Pigments" (R. S. Sohal, ed.), pp 303–316. Elsevier/North-Holland Biomedical Press, Amsterdam and New York.

Speit, G., Vogel, W., and Wolf, M. (1982). Characterization of sister chromatid exchange induction by hydrogen peroxide. *Environ. Mutagen.* **4**, 135–142.

Stohs, S. J., Lawson, T. A., Anderson, L., and Bueding, E. (1986). Effects of oltipratz, BHA, ADT and cabbage on glutathione metabolism, DNA damage and lipid peroxidation in old mice. *Mech. Ageing Dev.* **37**, 137–145.

Sullivan, J. L. (1982). Superoxide dismutase, longevity and specific metabolic rate. *Gerontology* **28**, 242–244.

Tolmasoff, J. M., Ono, T., and Cutler, R. G. (1980). Superoxide dismutase: Correlation with lifespan and specific metabolic rate in primate species. *Proc. Natl. Acad. Sci. USA* **77**, 2777–2781.

van der Schans, G. P., Bleichrodt, J. F., and Blok, J. (1973). Contribution of various types of damage to inactivation of a biologically-active double-stranded circular DNA by gamma-radiation. *Int. J. Radiat. Biol.* **23**, 133–160.

Wright, A. F., and Whalley, L. F. (1984). Genetics, aging and dementia. *Br. J. Psychiatry* **145**, 20–38.

Zs-Nagy, I., and Semsei, I. (1984). Centrophenoxine increases the rates of total and mRNA synthesis in the brain cortex of old rats: An explanation of its action in terms of the membrane hypothesis of aging. *Exp. Gerontol.* **19**, 171–178.

Chapter Seven

Additional Evidence Bearing on the DNA Damage Hypothesis of Aging

The DNA damage hypothesis of aging makes three further predictions (beyond those discussed in prior chapters). First, among species with differing life spans, the capacity to repair DNA damage should correlate with life span. Second, sublethal exposure of an animal to a DNA-damaging agent should shorten life span and accelerate at least some aspects of normal aging. Third, human genetic syndromes with features of premature aging should have elevated accumulation of DNA damage, excess intracellular DNA-damaging agents, and/or defective DNA repair. The evidence bearing on these predictions is discussed in this chapter.

I. Positive Correlation of Life Span with DNA Repair Capacity

A. Evidence for a Positive Correlation

Cutler (1972) suggested that the evolution of increased life span for mammals is the result of an increase in repair and/or protective processes for a common spectrum of damaging agents. This predicts that organisms with a higher repair capacity should have a longer life span than those with a lower one. If DNA damages are the cause of aging, then this increased repair capacity should include repair of DNA damage.

Work by Hart and Setlow (1974) has supported this concept. These workers determined the ability of skin fibroblasts of seven mammalian species to perform unscheduled DNA synthesis (a measure of repair synthesis) after UV irradiation. This repair synthesis reflects excision repair in which damaged regions are removed by nucleases, leaving single-strand gaps, which are then filled in by DNA polymerase and ligase (see Chapter 9, Sections II.A and II.B). The species studied were shrew, mouse, rat, hamster, cow, elephant, and human. The animals studied ranged in life span from 1.5 yr (shrew) to 95 yr (human). They found that both the initial rate and the extent of unscheduled DNA synthesis increased systematically with the expected life span of individuals of the different species. This positive correlation between DNA repair capacity and expected life span was striking and stimulated a series of 17 additional sets of experiments, which amplified this initial finding. We summarize the experiments below and tabulate them in Table X. In general, these further studies support the existence of a positive correlation between life span and DNA repair capacity.

Six studies involved mice. Ley *et al.* (1977) and Sutherland *et al.* (1980) showed that the low excision repair capacity of mouse cells compared with that of human cells is not only a property of skin fibroblasts *in vitro* but also is observed in skin *in vivo*. The other four mouse studies dealt with mouse strains of different longevities. Paffenholz (1978), using fibroblasts from three inbred mouse strains with life spans ranging from 300 to 900 days, found a positive correlation of UV-induced excision repair capacity and life span. Hall *et al.* (1981) compared the DNA repair capacities of lymphocytes of two mouse strains with life spans of about 300 and 900 days, respectively. Capacity for excision repair of UV-induced damage was observed to correlate positively with life span. However, Collier *et al.* (1982) found no correlation between UV-induced excision repair and longevity in embryonic cells derived from a congeneric pair of mouse strains, one long-lived and one short-lived. Hart *et al.* (1979) studied UV-induced DNA repair in the skin fibroblast of two closely related mouse species, one short-lived (3.4 yr) and one long-lived (8.2 yr). The rate of excision repair was 2.5-fold greater for cells of the latter, about the same as the ratio of life spans.

Two studies involved primates. Hart and Daniel (1980) measured the capacity of skin fibroblasts from six primate species to

Table X
Studies of the Correlation between Life Span and DNA Repair Capacity

Type of Repair	Cell Type	Animals Compared	Comment	Reference
UV excision repair	cultured skin fibroblasts	7 mammalian species	repair capacity increased systematically with life span	Hart and Setlow, 1974
UV excision repair	skin	mouse and human	repair capacity greater in human	Ley *et al.* 1977; Sutherland *et al.*, 1980
UV excision repair	embryo fibroblasts	3 mouse strains	repair capacity correlated with life span	Paffenholz, 1978
UV excision repair	lymphocytes	2 mouse strains	repair capacity correlated with life span	Hall *et al.*, 1981
UV excision repair	embryonic cells	2 mouse strains	no correlation between life span and repair capacity	Collier *et al.*, 1982
UV excision repair	skin fibroblasts	2 mouse species	repair capacity correlated with life span	Hart *et al.*, 1979
UV excision repair	skin fibroblasts	6 primate species	repair capacity correlated with life span	Hart and Daniel, 1980
UV excision repair	skin fibroblasts and lymphocytes	6–9 species of primate	both cell types showed a correlation of repair capacity with life span	Hall *et al.*, 1984
UV excision repair	mainly skin fibroblasts	21 species of mammal	repair capacity correlated with life span	Francis *et al.*, 1981
UV excision repair	epithelial cells of lens	5 species of mammal	repair capacity correlated with life span	Treton and Courtois, 1982
UV excision repair	hepatocytes	5 species of mammal	repair capacity correlated with life span	Maslansky and Williams, 1985
UV excision repair	fibroblasts mainly from lung	34 species	repair capacity correlated with life span if data on primates and bats excluded	Kato *et al.*, 1980

(*continued*)

Table X (*Continued*)

Type of Repair	Cell Type	Animals Compared	Comment	Reference
UV excision repair	turtle heart and trout gonadal and molly embryonic tissue	3 cold-blooded vertebrates	repair capacity very low; no correlation with life span	Woodhead *et al.*, 1980
UV photoreactivation	connective tissue fibroblasts	2 fish species	photoreactivation capacity correlated with life span	Regan *et al.*, 1982
O^6-methylguanine repair	liver	human and rat	repair capacity 10-fold greater in human	Montesano *et al.*, 1982
O^6-methylguanine repair	liver	human, monkey, and rat	human and monkey similar and 8–10-fold greater than rat	Hall *et al.*, 1985
O^6-methylguanine repair	chondrocytes	human and rabbit	repair capacity 5-fold greater in human	Lipman *et al.*, 1987

repair UV-induced damage. The data showed an excellent correlation between repair capacity and life span. Hall *et al.* (1984), in a double-blind study, examined UV-induced excision repair in fibroblasts cultured from punch biopsies from six primate species and in lymphocytes from the blood of nine primate species of widely different life spans. Both test systems, fibroblasts and lymphocytes, showed a good linear relationship between maximum life span and excision repair.

Four further studies used a variety of mammalian species. Francis *et al.* (1981) examined the amount of excision repair in response to UV-induced damage in cells, usually skin fibroblasts, from 21 different species of mammals. They measured both the number of repaired regions per 10^8 daltons of DNA and the patch size. Their data showed a correlation between the number of repaired regions and life span. Treton and Courtois (1982) measured unscheduled DNA synthesis in the epithelial cells of the lens of five species: rat, rabbit, cat, dog, and horse. The positive correlation between repair capacity and life span was excellent. Maslansky and Williams (1985) measured, by autoradiography, the amount of DNA repair synthesis induced by UV-irradiation in hepatocyte primary cultures derived from five species: mouse, hamster, rat, guinea pig, and rabbit. A positive correlation was found between the amount of DNA repair elicited by low UV fluences and species longevity.

Kato *et al.* (1980) investigated UV-induced unscheduled DNA synthesis in fibroblast cells from 34 species representing 11 orders. Most of the investigated cells were derived from lung tissue explants obtained from animals of unknown ages. This may have been a flaw in their procedure because repair capacity in mouse lung fibroblasts has been reported to decline with age (Kempf *et al.*, 1984). These authors found no correlation between repair and life span. According to Tice and Setlow (1985), two orders of species in the Kato *et al.* (1980) study destroy any positive correlation that might exist. These orders of species are the primates and Chiroptera (bats). In the data of Kato *et al.* (1980), the primates have a short life span and high repair synthesis, but this result contradicts the more extensive data of Hart and Daniel (1980). In the case of Chiroptera, the data of Kato *et al.* (1980) indicate long life span and low repair synthesis. However, Tice and Setlow (1985) argued that bats have a very different metabolic regime than the other animals considered. They pointed out that if both primates and Chiroptera are omitted from the data of Kato and coworkers there is a reasonable positive correlation between repair synthesis and life span among the remaining 22 data points.

Two studies used cold-blooded vertebrates. Woodhead *et al.* (1980) measured excision repair in response to UV-irradiation in cells from the Carolina box turtle, the rainbow trout, and the Amazon molly fish. The respective life spans of the species are 118–123 yr, up to 8 yr, and about 3 yr. The amount of excision repair measured in each case was small or nonexistent. This finding may be explained by the suggestion of Tice and Setlow (1985) that in the lower vertebrates another known repair process, photoreactivation (Chapter 9, Section II.F.1), may be more important than excision repair. Photoreactivation is a process that can restore the most frequent type of UV-induced DNA damage, the pyrimidine dimer, to its undamaged condition without the need for excising the dimer.

In support of the hypothesis that photoreactivation is important in lower vertebrates, Regan *et al.* (1982) have shown, for primary fibroblast cultures derived from the connective tissue of two closely related fish species, that there is fivefold more photoreactivating activity in cells from the species with the fivefold greater life span. This is an indication that life span correlates with DNA repair capacity (i.e., photoreactivation) in fish.

Three further investigations measured the ability to repair the altered base, O^6-methylguanine. Montesano *et al*. (1982) and Hall *et al*. (1985) found that the capacity to repair O^6-methylguanine damage in DNA of human and monkey liver is 8–10-fold greater than in rat liver. Lipman *et al*. (1987) found that the capacity of human chondrocytes to repair O^6-methylguanine damage is fivefold greater than for rabbit chondrocytes. Again, these results suggest that DNA repair capacity may correlate with life span in mammals.

Overall, the evidence indicates a good correlation between repair of DNA damages and life span in mammals. As pointed out by Tice and Setlow (1985), the expectation of a causal relationship between DNA repair capability and aging is based on the hypothesis that aging is a direct consequence of the detrimental accumulation of unrepaired DNA damages. They further noted that the more effective a cell is in preventing or repairing DNA damage, the more slowly it should age. Thus, the observations indicating that DNA repair capacity correlates with life span fulfill an important expectation of the DNA damage hypothesis of aging.

On the other hand, we should note that none of the studies reported in Table X involved repair of oxidative DNA damages. As we discussed in Chapter 6, Section III.A, oxidative DNA damages occur at a high rate in animals with a high metabolic rate and short life span, and at a low rate in animals with a low metabolic rate and long life span. For oxidative damages, we might not expect to observe a good correlation between repair capacity and longevity because high repair might be needed in metabolically active short-lived animals, merely to cope with the high incidence of DNA damage.

B. Possible Explanations for the Positive Correlation Found between Life Span and Excision Repair of UV-Induced Damages

Despite the correlation in most of the studies with mammals between capacity to repair UV-induced DNA damages and life span, UV damage is unlikely a primary cause of mammalian aging. UV-irradiation does not penetrate the skin and, hence, cannot be a factor in the aging of other organs (such as muscle and brain), which appear to play more crucial roles in aging of the whole animal (Chapter 5, Section II). What, then, is the significance of this correlation? Tice and Setlow (1985) suggested that many different types of repair activities may be coordinately expressed and that the findings for UV repair may be only an indication that other, presumably more important, repair systems are correlated with life span. Studies of the inherited human condition xeroderma pigmentosa (XP) provides supporting evidence for this idea. Thus, we now describe XP in some detail.

XP is an autosomal recessive disease. Individuals homozygous for the XP gene are hypersensitive to UV-irradiation and have a high incidence of skin cancers. XP cells are defective in the repair of damaged DNA containing pyrimidine dimers (cyclobutane pyrimidine dimers and pyrimidine [6-4] pyrimidine adducts) (Zelle and Lohman, 1979; Mitchell *et al.*, 1985). Fusion of cells from different individuals with XP has defined nine genetic complementation groups (A–I), implying that DNA repair in humans involves multiple gene products. So far, the molecular basis for the XP defects remains undefined. XP cells are not only defective in repair of pyrimidine dimers but are also defective in the removal of chemical adducts generated by DNA-damaging agents such as acetylaminofluorene (AAF), psoralen plus near UV light (PUVA), and benzpyrene (Amacher and Lieberman, 1977; Kaye *et al.*, 1980; Yang *et al.*, 1980; Wood *et al.*, 1988). These DNA-damaging agents introduce damages that cause major distortions in the double-helix structure. This implies that measurements of excision repair in response to UV-induced damage should reflect repair capacity for a range of damages, possibly including one or more types important in aging. Although the increased incidence of cancer in individuals with XP is dramatic in sun-exposed areas of the skin, other types of cancer are also elevated. Kraemer *et al.* (1984) estimated a 12-fold increase in occurrence of neoplasms in sites not exposed to UV-irradiation. This includes a disproportionate representation of malignant neoplasms of the brain and oral cavity. Thus, in XP there is presumably an inability to repair types of damages in addition to those caused by UV.

In Chapter 2, Section II.A, we discussed a type of natural DNA damage that appears to be very frequent, the apurinic–apyrimidimic site (AP site). This damage may arise by spontaneous hydrolysis of DNA or by the removal of oxidatively modified bases. AP sites, characteristically, are labile under alkaline conditions and give rise to DNA strand breaks. Hurt *et al.* (1983) showed that gaps or alkali-labile sites, or both, accumulate in the DNA of XP cells to a greater extent than in normal cells. Puvion-Dutilleul and Sarasin (1989) presented evidence for an abnormal weakness in the chromatin in a proportion (10–80%) of XP cells, which results in the breakage of some DNA fibers in alkaline conditions. These results suggest that XP cells may have a higher level of AP sites than normal cells.

AP endonucleases are enzymes that specifically attack AP sites as part of a repair process that removes these damages and replaces them by intact nucleotides (Chapter 9, Section II.B). Kuhnlein *et al.* (1976, 1978) and Lambert *et al.* (1983) showed that an AP endonuclease is defective in XP complementation groups A and D. Furthermore, the repair enzyme AP endonuclease I has been shown to compensate for the defect in XP-D cells suggesting that AP endonuclease I may be the defective gene product in these cells (Linn *et al.*, 1988). Individuals with cells in XP groups A and D have severe neurological abnormalities. These findings imply that normal individuals have the capacity to repair some intrinsic DNA damages, possibly AP sites, and that this capacity is particularly critical for achieving normal neuronal function. Andrews *et al.* (1978) and Robbins *et al.* (1983b) have postulated that defective DNA repair in individuals with XP causes premature neuron death. This suggests that, in humans, UV-induced damage may indeed be repaired by a pathway that is the same as, or coordinately regulated with, a DNA repair pathway important in determining neuron survival and therefore longevity.

To connect the studies we have summarized here to our arguments on DNA damage as the cause of aging, characteristics of accelerated aging need to be shown in individuals with XP. As noted previously (Chapter 6, Section IV), Martin (1978) compiled a list of 162 genetic syndromes in humans that show characteristics of accelerated aging. He used 21 separate criteria to establish a ranking. Among the top 10 syndromes, the highest (Down's syndrome [DS]) was positive by 15 of the 21 criteria and myotonic dystrophy was tenth, meeting 6 of the 21 criteria. XP fell just below this, meeting 5 of the 21 criteria for accelerated aging (see also Table XI, later). In addition, Andrews *et al.* (1978) considered that all XP patients develop premature aging of sun-exposed skin and that the neurological abnormalities of XP are the results of abnormal aging of the human nervous system.

Because none of the genetic syndromes in humans meet all 21 criteria of accelerated aging, Martin (1978) refers to those that meet some criteria as "segmental progeroid syndromes" to distinguish them from hypothetical "global progerias." If life span is determined by multiple protective and repair processes, loss of any of the more important processes might accelerate some aspects of aging, but not others. Thus, segmental progeroid syndromes may give clues about which processes of aging are significant. By this reasoning, the repair pathway(s) defective in individuals with XP, which are important in repairing UV-induced DNA damages, may be significant in determining life span in humans. This could explain the positive correlation between the capacity for UV-induced excision repair and the expected life span of individuals of various mammalian species.

II. *Experimental Acceleration of Aging by DNA-Damaging Agents*

If aging is the consequence of an accumulation of unrepaired DNA damages, then exposure of animals to a greater than normal level of an agent capable of damaging DNA might accelerate the aging process. On the other hand, with exposure from an external source, the type of damage, the distribution of damage by cell type, and the distribution of damage over time are likely to be very different than for naturally accumulating damage. The differences expected between external source damages and naturally accumulated damage can be illustrated by two examples. First, natural DNA damages, which occur over a lifetime, should reach high levels in nondividing cells but may not accumulate in dividing cells due to turnover (Chapter 5, Section I.C). In contrast, DNA damages introduced by high sublethal doses of total body irradiation administered over a short period would be expected to have a large effect on rapidly dividing cells and act by interfering with DNA replication. Second, if oxidative DNA damage is important (see Chapter 6), cells with a high level of oxidative metabolism, such as brain cells, should be more subject to natural DNA damages than most other cell types. However, whole body ionizing radiation should not necessarily produce this bias in the type of cell most affected. Similar arguments would also apply to DNA damages caused by administration of DNA-damaging chemicals. Thus, the effects of externally applied DNA-damaging agents should only partially, or poorly, mimic natural aging. We now discuss studies that measure the effect of ionizing radiation or chemical DNA-damaging agents on the acceleration of the aging process.

A. Ionizing Radiation

Numerous studies have shown a life-shortening effect of sublethal total body X-irradiation on mammals. Casarett (1964), in an early review of the subject, referred to 51 separate studies. In addition, he cited 17 of these studies in which the authors noted that animals dying prematurely after sublethal radiation exposure exhibited physical changes suggestive of premature senescence. In addition to the review of Casarett (1964), Alexander (1967) and Walburg (1975) have also reviewed and interpreted studies relating ionizing radiation to aging. We describe some of these studies below and give the major points of the three reviews.

Studies supporting the conclusion that ionizing radiation seems to accelerate natural aging are given first. Warren (1956), in a study of the deaths of 82,441 physicians reported in the period 1930–1954 inclusive, found that radiologists die on the average 5.2 yr earlier than do other physicians. The nonradiology specialists known to be exposed somewhat to radiation also showed definite shortening of life, but less than that of radiologists. Warren concluded that exposure to ionizing radiation is the predisposing factor in the shortening of life observed here. Analysis of the autopsy data suggested that the radiologists died younger from practically every cause of death and not merely from those manifestly related to radiation injury. He noted that this strongly suggests that radiologists are subject to some common factor that speeds up the aging process.

Upton (1957) showed that exposure of mice to ionizing radiation shortened their life span in proportion to the amount of radiation absorbed. He concluded that the life-shortening action of radiation involves the induction of neoplasia and non-neoplastic degenerative changes and mimics in many respects acceleration of the natural aging process.

Lindop and Rotblat (1962) irradiated mice with a single whole-body dose of X-rays at age 30 days and demonstrated a decrease in life span proportional to the dose. Upon postmortem examination of both the irradiated and control animals, it appeared that the life-shortening effect was due to an advancement of all causes of death, and that there was no real change in the distribution of the causes of death in irradiated animals compared with controls. Hence, they argued that X-irradiation induces premature aging.

Studies supporting the view that radiation-induced life-shortening does not closely mimic natural aging are given next. Hollingsworth *et al.* (1969) conducted neuromuscular tests of aging on Hiroshima subjects. The tests they used were designed to determine whether or not there were perceptible differences in the aging of atomic bomb survivors exposed to widely differing amounts of radiation. The results of the three tests carried out—strength of hand grip, reaction time in a light-extinction test, and the voltage required to detect vibration—clearly varied with age of the subjects. However, no changes definitely related to radiation exposure were detected.

Alexander and Connell (1960) found that the life span of mice was reduced by exposure to X-rays but that the radiation did not hasten typical changes seen in unirradiated aging animals, such as alterations in collagen. They concluded that the few facts available at that time did not indicate that the process is similar to spontaneous aging.

Work by Berech and Curtis (1964) showed a further difference between radiation-induced life-shortening and natural aging. Because various kidney functions decrease with age, these authors examined kidney function in both the aging and irradiated mouse to determine whether or not any similarities exist in the effects of radiation and aging. In general, no evidence indicated X-ray-induced premature aging of kidney function. For instance, female mouse kidneys showed a decline with respect to age in the ability to concentrate urine during a 24-hr water fast. X-irradiation early in life had no measurable long-term effect on the rate of aging of this process.

Lovelace and Davis (1963) showed that although the visual acuity of rhesus monkeys declines with age there is no effect of X-irradiation. Boer and Davis (1968) demonstrated that rhesus monkeys show remarkable changes in spontaneous behavior with age. Irradiated monkeys also showed behavioral changes, but these were in a different direction than their unirradiated counterparts. These investigators concluded that aging trends were not augmented by radiation effects.

Next we discuss three reviews on the subject of radiation-induced premature aging. Casarett (1964) described changes in four areas following irradiation in comparison with normal aging: actuarial, pathological, histopathological, and physiological. He concluded that the data indicate a strong qualitative resemblance between the permanent and late effects of life-shortening total-body irradiation and the manifestations of premature aging, although qualitative differences appear to exist and quantitative factors are not always equivalent or proportional. He further noted that although the manifestations of normal and radiological aging are similar there was a paucity of strictly defined and controlled information on the subject.

The review of Alexander (1967) summarized evidence leading to his conclusion that although radiation shortens average life span, the underlying processes responsible differ significantly from the cellular and subcellular changes associated with normal aging. He considered that the accumulation of DNA damage in postmitotic cells (e.g. nerve or muscle cells) was central to aging (see Chapter 1, Section III.A). He hypothesized that as these cells age breaks in DNA accumulate until the capacity for RNA synthesis decreases. However, he noted that the amount of radiation needed to produce this degree of fragmentation would be expected to be so high that it could not be given to mammals, because they would be killed owing to the destruction of essential stem cells in the bone marrow and the gut. Hence, this type of killing of postmitotic cells will not occur under the conditions of radiation-induced life-shortening of mammals.

Walburg (1975) published a critical re-examination of the hypothesis that radiation-induced life-shortening is equivalent to premature aging. He noted that the key test of the hypothesis is whether or not the time of all causes of death advances. According to Walburg, very few diseases were specifically identified as advanced in time, and only some neoplastic diseases and nephrosclerosis were accelerated by radiation. In addition, the available evidence reviewed by Walburg suggested an absence of radiation-induced acceleration of age changes in neuromuscular performance and behavior patterns. On the basis of the overall evidence, Walburg (1975) concluded that radiation does not result in a generalized advancement in the rate of aging. His own hypothesis was that radiation is a physical toxic agent that induces or accelerates some, but not all, specific diseases. At moderate to low doses, he concluded, radiation shortens life span principally, and perhaps exclusively, by induction or acceleration of neoplastic diseases.

It seems that despite the early enthusiasm for the idea that X-irradiation causes accelerated aging, more recent work indicates that the life-shortening effect of radiation only has limited similarity to normal aging. Although most of the attempts to induce accelerated aging were carried out with ionizing radiation, some experiments were also done with DNA-damaging chemicals. These are described next.

B. DNA-Damaging Chemicals

The effects of alkylating agents on life-shortening and aging have been studied. Alexander and Connell (1960) showed a life-shortening effect on mice with the nitrogen mustard Chlorambucil and the alkylating agent Myleran. Dunjic (1964) showed that Myleran reduces the life span of rats and noted that the causes of death apparently did not differ from those found in normal controls dying spontaneously. Conklin *et al.* (1963) subjected mice to mid-lethal doses of nitrogen mustard, triethylene melamine, or whole-body X-rays and then observed survivors throughout the remainder of their lives for effects on longevity and the incidence of late-occurring diseases. Life span was shortened by all treatments. They noted that the reduction in longevity was not attributable to the induction of neoplasia but was correlated with an increase in the age-specific death rate from diseases of all types. However, the increase varied in degree from one disease to another, and they concluded that the effects were not equivalent to premature aging in the simplest sense. Kodell *et al.* (1980) fed mice 2-acetylaminofluorene (2-AAF) at various concentrations. They found a significant dose-related life-shortening effect due to 2-AAF. While a substantial portion of the reduction in survival was attributable to bladder and liver tumors induced by 2-AAF, an additional effect was observed at the highest dose used, which did not appear to be specific to a particular cause of death whether neoplastic or otherwise.

Ohno and Nagai (1978) administered (7,12-dimethyl-benz(a) anthracene) (DMBA) neonatally to mice. The average life span of female mice was reduced by DMBA from 608 to 297 days. They noted that the DMBA-induced life-shortening appeared to be natural in that death was almost invariably preceded by menopause, graying of the coat, and a gradual loss of body weight (in that order). Neoplasms, although induced in considerable variety, were not a main cause of the DMBA-induced life-shortening. Furthermore, they reported that those mice that developed tumors died early, and no tumors were observed among those that lived beyond 184 days.

Both ionizing radiation and chemicals that mimic the effects of radiation probably exert their life-shortening effect primarily through DNA damage. However, as we pointed out in the introduction to Section II, the distribution of DNA damages in different tissues and over time when introduced from exogenous sources probably differs drastically from the natural distribution of DNA damages. We think that this difference is sufficient to explain the repeated observation that the pathological conditions associated with the life-shortening caused by exogenous DNA-damaging agents only partially, or poorly, mimic those of natural aging.

III. Genetic Syndromes in Humans with Features of Premature Aging

Many human genetic syndromes have been identified. A large number of these syndromes have one or more features of premature aging. On the DNA damage hypothesis of aging, syndromes showing premature aging should, at least in some cases, show evidence of excess DNA damage or deficiency in DNA repair. We have previously referred to the ranking by Martin (1978) of 162 genetic syndromes of humans according to their features of premature aging (Chapter 6, Section IV, and this chapter, Section I.B). In some of these 162 syndromes, cells of the affected individuals have been examined for the presence of excess DNA damage, the presence of excess intracellular DNA-damaging agents or for a deficiency in DNA repair. By Martin's ranking, Down's syndrome, which was discussed in Chapter 6, Section IV, ranked first with 15 of the 21 features of premature aging. Also, as discussed in Chapter 6, Section IV, cells from DS individuals have higher effective levels of DNA-damaging free radicals present and, furthermore, may be deficient in DNA repair. Therefore, DS conforms to the expectations of the DNA damage hypothesis of aging. A further four of the top 10 conditions in Martin's ranking (Cockayne's syndrome, Werner's syndrome, Hutchinson-Guilford progeria, ataxia telangiectasia) and six other syndromes showing some features of premature aging have also been found to have excessive levels of DNA damage or DNA-damaging agents or deficient levels of DNA repair. These 11 syndromes are listed in Table XI with their features of premature aging indicated. Two of them, DS and XP, have already been described (Chapter 6, Section IV, and Chapter 7, Section I.B). The other nine syndromes are described below.

Table XI

Genetic Syndromes with Features of Premature Aging and Possibly Increased DNA damage

Features of Aging	Syndromes[a]										
	DS	CS	WS	HGP	AT	HD	XP	DM	FA	AD	PD
Potential relevance to the "intrinsic mutagenesis" hypothesis of aging	X	X	X	X	X		X				
Increased frequencies of nonconstitutional chromosome aberrations	X		X		X		X				
Increased susceptibility to one or more types of neoplasms of relevance to aging	X		X		X		X				
Possibility of a defect in a stem-cell population or in the kinetics of stem-cell proliferation	X		X	X		X					
Premature graying or loss of hair or both	X		X	X	X	X					
Dementia or certain types of relevant degenerative neuropathology	X	X			X	X	X		X	X	X
May be susceptible to slow virus	X								X		

(*continued*)

Table XI (*Continued*)

Features of Aging	DS	CS	WS	HGP	AT	HD	XP	DM	FA	AD	PD
					Syndromes[a]						
Various types of amyloid depositions	X										
Increased depositions of lipofuscin pigments	X	X				X					
Diabetes mellitus	X	X	X		X	X		X	X		
Disorder of lipid metabolism		X		X							
Hypogonadism	X	X	X	X	X						
Autoimmunity	X										
Hypertension		X		X							
Degenerative vascular disease	X	X	X	X				X			
Osteoporosis		X	X	X							
Cataracts (may overlap with senile cataracts)	X	X	X					X			
Abnormalities of mitochondria											
Regional fibrosis		X	X	X	X		X		X		
Variations in amounts and/ or distributions of adipose tissue	X	X	X	X		X		X			
Miscellaneous aspects not listed above											

[a] Syndromes arranged from left to right in descending order with respect to numbers of features of premature aging in each syndrome. DS (Down's syndrome), CS (Cockayne's syndrome), WS (Werner's syndrome), HGP (Hutchinson-Guilford progeria), AT (ataxia telangiectasia), HD (Huntington's disease), XP (xeroderma pigmentosum), DM (diabetes mellitus), FA (Friedreich ataxia), AD (Alzheimer's disease), PD (Parkinson's disease). The data for all of the listed syndromes, except for HD, are from Martin (1978). The data on HD are from Farrer (1987).

A. Cockayne's Syndrome

Cockayne's syndrome (CS) is believed to be transmitted by an autosomal recessive gene as indicated by familial incidence and the history of consanguinity in some parents of affected children. Individuals with CS are characteristically dwarfed, with growth and development arrested at ages from several months to a few years.

The other prevalent symptoms include prominent photosensitivity, deafness, optic atrophy, intracranial calcifications, mental deficiency, large ears and nose, sunken eyes, long arms and legs, type II lipoproteinemia, and general appearance of premature aging (Cockayne, 1946; Guzzetta, 1972; Fujimoto *et al.*, 1969; Schmickel *et al.*, 1977). The 12 features of accelerated aging in CS noted by Martin (1978) are indicated in Table XI.

That individuals with CS are sensitive to sunlight suggested that they may be defective in the repair of UV-induced damages. Skin fibroblasts from these individuals proved to be significantly more sensitive to killing by UV than cells from normal individuals (Schmickel *et al.*, 1977; Wade and Chu, 1979; Deschavanne *et al.*, 1981; Marshall *et al.*, 1980). Cells from individuals with CS are also sensitive to the DNA-damaging chemicals NQO and AAF (Wade and Chu, 1978, 1979). CS is not associated with an increased rate of carcinogenesis. Cells from different individuals with CS can be fused in culture, allowing tests of genetic complementation to be carried out. By this test three genetic complementation groups have been identified (Tanaka *et al.*, 1981; Lehman, 1982), implying that defects in any one of at least three genes may give rise to the CS phenotype.

Of the 11 CS strains tested by Lehman (1982), two were assigned to complementation group A, eight to group B, and only one to group C. The strain in group C is derived from an individual who also had XP and was the sole known representative of XP-complementation group B. This implies that a single mutation can give rise to both conditions.

CS cells have an apparently normal ability to make incisions in DNA at sites of UV-induced damage (Friedberg *et al.*, 1979), to excise such damages, and to carry out DNA repair synthesis (Ahmed and Setlow, 1978; Lehman *et al.*, 1979; Wade and Chu, 1979). Mayne and Lehman (1982) have speculated about the nature of the repair defect in CS. Immediately after UV-irradiation of cells, there normally may be a rapid excision of UV-induced damages from actively transcribing regions, thereby preventing any significant inhibition of RNA synthesis. This supposed preferential DNA repair may be defective in CS cells. Such repair would be undetectable by conventional measurements because it represents only a small proportion of total excision repair. Recently, Mayne *et al.* (1988) reported evidence supporting this idea.

B. Werner's Syndrome

Werner's syndrome (WS) is inherited as an autosomal recessive trait and has its onset in persons between the ages of 15 and 20 yr (Epstein *et al.*, 1966; Zucker-Franklin *et al.*, 1968). As summarized by Epstein *et al.* (1966) and Brown (1987), individuals with WS generally appear normal during childhood but cease growth during their early teenage years. The hair turns prematurely gray and whitens. Other striking features are early cataract formation, skin that appears aged with a sclerodermatous appearance, a high-pitched voice, peripheral musculature atrophy, poor wound healing, chronic leg and ankle ulcers, hypogonadism, widespread atherosclerosis, soft tissue calcification, osteoporosis, and a high prevalence of diabetes mellitus. Neoplasms develop in about 10% of individuals, with a particularly high incidence of sarcomas and meningiomas. A common cause of death is complications of atherosclerosis in the fourth decade.

The 12 features of accelerated aging in WS noted by Martin (1978) are indicated in Table XI. WS is considered by some gerontologists to be a model of precocious aging. Cultured fibroblastlike cells from patients with WS have a reduced proliferative capacity (Martin *et al.*, 1970; Salk, 1982). Hoehn *et al.* (1975) reported that chromosomal aberrations occurred in high frequency in the cells of an individual with WS. Other studies showing that WS fibroblasts and lymphocytes have chromosome abnormalities and that the chromosomes of cultured fibroblasts have a variety of spontaneous translocations were reviewed by Salk (1982).

Fujiwara *et al.* (1977) investigated the DNA repair capacity and the rate of DNA replication in skin fibroblasts from five individuals afflicted with WS. These cells exhibited normal levels of sensitivity toward X-ray and UV-killing, normal levels of repair of X-ray-induced single-strand breaks, and normal levels of UV-induced repair synthesis. However, the elongation rate of DNA chains during replication was significantly slower in WS cells than in normal cells. Further studies on the levels of UV-induced repair synthesis in seven WS lines have also indicated normal repair levels (Higashikawa and Fujiwara, 1978). Nikaido *et al.* (1985) also found that WS cells were capable of repairing both UV and gamma-irradiation damage.

Gebhart *et al.* (1985) examined chromosomal instability in WS.

They found a 4–10-fold higher frequency of spontaneous chromosome breaks in peripheral blood samples from two individuals with WS than normal individuals. However, cells from these individuals did not have increased sensitivity to several chemical clastogens (agents that cause chromosomal breakage). The specific agents tested were diepoxybutane, isonicotinic acid hydrazide, 4-nitroquinoline-1-oxide and bleomycin. In addition, there was no increase in sister-chromatid exchange frequency in response to DNA-damaging agents in WS cells compared with normal cells. Stimulation of sister-chromatid exchange is a conventional test for DNA-damaging activity. Furthermore, there was no indication of a greater inhibition of proliferation by DNA-damaging agents in the WS cells compared with normal cells. In a recent paper, Stefanini *et al.* (1989) reported chromosome abnormalities in a WS individual but found normal repair capability with respect to DNA damages introduced by UV or by mono- and bifunctional alkylating agents.

In summary, evidence indicates reduced cellular proliferative capacity (Martin *et al.*, 1970; Salk, 1982) and an increase in spontaneous chromosome aberrations (Hoehn *et al.*, 1975; Salk, 1982; Gebhart *et al.*, 1985; Stefanini *et al.*, 1989) in WS cells compared with normal cells, suggesting increased DNA damage. However, the basis for this increased DNA damage does not appear to be reduced repair. This is indicated by the lack of increased sensitivity to UV- and X-ray-induced damages (Fujiwara *et al.*, 1977; Higashikawa and Fujiwara, 1978; Nikaido *et al.*, 1985) and to chemical clastogens (Gebhart *et al.*, 1985; Stefanini *et al.*, 1989). The apparent lack of a repair defect in WS suggests the alternative explanation that increased DNA damage is caused by a higher level of an endogenous DNA-damaging agent. The basic enzymatic or metabolic abnormality has not yet been established for WS. Fujiwara and Ichihashi (1982) presented evidence of abnormally enhanced synthesis and accumulation of sulfated glycosaminoglycan in the medium of WS fibroblasts. As one possibility for explaining the rapid cellular aging in WS, they suggested that increased sulfated glycosaminoglycan might affect the cells' DNA replicative potential. Furukawa and Bhovanandan (1982) reported that glycosaminoglycan components can inhibit DNA synthesis in isolated rat nuclei. Although the mechanism is unknown, it is possible that these components cause DNA damage.

C. Hutchinson-Guilford Progeria Syndrome

Hutchinson-Guilford progeria (HGP) is considered to be a disease model of certain aspects of accelerated aging. DeBusk (1972) and Brown (1987) have summarized the characteristic features of this condition. Individuals with HGP generally appear normal at birth, but by about age 1 yr, they show severe growth retardation. In the first few years of life, balding and loss of eyebrows and eyelashes commonly occur. As a result of widespread loss of subcutaneous tissue, the veins over the scalp become prominent. Pigmented age spots arise in the skin, which has a generally aged appearance. Individuals with HGP are very short, averaging 101 centimeters, and weigh no more than 11–14 kilograms as teenagers. They have a thin and high-pitched voice and usually do not develop to sexual maturity. Characteristically, they have prominent eyes, a beaked nose, a disproportionately large cranium, and a small jaw. Their large head, their balding, and their small face give them an extremely aged appearance. Individuals with HGP have normal to above-average intelligence. Over 80% of deaths are due to heart attacks or congestive heart failure, and the median age of death is 12 yr. The 10 features of accelerated aging in HGP listed by Martin (1978) are indicated in Table XI.

Cultured HGP fibroblasts exhibit poorer growth and generally earlier *in vitro* senescence than normal human cells (Goldstein, 1969; Epstein *et al.*, 1974). Also, they have been reported to have decreased ability to repair single-strand breaks following gamma-irradiation (Epstein *et al.*, 1973, 1974). Brown *et al.* (1976) detected a decrease in the repair of radiation-induced single-strand breaks in mid-passage fibroblasts from two individuals with HGP. Another experiment indicating a defect in repair of gamma-ray induced DNA damages was performed with adenovirus type 2 (Rainbow and Howes, 1977). Samples of irradiated and unirradiated virus were allowed to infect HGP and normal cells for 48 hr. At this time, the cells were examined for the presence of viral structural antigens by means of an immunofluorescent stain. Production of viral antigens by the irradiated virus was markedly reduced in the HGP cells compared with the normal cells, indicating a defective ability of HGP cells to repair the DNA damages in the virus.

The finding of a decreased DNA repair capability has not been confirmed by other investigators, however. Regan and Setlow (1974) and Bradley *et al.* (1976) reported normal single-strand break rejoining ability in HGP cell lines. Furthermore, Bender and Rary (1974) have reported normal levels of induced chromosomal aberrations in X-irradiated cultured lymphocytes from the HGP individual examined by Epstein *et al.* (1973). A higher level of aberrations might be expected if single-strand break rejoining were defective. The level of spontaneous chromosomal aberrations in lymphocyte and fibroblast cultures also was not elevated. Little (1976) has suggested that the repair loss in some HGP cell lines is a result of gene repression and/or of altered metabolism, but not of an intrinsic defect in DNA repair. Brown *et al.* (1980) obtained results indicating that heterogeneity for DNA repair capacity exists among HGP fibroblast strains. Because of this, they concluded that decreased DNA repair capacity is unlikely to be the basic genetic defect in HGP.

D. Ataxia Telangiectasia

As summarized by Friedberg (1985), ataxia telangiectasia (AT) is an autosomal recessive genetic disorder that affects many systems of the body, particularly the nervous system, the immune system, and the skin. Characteristically, profound cerebellar ataxia results in a staggering gait, severe lack of muscular coordination, and progressive mental retardation. The term telangiectasia refers to the marked dilation of small blood vessels, most easily observed in the eye and skin. There is also immune dysfunction and an increased incidence of neoplasms of the lymphoreticular system.

The eight features of accelerated aging in AT noted by Martin (1978) are indicated in Table XI. Cells from individuals with AT have a high frequency of spontaneous chromosomal aberrations, and X-irradiation markedly increases the number of chromosomal breaks in these cells (German, 1972; Bochkov *et al.*, 1974; Cohen *et al.*, 1975; Oxford *et al.*, 1975; Taylor *et al.*, 1976; Webb *et al.*, 1977). AT cells have a substantially increased sensitivity to ionizing radiation (Taylor *et al.*, 1975; Chen *et al.*, 1978; Paterson *et al.*, 1979; Arlett and Harcourt, 1978, 1980). The extreme sensitivity to X-rays is common to all AT cell lines so far examined (Cox *et al.*, 1978).

Chen *et al.* (1984) hybridized different AT cell lines and measured the frequency of radiation-induced chromosome aberrations to determine complementation. By this criterion seven A-T cell lines fell into four complementation groups. This study, as well as previous ones cited by the authors, indicates that AT, like XP and CS, can occur as the result of mutations in a number of different complementation groups, which probably correspond to different genes. AT cells are also sensitive to bleomycin (Lehman and Stevens, 1979; Taylor *et al.*, 1979; Cohen *et al.*, 1981), which is a radiomimetic chemical. Although many of the effects of bleomycin are similar to those of ionizing radiation, the chemical damages produced in DNA by bleomycin are less diverse than those produced by ionizing radiation. As reviewed by Muller and Zahn (1977), bleomycin causes release of thymine and, to a lesser extent, other bases in DNA. Both single- and double-strand breaks are produced in free DNA and in cellular DNA by bleomycin, and both are rejoined in mammalian cells. Coquerelle *et al.* (1987) have shown that in an AT fibroblast cell line, bleomycin-induced double-strand breaks were repaired less efficiently than in normal human fibroblasts. AT cells do not have increased sensitivity to UV and chemicals classified as UV-mimetic. This evidence indicates that individuals with AT are defective in repair of some type of damage, perhaps some kind of double-strand break, induced by ionizing radiation and bleomycin. While the clinical symptoms of AT have no rigorous molecular explanation, Thacker and Debenham (1988) presented evidence for the involvement of topoisomerase II, an enzyme that controls the topological state of DNA. In summary, AT individuals have a number of features of premature aging and are also apparently defective in repair of a type of DNA damage caused by ionizing radiation and bleomycin.

E. Huntington's Disease

Huntington's disease (HD) is an autosomal dominant condition characterized by chorea and dementia with considerable variation in age of symptom onset. Farrer (1987) pointed out that HD has many of the major stigmata of aging listed by Martin (1978). These include dementia and degenerative neuropathology, possible defect in stem-cell proliferation, diabetes mellitus, variations in amount of adipose tissue, premature graying or loss of hair, and marked lipofuscin deposition in cortical biopsies or postmortem histopathology (Table XI). Two additional pathogenic findings noted by Farrer (1987) were excessive degeneration in body organs and age-related abnormalities of homovanillic acid metabolism. HD, like XP and AT, is a primary neuronal degenerative disease. Such diseases are progressive and characterized by premature death of neurons in the absence of pathological evidence of a specific cause (Scudiero *et al.*, 1981).

Cultured lymphocyte lines from four patients with HD were reported to be abnormally sensitive to the lethal effects of X-rays but had normal survival after exposure to UV (Moshell *et al.*, 1980). The hypersensitivity to X-rays was interpreted as an inherited defect in DNA repair. Chen *et al.* (1981) measured radiation-induced chromosome aberrations in lymphocytes and clonogenic survival of lymphoblastoid cell lines from HD patients and controls. As a group, the HD cell lines were significantly more sensitive than the controls. Mc-Govern and Webb (1982) also reported that lymphocytes from HD patients were more sensitive to ionizing radiation than lymphocytes from control subjects.

Scudiero *et al.* (1981) have shown that fibroblast cell strains from six patients with HD were abnormally sensitive to killing by N-methyl-N'-nitro-N-nitrosoguanidine (MNNG). This hypersensitivity was interpreted as reflecting defective repair of MNNG-induced damage. Lymphoblastoid cell lines from HD patients are sensitive to the radiomimetic drug bleomycin as assayed by reduced cell viability and increased frequency of chromosome aberrations compared with normal controls (Imray and Kidson, 1983).

Moshell *et al.* (1980) have suggested that the defective repair of neuronal DNA damaged *in vivo* by endogenous chemicals causes the premature death of neurons. They considered this premature death of neurons to result in the neurological abnormalities of XP, AT, and HD. They further postulated that the DNA repair process(es) defective in these conditions are normally required to maintain the functional integrity of the nervous system.

F. Diabetes Mellitus

Diabetes mellitus (DM) is not only a common metabolic disorder in the elderly, but it is also considered to be one of the models of premature aging (Shagan, 1976; Kent, 1976; Cerami, 1985). Shagan (1976) noted that, among younger diabetics, a variety of degenerative diseases occur earlier and with greater severity than among their nondiabetic peers. These diseases include peripheral arterial disease, arteriosclerotic heart disease, stroke, senile cataract, osteoporosis, osteoarthritis, periodontal disease with premature loss of teeth, tendonitis, bursitis, and a loss of vibratory sensation in the extremities. Thus, Shagan (1976) concluded that in many ways the younger diabetic appears to age more rapidly than normal. According to Kent (1976), diabetes has adverse effects on virtually all systems of the human body. It hastens the onset of other degenerative diseases in a manner almost indistinguishable from aging; thus diabetes might be considered a special kind of accelerated aging.

As reviewed by Perlmutter *et al.* (1984) certain aspects of cognitive function decline in the course of normal aging. The ability to retrieve recently learned information deteriorates with age much more than the ability to register or encode new information. Perlmutter *et al.* (1984) carried out a study to determine whether older (>55 yr) patients with noninsulin-dependent (type II) diabetes are at increased risk of cognitive deterioration. The results indicated that cognitive function is inferior in the patients with type II diabetes compared with a comparably aged nondiabetic control group. Apparently, cognitive impairment in diabetes, as in normal aging, was due to a deficiency in memory retrieval rather than to an attentional or encoding deficit. Other evidence for involvement of the nervous system in diabetes, particularly the peripheral nervous system, has been reviewed by Locke and Tarsy (1985).

Most patients with diabetes are hyperglycemic (i.e., have a high blood glucose level). Bucala *et al.* (1984) have shown that glucose reacts nonenzymatically with DNA at a low but significant rate. The related compound glucose-6-phosphate (an intermediate in glucose metabolism) reacts much more rapidly with DNA, and has been shown to inactivate bacteriophage DNA and cause DNA strand scissions (Bucala *et al.*, 1984). These authors proposed that glucose-6-phosphate, and glucose at a much slower rate, forms adducts with the primary amino groups of DNA bases, and that these adducts inhibit template function. With time, these adducts undergo chemical rearrangement that can labilize the glycosidic bond between purine and ribose. This leads to depurination and, after still more time, strand scission.

The chain of events underlying the injuries that DM inflicts on blood vessels, eventually resulting in the destructive long-term consequences of the disease, is still undefined. However, morphological and functional abnormalities of vascular endothelium are well recognized in diabetes. Lorenzi *et al.* (1986) have presented evidence that a high glucose level induces DNA damage, probably single-strand breaks, in cultured human endothelial cells. Thus glucose-induced DNA damage may account for some of the features of premature aging in diabetes.

G. Friedreich Ataxia

Friedreich ataxia (FA), a degenerative neurological disease, is listed by Martin (1978) as having potential relevance to the pathobiology of aging. Evidence for cellular hypersensitivity to ionizing radiation in FA has been presented by Lewis *et al.* (1979) and Chamberlain and Lewis (1982). The latter authors suggested that defective DNA repair mechanisms may be involved in the pathogenesis of FA.

H. Alzheimer's Disease

Alzheimer's disease (AD) is a degenerative disorder of the central nervous system in humans. In developed countries, the prevalence ratio for AD ranges between 1.9 and 5.8 cases per 100 population aged ≥65 yr (Rocca *et al.*, 1986). It affects more than 2 million persons in the United States (Katzman, 1986). AD is characterized by progressive impairment of memory and intellectual function beginning in middle to late adult life. Victims of AD develop profound mental and physical disability that often requires institutional care. As reviewed by Kidson and Chen (1986), AD most often occurs as single sporadic cases in given families; however, in a minority of cases of AD, two or more affected relatives occur in the same family. Some kindreds have many affected individuals. This may reflect a genetic determination of the disease. Where a pattern consistent with Mendelian dominant inheritance is observed, a predominantly genetic basis is usually assumed; however, it is not clear if all cases of AD have a genetic component.

St. George-Hyslop *et al.* (1987) showed that in four families the inheritance of AD was due to a genetic defect in the long arm of chromosome 21. However, other evidence suggests that AD may be due to genetic defects elsewhere in the genome (Van Broeckhoven *et al.*, 1987; Schellenberg *et al.*, 1988). Martin (1978) listed AD as one of the 162 genetic syndromes of potential relevance to the pathobiology of aging. Although AD has only two of Martin's 21 features of accelerated aging (Table XI), AD is associated with DS, the genetic condition with the most features of premature aging. Most individuals with DS who are >40 yr of age also show the neuropathology of AD (Wright and Whalley, 1984; Mann, 1988). Other evidence also suggests that AD has features of premature aging. As reviewed by Sajdel-Sulkowska and Marotta (1985), the hallmark lesions in the AD brain are neuritic (senile) plaques and neurofibrillary tangles. These may be found in small numbers in the hippocampus as a consequence of normal aging, but their frequency increases markedly in both the neocortex and hippocampus in AD.

Chromosomal aberrations, which may arise from DNA damage, have been found at elevated levels in AD by some investigators (Bergener and Junklass, 1970; Ward *et al.*, 1979; Nordenson *et al.*, 1980; Buckton *et al.*, 1983), although not by others (Mark and Brun, 1973; Martin *et al.*, 1981; White *et al.*, 1981). Wright and Whalley (1984) reviewed this work as well as other early evidence concerning one type of chromosomal aberration, aneuploidy in AD. They concluded that lymphocyte aneuploidy does not correlate with the degree of dementia and is likely an index of age-related changes occurring independently of the neuropathological process. The evidence on chromosomal aberrations other than aneuploidy is ambiguous because some authors have reported increases in such chromosomal aberrations in AD (Bergener and Junklass, 1970; Nordenson *et al.*, 1980), whereas others have not been able to confirm these findings (Mark and Brun, 1973; Martin *et al.*, 1981; White *et al.*, 1981).

There have been a number of reports of reduced DNA repair capacity in AD. Kidson and Chen (1986) assessed lymphocytes (T cells) and lymphoblastoid cell lines (B cells) from a series of 15 unrelated individuals with AD for response to ionizing radiation. Sensitivity of DNA to ionizing radiation was assessed by cell survival, cell clonogenicity, and induced chromosome aberrations. Results obtained by these three methods gave reasonable agreement. They found a range of sensitivities of cells from different donors with AD. The majority showed significantly greater sensitivity than cells from age-matched control donors. Kidson and Chen (1986) also reported studies on one extensive pedigree with familial AD. They found that cells from all available affected individuals in the pedigree, as well as a proportion of their first degree relatives, had increased sensitivity to ionizing radiation. Robbins *et al.* (1983a, 1985) reported a significant increase in X-ray sensitivity in AD cells. However, other workers, using unstimulated lymphocytes (Smith *et al.*, 1987) or phytohemagglutinin-stimulated lymphocytes (Smith and Itzhaki, 1989), found no difference in sensitivity to gamma-irradiation between AD cells and those from age-matched normal individuals.

Scudiero *et al.* (1986) reported that AD cells showed a small but statistically significant hypersensitivity to the DNA-damaging alkylating agent MNNG. They also reported that a cell line from a patient with a dominantly inherited form of familial AD had a hypersensitivity similar to that of three sporadic AD lines. Li and Kaminskas (1985), using the technique of alkaline elution, found that when AD fibroblasts (from familial cases) were treated with MNNG, single-strand breaks were repaired less efficiently than in normal fibroblasts. On the other hand, they found that bleomycin-induced double-strand breaks were repaired equally efficiently by both types of cells. Robison *et al.* (1987, 1988) studied repair of methyl methanesulfonate- and MNNG-induced DNA damages in lymphoblasts from individuals with familial AD compared with lymphoblasts from normal individuals. Analysis of DNA repair was done by using alkaline elution and cesium chloride equilibrium density gradient centrifugation. AD lymphoblasts were found to be defective in repair of both methyl methanesulfonate- and MNNG-induced DNA damages. Similar results were also reported by Bartlett *et al.* (1988). Das (1986) found that the sensitivity of stimulated lymphocytes, as assayed by sister-chromatid exchange, was the same in AD and normal cells after mitomycin C or ethyl methanesulfonate treatment. Jones *et al.* (1989) measured DNA repair in lymphocytes as determined by alkaline elution in individuals with AD and normal individuals. The DNA-damaging agents used were methyl methanesulfonate and MNNG. Their results indicated that cell lines from affected patients repair significantly less damage than cell lines from healthy controls.

The diversity of agents, cell types, and assay methods used precludes easy comparison of these studies with each other. To summarize the present state of understanding, AD cells were reported to be defective in repair of X-ray damages by some investigators (Kidson and Chen, 1986; Robbins *et al.*, 1983a, 1985) but not by others (Smith *et al.*, 1987; Smith and Itzhaki, 1989). AD cells have been reported to be sensitive to MNNG by Scudiero *et al.* (1986), Li and Kaminskas (1985), Robison *et al.* (1987, 1988), Bartlett *et al.* (1988), and Jones *et al.* (1989). They are also sensitive to methyl methanesulfonate (Jones *et al.*, 1989). There are also reports of a lack of sensitivity to mitomycin C and ethyl methanesulfonate (Das, 1986) and bleomycin (Li and Kaminskas, 1985).

If some types of repair are defective in AD cells, the result might be an accelerated accumulation of particular DNA damages leading to an accelerated decline in transcription and gene expression followed by cellular degeneration. Sajdel-Sulkowska and Marotta (1984) compared the levels of mRNA in neurologically normal human brains (after a brief postmortem interval) with those from individuals diagnosed for AD. The average yield of RNA from control cortical specimens was 100 ± 8 μg of RNA per gram of tissue as compared with 55 ± 10 μg of RNA per gram of tissue from AD specimens. Sajdel-Sulkowska and Marotta (1985) also examined the translational activity of mRNA from AD cortices. In contrast to the high translational activity of control mRNA, the AD mRNA exhibited relatively lower activity. They concluded that their data support the hypothesis that there is a lower rate of protein synthesis in AD patients relative to age-matched controls. This lower rate of protein synthesis is related to a lowered concentration of functional RNA; more specifically, to a decreased concentration of ribosomal RNA and mRNA and to a lower specific translational activity of the mRNA. As discussed in Chapter 5, Section II.A.1, normal aging is also associated with reduced gene expression in the brain, and this decline may be the result of inhibition of mRNA transcription by DNA damage. Defective gene expression may lead to progressive cellular dysfunction. A distinct population of basal forebrain neurons, the nucleus basalis of Meynart, undergoes a profound and selective degeneration in AD (Whitehouse *et al.*, 1982). These neurons are a major source of cholinergic innervation of the cerebral cortex. Iverson *et al.* (1983) also found a variable loss (about 60% reduction, on average) in locus coeruleus cells when compared with controls of similar age. Robbins *et al.* (1983a) proposed that the death of neurons in AD is a consequence of the accumulation of unrepaired DNA damage in neurons resulting from intracellular metabolic and/or spontaneous hydrolytic reactions.

In conclusion, AD is characterized by progressive neuronal degeneration which is regarded as a feature of accelerated aging. Evidence suggests reduced repair of some types of DNA damage, a decline in gene expression in the brain, and cellular degeneration in a specific region of the brain. AD may result from several different genetic defects, including trisomy of chromosome 21, which also gives rise to DS. We speculate that in the region of the brain responsible for AD, DNA damage may be promoted by either a high level of H_2O_2 (when it occurs in individuals with DS) or by reduced DNA repair capability.

I. Parkinson's Disease

As reviewed by Robbins *et al.* (1985), Parkinson's disease (PD) is like AD in a number of respects. Like AD, PD is characterized by progressive degeneration of the central nervous system in the elderly. Both AD and PD are generally sporadic disorders. In AD, neurons in the cerebral cortex, basal forebrain, and locus ceruleus are progressively lost, whereas in PD, neurons in the substantia nigra, basal forebrain, and locus ceruleus are progressively lost. PD is one of the conditions listed by Martin (1978) as being of potential relevance to the pathobiology of aging (Table XI).

Robbins *et al.* (1983a) found that cell lines from six patients with PD were significantly more sensitive to X-rays than were normal cell lines. Sensitivity to UV-irradiation was normal in these patients. Robbins *et al.* (1985) also showed increased X-ray sensitivity in lymphoblastoid cell lines from eight individuals with PD. This suggested that in PD, as in AD, there may be defective repair of the type of damage caused by X-rays. They suggested that such a DNA repair defect could cause an abnormally rapid accumulation of spontaneously occurring DNA damage in PD and AD neurons *in vivo*, resulting in their premature death.

J. Other Premature Aging Syndromes

For five syndromes that had six or more features of premature aging listed by Martin (1978), we are aware of no direct evidence for higher than normal levels of DNA damage accumulation or lowered repair capacity. These include four syndromes having eight features of premature aging (Seip's syndrome, cervical lipodisplasia, Klinefelter's syndrome, and Turner's syndrome) and one with six features (myotonic dystrophy). In one of these five cases, Klinefelter's syndrome, some indirect evidence suggests a higher than normal level of DNA damage may be present. In Klinefelter's syndrome, there is a higher risk of malignant tumors (e.g., McNeil *et al.*, 1981). Also, cells of individuals with Klinefelter's syndrome, have an increased frequency of sister-chromatid exchanges after treatment with mitomycin C compared with cells from normal individuals (Yamagata *et al.*, 1989). This suggests a possible DNA repair defect.

K. Summary of Genetic Syndromes in Humans with Both Premature Aging and Evidence of Excess DNA Damage or DNA-Damaging Agents or Deficiency in DNA Repair

In Table XII, we list the evidence for excess DNA damage, DNA-damaging agents, or repair deficiency in 11 human genetic syndromes that have features of premature aging. All of these syndromes are segmental progerias in that they exhibit some, but not all, features of premature aging. In three syndromes, the features of premature aging may be generated by excess production of chemicals that cause DNA damage. Thus, oxidative damage in DS, glucose adducts in DM, and glycosaminoglycan damage in WS may be responsible for the features of premature aging. In three syndromes (AT, WS, and XP), the presence of increased damage was suggested either by increased chromosomal aberrations or by increased neoplasms. Reduced ability to repair damages appears to be a characteristic of eight syndromes. Cellular sensitivity to X-rays occurs in AD, AT, FA, HD, PD, and, possibly, HGP. UV sensitivity is a characteristic of XP and CS. This suggests that natural DNA damages, similar to damages caused by X-rays or UV, may produce features of premature aging. Because damages caused by oxidative reactions are similar to those caused by X-rays, this evidence is consistent with the idea, proposed in Chapter 6, that oxidative DNA damage is an important cause of aging.

As we have argued in Chapter 5, Section I.C, neurons should be particularly vulnerable to an increased incidence of DNA damage or reduced repair because neurons with lethally damaged DNA cannot be replaced by replication of undamaged cells. Among the 11 syn-

Table XII

Evidence for Elevated DNA Damage in Human Genetic Syndromes with Features of Premature Aging

Syndrome	Evidence for Spontaneous DNA Damage	Evidence for Repair Defect	Agents Identified as Possibly Causing Chromosomal Damage
Alzheimer's disease	evidence contradictory (see text)	cells hypersensitive to X-rays (1–3)[a] and MNNG (4–9) and MMS (6–9)	
Ataxia telangiectasia	increased chromosomal aberrations (10–15)	cells hypersensitive to X-rays (16–20) and bleomycin (21–23)	
Cockayne's syndrome		cells hypersensitive to UV (24–27), NQO, and AAF (25, 28)	
Diabetes mellitus			glucose (29, 30)
Down's syndrome	increased incidence of leukemia (31)	cells hypersentive to X-rays (32–34); reduced ability to repair X-ray-induced single-strand breaks (35)	hydrogen peroxide, hydroxyl radical, and singlet oxygen (36, 37)
Friedreich ataxia		cells hypersensitive to X-rays (38, 39)	
Huntington's disease		cells hypersensitive to X-rays (40–42), MNNG (43), and bleomycin (44)	
Hutchinson-Guilford progeria		reduced ability to repair X-ray-induced single-strand breaks under some conditions (45–48), but not under other conditions (49, 50)	
Parkinson's Disease		cells hypersensitive to X-rays (51, 52)	
Werner's syndrome	increased chromosomal breaks and aberrations (53–56)		glycosaminoglycans (57, 58)
Xeroderma pigmentosum	increased neoplasms at sites not exposed to UV (59)	reduced ability to repair pyrimidine dimers (60, 61) and damages caused by AAF, PUVA, and benzpyrene (62–65)	

[a] Numbers indicate the following references: (1) Kidson and Chen (1986), (2) Robbins *et al.* (1983), (3) Robbins *et al.* (1985), (4) Scudiero *et al.* (1986), (5) Li and Kaminskas (1985), (6) Robison *et al.* (1987),

dromes with evidence of both elevated DNA damage and premature aging (Table XII), nine are characterized by neuropathology (AD, AT, CS, DS, DM, FA, HD, PD, and XP.

IV. Conclusions

In Section I of this chapter, we reviewed evidence that, among mammalian species, life span correlates well with DNA repair capacity. In Section II, we reviewed evidence that treatment of mammals with exogenous DNA-damaging agents induces life-shortening with some similarity to normal aging. In Section III, we described 11 genetic syndromes of humans in which evidence indicates both features of premature aging and an increased incidence of DNA damage. In nine of these syndromes, evidence also suggested neuropathology. Overall, the findings in the three sections are in agreement with the DNA damage hypothesis of aging.

References

Ahmed, F. E., and Setlow, R. B. (1978). Excision repair in ataxia telangiectasia, Fanconi's anemia, Cockayne syndrome, and Bloom's syndrome, after treatment with ultraviolet radiation and N-acetoxy-2-acetylaminofluorene. *Biochim. Biophys. Acta* **521,** 805–817.

Alexander, P. (1967). The role of DNA lesions in processes leading to aging in mice. *Symp. Soc. Exp. Biol.* **21,** 29–50.

Alexander, P., and Connell, D. I. (1960). Shortening of the life span of

) Robison *et al.* (1988), (8) Bartlett *et al.* (1988), (9) Jones *et al.* (1989), (10) German (1972), (11) Bochkov *et al.* 974), (12) Cohen *et al.* (1975), (13) Oxford *et al.* (1975), (14) Taylor *et al.* (1976), (15) Webb *et al.* (1977), 6) Taylor *et al.* (1975), (17) Chen *et al.* (1978), (18) Paterson *et al.* (1979), (19) Arlett and Harcourt (1980), 0) Cox *et al.* (1978), (21) Lehman and Stevens (1979), (22) Taylor *et al.* (1979), (23) Cohen *et al.* (1981), 4) Schmickel *et al.* (1977), (25) Wade and Chu (1979), (26) Deschevanne *et al.* (1981), (27) Marshall *et al.* 980), (28) Wade and Chu (1978), (29) Bucala *et al.* (1984), (30) Lorenzi *et al.* (1986), (31) Miller (1966), 2) Sasaki *et al.* (1970), (33) Countryman *et al.* (1977), (34) Preston (1981), (35) Athanasiou *et al.* (1980), 6) Balazs and Brookshank (1985), (37) Elroy-Stein et al. (1986), (38) Lewis *et al.* (1979), (39) Chamberlain d Lewis (1982), (40) Moshell *et al.* (1980), (41) Chen *et al.* (1981), (42) McGovern and Webb (1982), 3) Scudiero *et al.* (1981), (44) Imray and Kidson (1983), (45) Epstein *et al.* (1973), (46) Epstein *et al.* 974), (47) Brown *et al.* (1976), (48) Rainbow and Howes (1977), (49) Regan and Setlow (1974), (50) Bradley *al.* (1976), (51) Robbins *et al.* (1983a), (52) Robbins *et al.* (1985), (53) Hoehn *et al.* (1975), (54) Salk (1982), 5) Gebhart *et al.* (1985), (56) Stefanini *et al.* (1989), (57) Fujiwara and Ichihashi (1982), (58) Furukawa and hovanandan (1982), (59) Kraemer *et al.* (1984), (60) Zelle and Lohman (1979), (61) Mitchell *et al.* (1985), 2) Amacher and Lieberman (1977), (63) Kaye *et al.* (1980), (64) Yang *et al.* (1980), (65) Wood *et al.* (1988).

mice by irradiation with X-rays and treatment with radiomimetic compounds. *Radiat. Res.* **12,** 38–48.

Amacher, D. E., and Lieberman, M. W. (1977). Removal of acteylaminofluorene from the DNA of control and repair-deficient human fibroblasts. *Biochem. Biophys. Res. Commun.* **74,** 285–290.

Andrews, A. D., Barrett, S. F., and Robbins, J. H. (1978). Xeroderma pigmentosum neurological abnormalities correlate with colony forming ability after ultraviolet radiation. *Proc. Natl. Acad. Sci. USA* **75,** 1984–1988.

Arlett, C. F., and Harcourt, S. A. (1978). Cell killing and mutagenesis in repair defective human cells. *In* "DNA Repair Mechanisms" (P. C. Hanawalt, E. C. Friedberg, and C. F. Fox, eds.), pp. 633–636. Academic Press, New York.

Arlett, C. F., and Harcourt, S. A. (1980). Survey of radiosensitivity in a variety of human cell strains. *Cancer Res.* **40,** 926–932.

Athanasiou, K., Sideris, E. G., and Bartsocas, C. (1980). Decreased repair of X-ray induced DNA single-strand breaks in lymphocytes in Down syndrome. *Pediatr. Res.* **14,** 336–338.

Balazs, R., and Brookshawk, B. W. (1985). Neurochemical approaches to the pathogenesis of Down's Syndrome. *J. Ment. Defic. Res.* **29,** 1–14.

Bartlett, J. D., Hartshorn, J. N., and Robison, S. H. (1988). Alzheimer's disease patient's monocytes and T-lymphocytes display decreased DNA repair efficiencies after exposure to alkylating agents. *J. Cell Biochem.* **12A,** 307.

Bender, M. A., and Rary, J. M. (1974). Spontaneous and X-ray-induced chromosomal aberrations in progeria. *Radiat. Res.* **59,** 181a.

Berech, J., and Curtis, H. J. (1964). The role of age and X-irradiation on kidney function in the mouse. *Radiat. Res.* **22,** 95–105.

Bergener, M., and Junklass, F. K. (1970). Genetische befunde bei morbus Alzheimer und seniler demenz. *Gerontol. Clin.* **12,** 71–75.

Bochkov, N. P., Lopukhin, Y. M., Kuleshov, N. P., and Kovalchuk, L. V. (1974). Cytogenetic study of patients with ataxia-telangiectasia. *Humangenetik* **24,** 115–128.

Boer, A. P., and Davis, R. T. (1968). Age changes in the behavior of monkeys induced by ionizing radiations. *J. Gerontol.* **23,** 337–342.

Bradley, M. O., Erickson, L. C., and Kohn, K. W. (1976). Normal DNA strand rejoining and absence of DNA crosslinking in progeroid and aging human cells. *Mutat. Res.* **37,** 279–292.

Brown, W. T. (1987). Genetic aspects of aging in humans. *In* "Review of Biological Research in Aging," Vol. 3 (M. Rothstein, ed.), pp. 77–91. Alan R. Liss, Inc., New York.

Brown, W. T., Epstein, J., and Little, J. B. (1976). Progeria cells are stimulated to repair DNA by cocultivating with normal cells. *Exp. Cell. Res.* **97,** 291–296.

Brown, W. T., Ford, J. P., and Gershey, E. L. (1980). Variation of DNA repair capacity in progeria cells unrelated to growth conditions. *Biochem. Biophys. Res. Commun.* **97**, 347–353.

Bucala, R., Model, P., and Cerami, A. (1984). Modification of DNA by reducing sugars: A possible mechanism for nucleic acid aging and age-related dysfunction in gene expression. *Proc. Natl. Acad. Sci. USA* **81**, 105–109.

Buckton, K. E., Whalley, L. J., Lee, M., and Christie, J. E. (1983). Chromosome changes in Alzheimer's presenile dementia. *J. Med. Genet.* **20**, 46–51.

Casarett, G. W. (1964). Similarities and contrasts between radiation and time pathology. *Adv. Gerontol. Res.* **1**, 109–163.

Cerami, A. (1985). Hypothesis. Glucose as a mediator of aging. *J. Am. Geriatrics Soc.* **33**, 626–634.

Chamberlain, S., and Lewis, P. D. (1982). Studies of cellular hypersensitivity to ionizing radiation in Friedreich's ataxia. *J. Neurol. Neurosurg. Psychiatry* **45**, 1136–1138.

Chen, P., Kidson, C., and Imray, F. P. (1981). Huntington's disease: Implications of associated cellular radiosensitivity. *Clin. Genet.* **20**, 331–336.

Chen, P., Imray, F. P., and Kidson, C. (1984). Gene dosage and complementation analysis of ataxia telangiectasia lymphoblastoid cell lines assayed by induced chromosome aberrations. *Mutat. Res.* **129**, 165–172.

Chen, P. C., Lavin, M. F., Kidson, C., and Moss, D. (1978). Identification of ataxia telangiectasia heterozygotes, a cancer prone population. *Nature* **274**, 484–486.

Cockayne, E. A. (1946). Dwarfism with retinal atrophy and deafness. *Arch. Dis. Child.* **21**, 52–54.

Cohen, M. M., Shahan, M., Dagan, J., Shmueli, E., and Kohn, G. (1975). Cytogenetic investigation in families with ataxia telangiectasia. *Cytogenet. Cell. Genet.* **15**, 338–356.

Cohen, M. M., Simpson, S. J., and Pazos, L. (1981). Specificity of bleomycin-induced cytotoxic effects on ataxia telangiectasia lymphoid cell lines. *Cancer Res.* **41**, 1817–1823.

Collier, I. E., Popp, D. M., Lee, W. H., and Regan, J. D. (1982). DNA repair in a congeneic pair of mice with different longevities. *Mech. Ageing Dev.* **19**, 141–146.

Conklin, J. W., Upton, A. C., Christenberry, K. W., and McDonald, T. P. (1963). Comparative late somatic effects of some radiomimetic agents and x-rays. *Radiat. Res.* **19**, 156–168.

Coquerelle, T. M., Weibezahn, K. F., and Lucke-Huhle, C. (1987). Rejoining of double-strand breaks in normal human and ataxia-telangiectasia fibroblasts after exposure to ^{60}Co gamma rays, ^{241}Am alpha particles or bleomycin. *Int. J. Radiat. Biol.* **51**, 209–218.

Countryman, P. I., Heddle, J. A., and Crawford, E. (1977). The repair of X-ray-induced chromosomal damage in trisomy 21 and normal lymphocytes. *Cancer Res.* **37**, 52–58.

Cox, R., Hosking, G. P., and Wilson, J. (1978). Ataxia telangiectasia: The evaluation of radiosensitivity in cultured skin fibroblasts as a diagnostic test. *Arch. Dis. Child.* **53**, 386–390.

Cutler, R. G. (1972). Transcription of reiterated DNA sequence classes throughout the lifespan of the mouse. *Gerontol. Res.* **4**, 219–321.

Das, R. K. (1986). Mitomycin C and ethyl methanesulphonate-induced sister-chromatid exchanges in lymphocytes from individuals with Alzheimer's presenile dementia. *Mutat. Res.* **173**, 127–130.

DeBusk, F. L. (1972). The Hutchinson-Guilford progeria syndrome. *J. Pediatri.* **80**, 697–724.

Deschavanne, P. J., Diatloff-Zito, C., Maciera-Coelho, A., and Malaise, E.-P. (1981). Unusual sensitivity of two Cockayne's syndrome cell strains to both UV and gamma irradiation. *Mutat. Res.* **91**, 403–406.

Dunjic, A. (1964). Shortening of the span of life of rats by 'Myleran'. *Nature* **203**, 887–888.

Elroy-Stein, O., Bernstein, Y., and Groner, Y. (1986). Overproduction of human Cu/Zn superoxide dismutase in transfected cells: extenuation of paraquat-mediated cytotoxicity and enhancement of lipid peroxidation. *EMBO J.* **5**, 615–622.

Epstein, C. J., Martin, G. M., Schultz, A. L., and Motulsky, A. (1966). Werner's syndrome. *Medicine (Baltimore)* **45**, 177–221.

Epstein, J., Williams, J. R., and Little, J. B. (1973). Deficient DNA repair in human progeroid cells. *Proc. Natl. Acad. Sci. USA* **70**, 977–981.

Epstein, J., Williams, J. R., and Little, J. B. (1974). Rate of DNA repair in progeric and normal fibroblast. *Biochem. Biophys. Res. Commun.* **59**, 850–857.

Farrer, L. A. (1987). Genetic neurodegenerative disease models for human aging. *In* "Review of Biological Research in Aging," Vol. 3 (M. Rothstein, ed.), pp. 163–189. Alan R. Liss, New York.

Francis, A. A., Lee, W. H., and Regan, J. D. (1981). The relationship of DNA excision repair of ultraviolet induced lesions to the maximum lifespan of mammals. *Mech. Ageing Develop.* **16**, 181–189.

Friedberg, E. C. (1985). "DNA Repair." W. H. Freeman, New York.

Friedberg, E. C., Ehmann, U. K., and Williams, J. I. (1979). Human diseases associated with defective DNA repair. *Adv. Rad. Biol.* **8**, 85–174.

Fujimoto, W., Greene, M., and Seegmillar, J. (1969). Cockayne's syndrome: Report of a case with hyperlipoproteinemia hyperinsulinemia, renal disease and normal growth hormone. *J. Pediat.* **75**, 881–884.

Fujiwara, Y., and Ichihashi, M. (1982). Glycosaminoglycan synthesis in untransformed and transformed Werner syndrome fibroblasts: A preliminary report. *Adv. Exp. Med. Biol.* **190**, 613–626.

Fujiwara, Y., Higashikawa, T., and Tasumi, M. (1977). A retarded rate of DNA replication and normal level of DNA repair in Werner's syndrome fibroblasts in culture. *J. Cell. Physiol.* **92**, 365–374.

Furukawa, K., and Bhovanandan, V. P. (1982). Influence of glycosaminoglycans on endogenous DNA synthesis in isolated normal and cancer cell nuclei: Differential effect of heparin. *Biochim. Biophys. Acta.* **697**, 344–352.

Gebhart, E., Schinzel, M., and Ruprecht, K. W. (1985). Cytogenetic studies using various clastogens in two patients with Werner syndrome and control individuals. *Hum. Genet.* **70**, 324–327.

German, J. (1972). Genes which increase chromosomal instability in somatic cells and predispose to cancer. *In* "Progress in Medical Genetics" (A. Steinberg and A. Bearn, eds.), pp. 61–101. Grune and Stratton, New York.

Goldstein, S. (1969). Lifespan of cultured cells in progeria. *Lancet* **1**, 424.

Guzzetta, F. (1972). Cockayne-Neill-Dingwall syndrome. *In* "Handbook of Clinical Neurology," Vol. 13 (P. J. Vinken and G. W. Bruyn, eds.), pp. 431–440. Elsevier/North Holland, Amsterdam.

Hall, J., Bresil, H., and Montesano, R. (1985). O^6-alkylguanine DNA alkyltransferase activity in monkey, human and rat liver. *Carcinogenesis* **6**, 209–211.

Hall, K. Y., Bergmann, K., and Walford, R. L. (1981). DNA repair, H-2, and aging in NZB and CBA mice. *Tissue Antigens* **16**, 104–110.

Hall, K. Y., Hart, R. W., Benirshke, A. K., and Walford, R. L. (1984). Correlation between ultraviolet induced DNA repair in primate lymphocytes and fibroblasts and species maximum achievable life span. *Mech. Aging Dev.* **24**, 163–173.

Hart, R. N., and Setlow, R. B. (1974). Correlation between deoxyribonucleic acid excision-repair and lifespan in a number of mammalian species. *Proc. Natl. Acad. Sci. USA* **71**, 2169–2173.

Hart, R. W., and Daniel, F. B. (1980). Genetic stability *in vitro* and *in vivo*. *Adv. Pathobiol.* **7**, 123–141.

Hart, R. W., Sacher, G. A., and Hoskins, T. L. (1979). DNA repair in a short- and a long-lived rodent species. *J. Gerontol.* **34**, 808–817.

Higashikawa, T., and Fujiwara, Y. (1978). Normal level of unscheduled DNA synthesis in Werner's syndrome fibroblasts in culture. *Exp. Cell Res.* **113**, 438–441.

Hoehn, H., Bryant, E. M., Au, K., Norwood, T. H., Boman, H., and Martin, G. M. (1975). Varigated translocation mosaicism in human skin fibroblast cultures. *Cytogenet. Cell Genet.* **15**, 282–298.

Hollingsworth, D. R., Hollingsworth, J. W., Bogitch, S., and Keehn, R. J. (1969). Neuromuscular tests of aging Hiroshima subjects. *J. Gerontol.* **24**, 276–291.

Hurt, M. M., Beaudet, A. L., and Moses, R. E. (1983). Stable low molecular weight DNA in xeroderma pigmentosum cells. *Proc. Natl. Acad. Sci. USA* **80**, 6987–6991.

Imray, F. P., and Kidson, C. (1983). Responses of Huntington's disease and ataxia telangiectasia lymphoblastoid cells to bleomycin. *Chem. Biol. Interact.* **47**, 325–336.

Iverson, L. L., Rossor, M. N., Reynolds, G. P., Hills, R., Roth, M., Mountjoy, C. Q., Foote, S. L., Morrison, J. H., and Bloom, F. E. (1983). Loss of pigmented dopamine-beta-hydroxylase positive cells from locus coeruleus in senile dementia of Alzheimer's type. *Neurosci. Lett.* **39**, 95–100.

Jones, S. K., Nee, L. E., Sweet, L., Polinsky, R. J., Bartlett, J. D., Bradley, W. G., and Robinson, S. H. (1989). Decreased DNA repair in familial Alzheimer's disease. *Mutat. Res.* **219**, 247–255.

Kato, H., Harada, M., Tsuchiya, K., and Moriwaki, K. (1980). Absence of correlation between DNA repair in ultraviolet irradiated mammalian cells and lifespan of the donor species. *Jap. J. Genet.* **55**, 99–108.

Katzman, R. (1986). Alzheimer's disease. *N. Eng. J. Med.* **314**, 964–973.

Kaye, J., Smith, C. A., and Hanawalt, P. C. (1980). DNA repair in human cells containing photoadducts of 8-methyoxy-psoralen or angelicin. *Cancer Res.* **40**, 696–702.

Kempf, C., Schmitt, M., Danse, J.-M., and Kempf, J. (1984). Correlation of DNA repair synthesis with aging in mice, evidenced by quantitative autoradiography. *Mech. Ageing Dev.* **26**, 183–194.

Kent, S. (1976). Is diabetes a form of accelerated aging? *Geriatrics* **31(Nov.)**, 140–151.

Kidson, C., and Chen, P. (1986). DNA damage, DNA repair and the genetic basis of Alzheimer's disease. *Prog. Brain Res.* Chapter **18**, 291–301.

Kodell, R. L., Farmer, J. H., and Littlefield, N. A. (1980). Analysis of life-shortening effects in female BALB/C mice fed 2-acetylaminofluorene. *J. Environ. Pathol. Toxicol.* **3**, 69–87.

Kraemer, K. H., Lee, M. M., and Sotto, J. (1984). DNA repair protects against cutaneous and internal neoplasia: Evidence from xeroderma pigmentosum. *Carcinogenesis* **5**, 511–514.

Kuhnlein, U., Penhoet, E. E., and Linn, S. (1976). An altered apurinic DNA endonuclease activity in group A and group D xeroderma pigmentosum fibroblasts. *Proc. Natl. Acad. Sci. USA* **73**, 1169–1173.

Kuhnlein, U., Lee, B., Penhoet, E. E., and Linn, S. (1978). Xeroderma pigmentosum fibroblasts of the D group lack an apurinic DNA endonuclease species with a low apparent Km. *Nucleic Acids Res.* **5**, 951–960.

Lambert, M. W., Lambert, W. C., and Okorodudu, A. O. (1983). Nuclear DNA endonuclease activities on partially apurinic/apyrimidinic DNA in normal human and xeroderma pigmentosum lymphoblastoid and mouse melanoma cells. *Chem. Biol. Interact.* **46**, 109–120.

Lehman, A. R. (1982). Three complementation groups in Cockayne's syndrome. *Mutat. Res.* **106**, 347–356.

Lehman, A. R., and Stevens, S. (1979). The response of ataxia telangiectasia cells to bleomycin. *Nucleic Acids Res.* **6**, 1953–1960.

Lehman, A. R., Kirk-Bell, S., and Mayne, L. (1979). Abnormal kinetics of DNA synthesis in ultraviolet light-irradiated cells from patients with Cockayne's syndrome. *Cancer Res.* **39**, 4237–4241.

Lewis, P. D., Corr, J. B., Arlett, C. F., and Harcourt, S. A. (1979). Increased sensitivity to gamma radiation of skin fibroblasts in Friedreich's ataxia. *Lancet* **2**, 474–475.

Ley, R. D., Sedita, B. A., Grube, D. D., and Fry, R. J. M. (1977). Induction and persistence of pyrimidine dimers in the epidermal DNA of two strains of hairless mice. *Cancer Res.* **37**, 3243–3248.

Li, J. C., and Kaminskas, E. (1985). Deficient repair of DNA lesions in Alzheimer's disease fibroblasts. *Biochem. Biophys. Res. Commun.* **129**, 733–738.

Lindop, P. J., and Rotblat, J. (1962). Induction of aging by radiation. *In* "Biological Aspects of Aging" (N. W. Shock, ed.), pp. 216–221. Columbia University Press, New York.

Linn, S., Kim, J., Choi, S.-Y., Keeney, S., Randahl, H., Tomkinson, A. E., Nishida, C., Syvaoja, S., and Krauss, S. W. (1988). The enzymology of mammalian DNA excision repair. *J. Cell Biol.* **107**, 448a.

Lipman, J. M., Sokoloff, L., and Setlow, R. B. (1987). DNA repair by articular chondrocytes. V. O^6-methylguanine-acceptor protein activity in resting and cultured rabbit and human chondrocytes. *Mech. Ageing Dev.* **40**, 205–213.

Little, J. B. (1976). Relationship between DNA repair capacity and cellular aging. *Gerontology* **22**, 28–55.

Locke, S., and Tarsy, D. (1985). The nervous system and diabetes. *In* "Joslin's Diabetes Mellitus," 12th ed. (A. Marble, L. P. Krall, R. Bradley, A. R. Christlieb, and J. S. Soeldner, eds.), pp. 665–685. Lea and Febiger, Philadelphia.

Lorenzi, M., Montesano, D. F., Toledo, S., and Barrieux, A. (1986). High glucose induces DNA damage in cultured human endothelial cells. *J. Clin. Invest.* **77**, 322–325.

Lovelace, W. E., and Davis, R. T. (1963). Minimum-separable visual acuity of rhesus monkeys as a function of aging and whole body radiation with X-rays. *J. Genet. Psychol.* **103**, 251–257.

Mann, D. M. A. (1988). The pathological association between Down syndrome and Alzheimer disease. *Mech. Ageing Dev.* **43**, 99–136.

Mark, J., and Brun, A. (1973). Chromosomal deviations in Alzheimer's disease compared to those in senescence and senile dementia. *Gerontol. Clin.* **15**, 253–258.

Marshall, R. R., Arlett, C. F., Harcourt, S. A., and Broughton, B. A. (1980). Increased sensitivity of cell strains from Cockayne's syndrome to sister chromatid-exchange induction and cell killing by UV light. *Mutat. Res.* **69**, 107–112.

Martin, G. M. (1978). Genetic syndromes in man with potential relevance to the pathobiology of aging. *In* "Genetic Effect on Aging. Birth Defects, Original Article Series" (D. Bergsma and D. E. Harrison, eds.), pp. 5–39. The National Foundation March of Dimes, New York.

Martin, G. M., Sprague, C. A., and Epstein, C. J. (1970). Replicative lifespan of cultivated human cells. Effects of donor's age, tissue and genotype. *Lab. Invest.* **23**, 86–92.

Martin, J. M., Kellett, J. M., and Kahn, J. (1981). Aneuploidy in cultured human lymphocytes. II. A comparison between senescence and dementia. *Age Ageing* **10**, 24–28.

Maslansky, C. J., and Williams, G. M. (1985). Ultraviolet light-induced DNA repair synthesis in hepatocytes from species of differing longevities. *Mech. Ageing Dev.* **29**, 191–203.

Mayne, L. V., and Lehmann, A. R. (1982). Failure of RNA synthesis to recover after UV irradiation: An early defect in cells from individuals with Cockayne's syndrome and xeroderma pigmentosum. *Cancer Res.* **42**, 1473–1478.

Mayne, L. V., Mullenders, L. H. F., and van Zeeland, Z. (1988). Cockayne's syndrome: A UV sensitive disorder with a defect in the repair of transcribing DNA but normal overall excision repair. *In* "Mechanisms and Consequences of DNA Damage Processing" (E. C. Friedberg and P. C. Hanawalt, eds.), pp. 349–353. Alan R. Liss, New York.

McGovern, D., and Webb, T. (1982). Sensitivity to ionizing radiation of lymphocytes from Huntington's chorea patients compared to controls. *J. Med. Genet.* **19**, 168–174.

McNeil, M. M., Leong, A. S-Y., and Sage, R. E. (1981). Primary mediastinal embryonal carcinoma in association with Klinefelter's syndrome. *Cancer* **47**, 343–345.

Miller, R. (1966). Relation between cancer and congenital defects in man. *N. Eng. J. Med.* **275**, 87–93.

Mitchell, D. L., Haipek, C. A., and Clarkson, J. M. (1985). (6-4) Photoproducts are removed from the DNA of UV-irradiated mammalian cells more efficiently than cyclobutane pyrimidine dimers. *Mutat. Res.* **143**, 109–112.

Montesano, R., Bresil, H., Likhachev, A., vonBahr, C., Roberfroid, M., and Pegg, A. E. (1982). Removal from DNA of O^6-methylguanine (O^6-MeG) by human liver fractions. *Proc. Amer. Assoc. Cancer Res.* **23**, 11.

Moshell, A. N., Barrett, S. F., Tarone, R. E., and Robbins, J. H. (1980). Radiosensitivity in Huntington's disease: Implications for pathogenesis and presymptomatic diagnosis. *Lancet* **1**, 9–11.

Muller, W. E. G., and Zahn, R. K. (1977). Bleomycin an antibiotic that removes thymine from double-stranded DNA. *Prog. Nucleic Acids Res. Mol. Biol.* **20**, 21–57.

Nikaido, O., Nishida, T., and Shima, A. (1985). Cellular mechanisms of aging in the Werner syndrome. *Adv. Exp. Med. Biol.* **190**, 421–438.

Nordenson, I., Adolfsson, R., Beckman, G., Bucht, G., and Winblad, B. (1980). Chromosomal abnormalities in dementia of the Alzheimer type. *Lancet* **1**, 481–482.

Ohno, S., and Nagai, Y. (1978). Genes in multiple copies as the primary cause of aging. *In* "Genetic Effects of Aging" (D. Bergsma, D. E. Harrison, and N. W. Paul, eds.), pp. 501–514. Alan R. Liss, New York.

Oxford, J. M., Harnden, D. G., Parrington, J. M., and Delhanty, J. D. A. (1975). Specific chromosomal aberrations in ataxia telangiectasia. *J. Med. Genet.* **12**, 251–262.

Paffenholz, V. (1978). Correlation between DNA repair of embryonic fibroblasts and different life span of 3 inbred mouse strains. *Mech. Ageing Dev.* **7**, 131–150.

Paterson, M. C., Anderson, A. K., Smith, B. P., and Smith, P. J. (1979). Enhanced radiosensitivity of cultured fibroblasts from ataxia telangiectasia heterozygotes manifested by defective colony-forming ability and reduced DNA repair replication after hypoxic gamma-irradiation. *Cancer Res.* **39**, 3725–3734.

Perlmutter, L. C., Hakami, M. K., Hodgson-Harrington, C., Ginsberg, J., Kataz, J., Singer, D. E., and Nathan, D. M. (1984). Decreased cognitive function in aging non-insulin dependent diabetic patients. *Am. J. Med.* **77**, 1043–1048.

Preston, R. J. (1981). X-ray-induced chromosome aberrations in Down lymphocytes and explanation of their increased sensitivity. *Environ. Mutagen.* **3**, 85–89.

Puvion-Dutilleul, F., and Sarasin, A. (1989). Chromatin and nucleolar changes in xeroderma pigmentosum cells resemble aging-related nuclear events. *Mutat. Res.* **219**, 57–70.

Rainbow, A. J., and Howes, M. (1977). Decreased repair of gamma ray damaged DNA in progeria. *Biochem. Biophys. Res. Commun.* **74**, 714–719.

Regan, J. D., and Setlow, R. B. (1974). DNA repair in progeroid cells. *Biochem. Biophys. Res. Commun.* **59**, 858–864.

Regan, J. D., Carrier, W. L., Samet, C., and Olla, B. L. (1982). Photoreactivation in two closely related marine fishes having different longevities. *Mech. Ageing Dev.* **18**, 59–66.

Robbins, J. H., Otsuka, F., Tarone, R. E., Polinsky, R. J., Brumback, R. A., Moshell, A. N., Nee, L. E., Ganges, M. B., and Cayeux, S. J. (1983a).

Radiosensitivity in Alzheimer disease and Parkinson disease. *Lancet* **1**, 468–469.

Robbins, J. H., Polinsky, R. J., and Moshell, A. N. (1983b). Evidence that lack of deoxyribonucleic acid repair causes death of neurons in xeroderma pigmentosum. *Ann. Neurol.* **13**, 682–684.

Robbins, J. H., Otsuka, F., Tarone, R. E., Polinsky, R. J., Brumback, R. A., and Nee, L. E. (1985). Parkinson's disease and Alzheimer's disease: Hypersensitivity to X-rays in cultured cell lines. *J. Neurol. Neurosurg. Psychiatry* **48**, 916–923.

Robison, S. H., Munzer, S. J., Tandan, R., and Bradley, W. G. (1987). Alzheimer's disease cells exhibit defective repair of alkylating agent-induced DNA damage. *Ann. Neurol.* **21**, 250–258.

Robison, S. H., Jones, S. K., Nee, L. E., Polinsky, R. J., and Bradley, W. G. (1988). Decreased repair of alkylating agent induced damage in familial Alzheimer's disease. *J. Cell. Biochem.* **12A**, 355.

Rocca, W. A., Amaducci, L. A., and Schoenberg, B. S. (1986). Epidemiology of clinically diagnosed Alzheimer's disease. *Ann. Neurol.* **19**, 415–424.

Sajdel-Sulkowska, E. M., and Marotta, C. A. (1984). Alzheimer's disease brain: Alterations in RNA levels and in a ribonuclease-inhibitor complex. *Science* **225**, 947–949.

Sajdel-Sulkowska, E. M., and Marotta, C. A. (1985). Functional messenger RNA from the postmortem human brain: Comparison of aged normal with Alzheimer's disease. *In* "Molecular Biology of Aging: Gene Stability and Gene Expression" (R. S. Sohal, L. S. Birnbaum, and R. G. Cutler, eds.), pp. 243–256. Raven Press, New York.

Salk, D. (1982). Werner syndrome: A review of recent research with an analysis of connective tissue metabolism, growth control of cultured cells, and chromosomal aberrations. *Hum. Genet.* **62**, 1–20.

Sasaki, M. S., Tonomura, A., and Matsubara, S. (1970). Chromosome constitution and its bearing on the chromosomal radiosensitivity in man. *Mutat. Res.* **10**, 617–633.

Schellenberg, G. D., Bird, T. D., Wijsman, E. M., Moore, D. K., Boehnke, M., Bryant, E. M., Lampe, T. H., Nochlin, D., Sumi, S. M., Deep, S. S., Beyreuther, K., and Martin, G. M. (1988). Absence of linkage of chromosome 21q21 markers to familial Alzheimer's disease. *Science* **241**, 1507–1510.

Schmickel, R. D., Chu, E. H. Y., Trosko, J. E., and Chang, C.-C. (1977). Cockayne syndrome: A cellular sensitivity to ultraviolet light. *Pediatrics* **60**, 135–139.

Scudiero, D. A., Meyer, S. A., Clatterbuck, B. E., Tarone, R. E., and Robbins, J. H. (1981). Hypersensitivity to N'-methyl-N'-nitro-N-nitrosoguanidine in fibroblasts from patients with Huntington disease, familial dysautonomia, and other primary neuronal degenerations. *Proc. Natl. Acad. Sci. USA* **78**, 6451–6455.

Scudiero, D. A., Polinsky, R. J., Brumback, R. A., Tarone, R. E., Nee, L. E., and Robbins, J. H. (1986). Alzheimer disease fibroblasts are hypersensitive to the lethal effects of a DNA-damaging chemical. *Mutat. Res.* **159**, 125–131.

Shagan, B. P. (1976). Is diabetes a model for aging? *Med. Clin. N. Am.* **60**, 1209–1230.

Smith, T. A. D., and Itzhaki, R. F. (1989). Radiosensitivity of lymphocytes from patients with Alzheimer's disease. *Mutat. Res.* **217**, 11–17.

Smith, T. A. D., Neary, D., and Itzhaki, R. F. (1987). DNA repair in lymphocytes from young and old individuals and from patients with Alzheimer's disease. *Mutat. Res.* **184**, 107–112.

St. George-Hyslop, P. H., Tanzi, R. E., Polinsky, R. J., Haines, J. L., Nee, L., Watkins, P. C., Myers, R. H., Feldman, R. G., Pollen, D., Drachman, D., Growden, J., Bruni, A., Foncin, J.-F., Salmon, D., Frommelt, P., Amaducci, L., Sorbi, S., Piacentini, S., Stewart, G. D., Hobbs, W. J., Conneally, P. M., and Gusella, J. F. (1987). The genetic defect causing familial Alzheimer's disease maps on chromosome 21. *Science* **235**, 885–889.

Stefanini, M., Scappaticci, S., Lagomarsini, P., Borroni, G., Berardesca, E., and Nuzzo, F. (1989). Chromosome instability in lymphocytes from a patient with Werner's syndrome is not associated with DNA repair defects. *Mutat. Res.* **219**, 179–185.

Sutherland, B. M., Harber, L. C., and Kochevar, I. E. (1980). Pyrimidine dimer formation and repair in human skin. *Cancer Res.* **40**, 3181–3185.

Tanaka, K., Kawai, K., Kumahara, Y., Ikenaga, M., and Okada, Y. (1981). Genetic complementation groups in Cockayne syndrome. *Somat. Cell Genet.* **7**, 445–455.

Taylor, A. M. R., Harnden, D. G., Arlett, C. F., Harcourt, S. A., Lehmann, A. R., Stevens, S., and Bridges, B. A. (1975). Ataxia telangiectasia: A human mutation with abnormal radiation sensitivity. *Nature* **258**, 427–429.

Taylor, A. M. R., Metcalfe, J. A., Oxford, J. M., and Harnden, D. G. (1976). Is chromatid-type damage in ataxia telangiectasia after irradiation at G_o a consequence of defective repair? *Nature* **260**, 441–443.

Taylor, A. M. R., Rosney, C. M., and Campbell, J. B. (1979). Unusual sensitivity of ataxia telangiectasia cells to bleomycin. *Cancer Res.* **39**, 1046–1050.

Thacker, J., and Debenham, P. G. (1988). The molecular basis of radiosensitivity in the human disorder ataxia-telangiectasia. *In* "Mechanisms and Consequences of DNA Damage Processing" (E. C. Friedberg and P. C. Hanawalt, eds.), pp. 361–369. Alan R. Liss, New York.

Tice, R. R., and Setlow, R. B. (1985). DNA repair and replication in aging organisms and cells. *In* "Handbook of the Biology of Aging" (C. E. Finch and E. L. Schneider, eds.), pp. 173–224. Van Nostrand Reinhold, New York.

Treton, J. A., and Courtois, Y. (1982). Correlation between DNA excision repair and mammalian lifespan in lens epithelial cells. *Cell Biol. Internat. Rpts.* **6**, 253–260.

Upton, A. C. (1957). Ionizing radiation and the aging process. *J. Gerontol.* **12**, 306–313.

Van Broeckhoven, C., Genthe, A. M., Vandenberghe, A., Horsthemke, B., Backhovens, H., Raeymaekers, P., Van Hul, W., Wehnert, A., Gheuens, J., Cras, P., Bruyland, M., Martin, J. J., Salbaum, M., Multhaup, G., Masters, C. L., Beyreuther, K., Gurling, H. M. D., Mullan, M. J., Holland, A., Barton, A., Irving, N., Williamson, R., Richards, S. J., and Hardy, J. A. (1987). Failure of familial Alzheimer's disease to segregate with the A4-amyloid gene in several European families. *Nature* **329**, 153–157.

Wade, M. H., and Chu, E. H. Y. (1978). Effects of DNA damaging agents on cultured fibroblasts derived from patients with Cockayne syndrome. In "DNA Repair Mechanisms" (P. C. Hanawalt, E. C. Friedberg, and C. F. Fox, eds.), pp. 667–670. Academic Press, New York.

Wade, M. H., and Chu, E. H. Y. (1979). Effects of DNA damaging agents on cultured fibroblasts derived from patients with Cockayne syndrome. *Mutat. Res.* **59**, 49–60.

Walburg, H. E., Jr. (1975). Radiation-induced life-shortening and premature aging. *Adv. Radiat. Biol.* **5**, 145–179.

Ward, B. E., Cook, R. H., Robinson, A., and Austin, J. H. (1979). Increased aneuploidy in Alzheimer disease. *Am. J. Med. Genet.* **3**, 137–144.

Warren, S. (1956). Longevity and causes of death from irradiation in physicians. *JAMA* **162**, 464–467.

Webb, T., Harnden, D. G., and Harding, M. (1977). The chromosome analysis and susceptibility to transformation by simian virus 40 of fibroblasts from ataxia telangiectasia. *Cancer Res.* **37**, 997–1002.

White, B. J., Crandell, C., Goudsmit, J., Morrow, C. H., Alling, D. W., Gajdusek, D. C., and Tijio, J. H. (1981). Cytogenetic studies of familial and sporadic Alzheimer's disease. *Am. J. Med. Genet.* **10**, 77–89.

Whitehouse, P. J., Price, D. L., Struble, R. G., Clark, A. W., Coyle, J. T., and DeLong, M.R. (1982). Alzheimer's disease and senile dementia: Loss of neurons in the basal forebrain. *Science* **215**, 1237–1239.

Wood, R. D., Robins, P., and Lindahl, T. (1988). Complementation of the xeroderma pigmentosum DNA repair defect in cell free extracts. *Cell* **53**, 97–106.

Woodhead, A. D., Setlow, R. B., and Grist, E. (1980). DNA repair and longevity in three species of cold-blooded vertebrates. *Exp. Gerontol.* **15**, 301–304.

Wright, A. F., and Whalley, L. F. (1984). Genetics, aging and dementia. *Br. J. Psychiatry* **145**, 20–38.

Yamagata, Z., Iijima, S., Takeshita, T., Ariizumi, C., and Higurashi, M. (1989). Mitomycin-C-induced sister-chromatid exchanges and cell-cycle kinetics in lymphocytes from patients with Klinefelter syndrome. *Mutat. Res.* **212**, 263–268.

Yang, L. L., Maher, V. M., and McCormick, J. J. (1980). Error-free excision of the cytotoxic and mutagenic N2-deoxyguanosine DNA adduct formed in human fibroblasts by (±)7 beta,8 alpha-dihydroxy-9 alpha,10 alpha-epoxy-7,8,9,10- tetrahydrobenzo(a)pyrene. *Proc. Natl. Acad. Sci. USA* **77**, 5933–5937.

Zelle, B., and Lohman, P. H. M. (1979). Repair of UV-endonuclease-susceptible sites in the seven complementation groups of xeroderma pigmentosum A through G. *Mutat. Res.* **62**, 363–368.

Zucker-Franklin, D., Rifkind, H., and Jacobson, H. G. (1968). Werner's syndrome, an analysis of ten cases. *Geriatrics* **23(Aug.)**, 123–135.

Aging in Nonmammalian Organisms, with Comparisons to Aspects of Aging in Mammals

Aging has been studied in a number of nonmammalian species. These include species of bacteria, protozoa, fungi, nematodes, insects, and higher plants. For many species, aging seems to be primarily due to DNA damage, as it is in mammals. In some species, aging is circumvented by a replacement strategy, as it is for some tissues in mammals such as the hemopoietic stem cells (Chapter 5, Section I.C). However, for some species, aging is primarily due to factors that are of lesser importance in mammalian aging. These are the factors of mutation and of wear and tear. We also discuss the contributions of mutation and wear and tear to aging in mammals.

I. *Aging and DNA Damage in Single-Celled Organisms*

A. Bacteria

It has often been speculated that dividing cells, such as prokaryotic bacteria, are essentially immortal. Cutler (1972) hypothesized, however, that no genetic material is immortal. To test the concept of a finite life span of the genome, he carried out experiments in which cells of the bacterium *E. coli* were attached to a nitrocellulose filter. They were then allowed to grow and form daughter cells. Those daughter cells, which were unattached, could be eluted into the media. The rate of duplication of cells originally attached declined steadily with time until at 6 days no further replication occurred. These findings were interpreted as evidence for the aging of the genome due to accumulation of mutations; however, an explanation in terms of DNA damages is equally reasonable because both kinds of genetic "error" would increase with time.

We think that bacterial populations respond to the problem of DNA damage in a fashion similar to the response of populations of rapidly dividing somatic cells in mammals. As we suggested in Chapter 5, Section I.C, cell populations such as the hemopoietic cells of the bone marrow may maintain themselves by a strategy of replacement, where lethally damaged cells are replaced by replication of undamaged ones. Such a population, whether of bacteria or of mammalian cells, will persist so long as the incidence of unrepaired lethal damages is low enough at each generation to permit replacement of losses. Despite the attrition, aging of the population as a whole may not be apparent.

B. Protozoa

Among the simplest eukaryotic organisms in which aging occurs are strains of paramecium and other ciliates. *Paramecium tetraurelia* is a eukaryote that undergoes both asexual and sexual reproduction. Asexual or clonal reproduction occurs by binary fission involving a mitosislike behavior of the chromosomes, which resembles somatic cell division in higher animals. Asexual reproduction is initiated after the completion of autogamy (a form of self-fertilization) or conjugation (sexual reproduction). In the asexual phase there is a gradual loss of vitality, or clonal aging, over successive generations (Smith-Sonneborn, 1987). After either autogamy or conjugation, a new, revitalized asexual phase is initiated. *Paramecium tetraurelia* has a life span of about 200 cell divisions since the previous fertilization (Smith-Sonneborn, 1987).

Aging in these organisms may reflect a net accumulation of DNA damage during the asexual phase, and rejuvenation may result from efficient DNA repair in the sexual phase (Martin, 1977). Evidence in support of the idea that DNA damage causes aging in *P. tetraurelia* was reported by Smith-Sonneborn (1979). The experiments depended on using photoreactivation, a process that corrects pyrimidine dimers, a major damage caused by UV (Chapter 9, Section II.F.1). She found that when *P. tetraurelia* was UV-irradiated without photoreactivation, clonal life span was reduced. However, when *P. tetraurelia* was both UV-irradiated and then subjected to photoreactivation, clonal life span was not only restored but was substantially increased. These findings suggest that DNA repair enzymes were induced by the treatment with UV. After the pyrimidine dimers were removed by photoreactivation, the higher repair capability could more effectively overcome other unidentified DNA damages that normally limit the life span.

Holmes and Holmes (1986) used an alkaline elution procedure to measure DNA damages in *P. tetraurelia* cells that had grown asexually for various numbers of generations. This procedure detects apurinic–apyrimidimic sites (AP sites), strand breaks, or strand gaps but does not distinguish between these types of damages. The results obtained indicated that, as the cells aged through approximately 16, 60, and 108 asexual generations, DNA damages accumulated. Holmes and Holmes (1986) suggested that accumulation of DNA damage may be the basic cause of aging in paramecium. In Chapter 5, Section I.C, we argued that in a rapidly dividing population of cells in which the rate of occurrence of unrepaired lethal DNA damage is low compared with cell generation time, the damaged cells can be replaced by the duplication of undamaged ones. Such cell populations may not experience the progressive impairments of function that define aging. Because *P. tetraurelia* populations age even though they are rapidly dividing, we need to explain in this particular case how DNA damages may accumulate.

In *P. tetraurelia,* two functionally different nuclei coexist: the macronucleus and the micronucleus. The former is comparable to the somatic cell nucleus and the latter to a germ cell nucleus of higher organisms. Micronuclear genes are apparently not expressed during vegetative growth as shown in the related protozoan *Tetrahymena thermophila* (Mayo and Orias, 1981). Within two cell cycles subsequent to either automixis or conjugation, new macronuclei develop from the micronuclei and this is accompanied by rejuvenation of the cell line. As the new macronuclei develop, the old macronuclei fragment and their replication declines (Berger, 1973). The new macronucleus is highly polyploid, with about 800 copies of the haploid genome (Berger, 1973). It is considered to be responsible for the phenotype of the cell. Because the measurements of strand breaks by

Holmes and Holmes (1986) were with a mixture of both macronuclear and micronuclear DNA, most of the accumulated damages
detected probably were in the majority macronuclear DNA. Therefore, the aging of the asexual line may be due to accumulated damage
in the macronuclear DNA.

We speculate that the approximately 800 genome copies in a
new macronucleus provide a buffer against the consequences of
DNA damage. At each generation after macronuclear regeneration,
damage may occur in the macronuclear DNA. A DNA damage may
lead to failure of a genome to replicate or function properly. Damaged genomes that fail to replicate may nevertheless survive cell
division and be carried along in one of the two daughter cells. Thus,
DNA damages will accumulate, and there will be a progressive attrition of intact functional copies of the macronuclear genome. We
think this attrition may account for the progressive aging of *P. tetraurelia* over many generations.

Smith-Sonneborn (1987) suggested, on the basis of the work by
Holmes and Holmes (1986), that the UV-stimulated increase in life
span observed earlier by her (Smith-Sonneborn, 1979) could be due
to enhanced repair of AP sites. This is consistent with the evidence
reviewed in Chapter 7, Section I.B, that UV-induced damages and
AP sites can be repaired by common repair functions.

In Chapter 5, Section II.A, we presented evidence that those
mammalian tissues that accumulate DNA damages with age also
experience an age-associated decline in gene expression. Klass and
Smith-Sonneborn (1976) found that as *P. tetraurelia* undergoes
clonal aging the amount of macronuclear DNA decreases and RNA
synthesis declines. The decline in RNA synthesis may reflect the loss
of functional macronuclear DNA because of DNA damage.

A series of experiments has explored whether the factors
that determine aging in *P. tetraurelia* reside in the cytoplasm or the
macronucleus. These experiments showed that the determinants of
aging are located in the macronucleus and not the cytoplasm, consistent with the hypothesis that DNA damage is the cause of aging in
these organisms. One set of experiments, to determine whether or
not cytoplasm was important in clonal aging, was performed by
Aufderheide (1984). Using a microinjection protocol, he transferred
cytoplasm from young cells into old cells to see if the mean age of
death of the injected cell lines could be increased compared with

uninjected controls and sham-injected controls. Similarly, cytoplasm from old cells was transferred into young cells to see if the mean age of death of the injected cell lines would decrease. In neither case was there any statistically significant change in mean ages at death. Aufderheide (1984) concluded on the basis of his own results and previous work of others that a cytoplasmic effect on vegetative clonal aging in *P. tetraurelia* cannot be demonstrated.

A role of the macronucleus in determining clonal aging was suggested by the finding that mutations that reduced clonal life span to 1/3–1/4 of the wild-type life span all had a common phenotypic defect: a frequent inability to properly divide the macronucleus during cell division (Aufderheide and Schneller, 1985). In further experiments to directly measure the contribution of the macronucleus to proliferative potential, Aufderheide (1987) used a nuclear transplantation procedure. He injected macronuclei from young or old wild-type cells into genetically marked host cells of a standard clonal age. The preexisting macronucleus in each of these cells remained in the cell. He then compared the clonal life span of transformed hybrids with the life span of injected but untransformed lines (injection controls). Donated young macronuclei substantially prolonged the life span of the hybrid cell lines over that of the injection controls. The mean age at death in three different experiments for the transformed lines was almost double that of the controls. Old donor macronuclei caused only a small increase in postinjection proliferation of the hybrids. These results indicate that as macronuclei age their ability to support subsequent cell growth and division declines. These results also indicate that the proliferation potential of the donor macronucleus is not changed by its transplanation into a host cell of a different clonal age. The macronucleus thus appears to "remember its age" after transplantation. These results, coupled with an absence of any detectable cytoplasmic effects on aging during vegetative growth, argue in favor of a macronuclear determination of the proliferative potential of a clonal cell line. Aufderheide (1987) postulated that the aging process is associated with progressive and irreversible changes in the genome of the macronucleus. These findings, together with the evidence described earlier for the accumulation of DNA damage with clonal age and for the prolongation of clonal life span by induced DNA repair, suggest that aging in *P. tetraurelia* is due to DNA damage.

II. *Aging in Multicellular Organisms*

A. Filamentous Fungi

Clonal deterioration leading to cellular death occurs regularly in filamentous fungi after prolonged asexual reproduction. This phenomenon has been studied extensively in *Podospora anserina* and *Neurospora crassa* (see review by Bertrand, 1983). The arrest of growth is the final manifestation of a succession of events that ultimately results in death of the hyphal filaments at the frontier of the culture. Extensive studies led to the conclusion that senescence in filamentous fungi is triggered by mutational events that affect the mitochondrial genome. It appears that mitochondria with defective genomes selectively multiply during senescence and slowly displace normal mitochondria as senescence progresses. This results in vegetative death as the proportion of normal mitochondrial genomes declines to values less than the absolute minimum necessary to sustain essential cellular functions. As reviewed by Bertrand (1983), this model explains the observation in *Podospora* that when senescent clones are crossed as female (protoperithecial) parents with normal clones, normal as well as senescent progeny are recovered. In such crosses, the four ascospore progeny (meiotic products) from any given ascus are either all normal or all senescent. The senescent female can transmit either a senescent or a nonsenescent element through meiosis, implying that the aged mycelium of *Podospora* contains a mixture of the mutant and nonmutant forms of the genetic factor responsible for vegetative death. Senescence is not transmitted through meiosis by the male parent. This indicates that the genetic element responsible for senescence is located in the cytoplasm of the cell. Since the early work outlined here, the molecular nature of the mitochondrial mutations that cause senescence have been elucidated. As reviewed by Bertrand (1983), senescence may arise by a variety of different structural changes in mitochondrial DNA. A detailed genetic and molecular analysis in one race of *P. anserina* has revealed that senescence is correlated with the liberation and amplification of a mitochondrial plasmid (p1DNA or α-SEN DNA). In juvenile cultures, this DNA sequence is an integral part of the mitochondrial DNA representing the first intron of the gene coding for subunit 1 of cytochrome c oxidase. During aging, the intron is liberated and amplified. In senescent mycelia it exists as a covalently closed-circular DNA "mobile intron" (Osiewacz and Esser, 1984; Cummings *et al.*, 1985).

Liberation of senescence plasmids from mitochondrial DNA may be stimulated by DNA damage. Munkres and Rana (1978) found that a dietary supply of free radical scavengers effectively postpones the onset of senescence but does not slow the rate of deterioration once senescence has started. According to Bertrand (1983), these results indicate that the onset of senescence might be triggered by free radicals, but they also imply that free radicals are not directly involved in the degenerative process leading to vegetative death. The latter conclusion was based on the above observation that free radical scavengers do not slow the rate of deterioration once senescence has started.

Mitochondrial DNA receives about 16-fold more oxidative damage on a per nucleotide basis than nuclear DNA in mammalian cells (Richter *et al.*, 1988) and a similar ratio may apply to fungi. Munkres (1985) has speculated that there may be features of the mitochondrial chromosome structure that render the DNA susceptible to site-specific free radical-mediated scission of the plasmids. In any case, senescence plasmids may tend to arise from mitochondrial DNA because of the greater vulnerability of these DNA elements to oxidative damage than chromosomal DNA.

In concluding his review, Bertrand (1983) commented that although senescence and vegetative death in fungi are now relatively well understood, fungal systems have not been validated yet as models for aging and senescence in higher organisms. One of the striking contrasts between fungal senescence and mammalian senescence is the ability of senescent maternal fungi to pass the character of senescence through meiosis to their sexual offspring, whereas there is no evidence of this in mammals.

B. Nematodes

Two nematode species used to study aging are *Caenorhabditis elegans* and *Turbatrix aceti. Caenorhabditis elegans* has a short life cycle: reproduction normally starts on day 3 or 4, ceases by day 14, and by day 25 it dies (Zuckerman, 1983). Thus, for aging studies, all the symptoms of senescence are compressed into a short period. In addition, *C. elegans* has a small fixed number of cells (about 830 at maturity) and differentiated organ systems. To test the idea that longevity is linked to repair capacity, Hartman *et al.* (1988) measured the sensitivities to DNA-damaging agents of several inbred strains of *C. elegans* with life spans ranging from 13 to 30.9 days. These inbred lines were obtained as recombinational variants from an initial cross between two different wild-type strains (Johnson and Wood, 1982). The life spans of the two wild-type parents were 17 and 19 days, respectively. The progeny of the cross were allowed to self-fertilize for 19 generations. The DNA-damaging agents used by Hartman *et al.* (1988) were UV-irradiation, gamma-irradiation and methyl meth-anesulfonate. Sensitivities at several stages of the developmental cycle were tested. They found no significant correlations between mean life span and sensitivity to the lethal effects of these agents. Excision repair of two UV-irradiation-induced photoproducts was also measured. Long-lived strains were no more repair competent than shorter-lived strains. The authors concluded on the basis of this evidence that DNA repair plays at best a minor role in the aging process of *C. elegans*. However, there are alternative explanations for this evidence. Zuckerman and Geist (1983) have noted that any condition that inhibits growth in nematodes, including starvation and environmental factors, induces increased longevity; therefore, any genetic allele that directly or indirectly reduces growth may increase life span. It is not surprising, given these complications, that in-creased repair capacity was not linked to increased longevity in the particular inbred lines studied. Johnson (1987) has pointed out that a defect in any organ or cellular system might lead to a premature death that would be completely unrelated to the mechanisms responsible for normal aging. Thus, measurements of repair levels in short-lived genotypes also may have no bearing on the role of DNA repair in the aging process. In conclusion, the above experiments, which find no correlation between life span and repair capacity, do not critically address the question of whether or not DNA damage causes aging in nematodes.

Analysis of DNA from *C. elegans* by Klass *et al.* (1983) demon-strated a number of significant age-correlated changes. The number of single-strand breaks as assayed by an *in vitro* procedure using *E. coli* DNA polymerase I increased significantly with age. They also observed an exponential increase in the amount of 5-methylcytosine in the DNA as the worm matured and aged. Furthermore, transcriptional capacity of DNA isolated from older worms was reduced when assayed in a HeLa cell *in vitro* transcription system. A biological assay to determine age-correlated changes in the DNA of sperm showed that a decline in the capacity of the sperm to support zygotic development occurred as the age of the male increased. These results indicate that significant age-correlated alterations occur in the DNA of the nematode.

DNA damage accumulates when the rate of occurrence of damage exceeds the rate of repair. Accumulation of damage leads to a decline in gene expression, including, possibly, expression of genes that encode repair enzymes (Chapter 9, Section III.A). Targovnik *et al.* (1984) measured excision repair capacity after UV-irradiation in young and old *T. aceti*. They assayed repair synthesis activity and the actual removal of pyrimidine dimers from the genome. The data consistently indicated a decline in DNA excision repair capacity with age in the nematode. Targovnik *et al.* (1985) also measured the capacity to excise thymine glycol in young and old nematodes after exposure to ionizing radiation. They found that the young nematodes were strikingly more capable of this type of DNA repair than old nematodes.

If endogenously produced oxidative radicals are important DNA-damaging agents in nematodes, then antioxidants might enhance their longevity. Vitamin E, a strong antioxidant, has been reported to increase life span of the nematodes *Caenorhabditis brigsae* (Epstein and Gershon, 1972), *C. elegans* (Zuckerman and Geist, 1983), and *T. aceti* (Kahn and Enesco, 1981; Kahn-Thomas and Enesco, 1982). In two of these studies (Kahn-Thomas and Enesco, 1982; Zuckerman and Geist, 1983), a positive correlation was demonstrated among enhanced growth, development, and life span increases.

In conclusion, the accumulation of DNA damages and decline in transcription with age in nematodes suggest that DNA damage may cause aging in these organisms. That an antioxidant increases life span in nematodes suggests that oxidative DNA damage, specifically, may be important.

C. Insects

A number of studies on the effects of oxidative free radicals on aging have been done with *Musca domestica*, the common housefly.

In Chapter 6, Section III, we discussed the inverse correlation between metabolic activity and life span in different mammalian species and observed that this relationship may reflect the higher rate of oxidative DNA damage associated with higher metabolic rate. Several studies with adult houseflies (Sohal, 1976; 1981; Sohal and Buchan, 1981) have indicated that life span can be prolonged by reducing flight activity (and consequently oxygen consumption and metabolic rate). These observations suggest that decreased metabolic rate and, thus, decreased oxidative damage, may prolong life span. To test this idea, Sohal *et al.* (1986) attempted to correlate variations in the life span of houseflies belonging to the same cohort with antioxidant defenses and the products of free radical reactions. All houseflies lose flight ability prior to death. Therefore, in an aging population, short-lived flies can be identified as "crawlers," as distinguished from their longer-lived cohorts, the "fliers." The average life span of crawlers is only about 2/3 that of the fliers. The fliers were found to contain higher levels of superoxide dismutase and catalase, enzymes that remove reactive oxygen (Fig. 9, Chapter 6, Section I), and a higher level of the antioxidant glutathione. In addition, the concentration of H_2O_2 was lower in the fliers. The concentration of thiobarbaturic acid, a product of free radical reactions commonly found to accumulate with age, was also lower in these flies. These observations suggest that oxidative damage promotes aging in house flies, as it apparently does in mammals.

Panno and Nair (1984) showed that chromatin condensation occurs in the brain cells of the housefly with age. They also reviewed evidence that chromatin condensation is associated with reduced transcription. Panno and Nair (1986) examined the time course of chromatin condensation in brain nuclei from two groups of adult male flies: one reared under conditions of low physical activity (long-lived) and one reared under conditions of high physical activity (short-lived). The rate of chromatin condensation in the low activity group was very much less than that observed in the high activity group. In summary, the studies presented above suggest that lower physical activity leads to lower metabolic rates, lower levels of oxidative damage, lower rates of condensation of brain chromatin, a longer retention of high levels of transcription in brain cells, and increased fly longevity.

Recently, Newton *et al.* (1989a, 1989b) studied the effect of age on endogenous DNA single-strand breakage in the housefly. DNA single-strand breaks were not found to accumulate as the flies aged from 3 to 23 days and the degree of endogenous single-strand breaks was found to be unrelated to physiological age. The authors concluded that their results do not support the view that DNA single-strand breaks are a causal factor in aging in the housefly. As reviewed above, however, aging in the housefly may be associated with oxidative damage. If this damage is at the level of DNA, then oxidatively altered bases (which make up over 90% of oxidative DNA damages) may be more important in aging than single-strand breaks (which represent only about 2–4% of oxidative damages) (see Chapter 6, Section I, for further discussion).

Aging in the fruitfly, *Drosophila,* was reviewed by Miquel *et al.* (1979). *Drosophila* does not have any dividing cells in its somatic tissues. These authors concluded that aging in *Drosophila* is accompanied by a gradual disorganization of its postmitotic cells, probably resulting from deleterious reactions associated with oxidative metabolism. They also compared aging in *Drosophila* to aging in the mouse. In contrast to *Drosophila,* aging in the mouse is complicated by the presence of both fast-turnover cells and slow-turnover cells. They theorized that, in view of the normal fine structure of most dividing (fast-turnover) cells in old mice and other animals, it is unlikely that these cells play the main role in the multifaceted deterioration of metazoan aging. They summarized work indicating that some postmitotic tissues of the mouse have age-related structural senescent deterioration similar to that found in *Drosophila* tissues. Thus, the cytoplasmic changes in fixed postmitotic cells of *Drosophila* and the mouse suggest that these cells are the target of primary senescent deterioration.

Herman *et al.* (1971) studied the aging *Drosophila* brain. Young and old *Drosophila* were compared with respect to behavioral changes and cytological changes in the brain. Negative geotaxis and mating were used as indices of behavior. These began to decline at ages 28–35 days and very low values were obtained at 70–85 days. Other general behavioral differences between young and old flies included the following features. The young animals continuously fly or move around, their movements are well coordinated, and they rapidly escape when the vial containing them is open. Most of the old *Drosophila* have little, if any, ability to fly but can walk or hop when disturbed. They also are poorly coordinated, have difficulty righting themselves, and ordinarily do not escape when the vial is open. A variety of degenerative changes were observed in the giant neurons of the brain by both light and electron microscopy in 70–100-day-old *Drosophila*. These observations suggest that in *Drosophila*, as in mammals (Chapter 5, Section II.B.1), a decline in structure and function of neurons is correlated with the aging process.

The life span of *Drosophila* is increased by the antioxidants α-tocopherol (vitamin E) (Miquel *et al.*, 1973), L-thiazolidine-4-carboxylate (Miquel and Economos, 1979), and propylgallate (Ruddle *et al.*, 1988). This suggests that oxidative damage may contribute to aging in *Drosophila*.

D. Plants

In plants, sexual reproduction is generally the norm, with gamete fusion as the prelude to embryogenesis and seed formation. However plants, unlike most animals, possess the additional capacity to regenerate complete new individuals of similar genetic complement from vegetative parts. Meristematic buds or excised pieces of tissue can be induced to do this when placed in the appropriate environment. Plant vegetative cell lines apparently can be maintained indefinitely (Osborne, 1985). These lines of vegetative cells probably maintain themselves by a strategy of replacement, which we discussed for bacterial cells in Section A, above. Nevertheless, in some plant tissues senescence occurs, and this may be associated with DNA damage. Cheah and Osborne (1978) showed that in dry seeds, fragmentation of nuclear DNA occurs *in vivo* during embryo senescence. They suggested that the loss of DNA integrity could be the source of chromosomal aberrations and impaired transcription observed when seeds of low viability germinate.

Most forest trees live at least 100 yr, many of them >300 yr, and a few >1,000 yr. Clonal tree species may occupy a site for thousands of years (Harper and White, 1974). Most of the tree is dead, with only a thin shell of dividing cells (cambium) around the trunk and in the leaves. Cutler (1976) has noted that a tree actually represents a free-living clone of cells in which selective removal of irreversible accumulated damage is constantly occurring. He noted that one would not expect to find old cells in a tree any more than one would find old cells in a growing culture of bacteria.

Loehle (1988) analyzed the partitioning of energy between investments in defense and investments in growth in woody plants. This analysis indicated that life span of individual trees is determined by energy investment in protective measures such as thick bark and defensive chemicals. Longevity of angiosperms, but not gymnosperms, was correlated with increased investment in defense as measured by volumetric heat content of the wood. Longevity of gymnosperms was predicted by resistance to wood decay. Thus, longevity of the individual tree appears to be determined by the resistance to destruction of its nonliving woody tissues.

III. Similarities of Aging in Nonmammalian Organisms and in Mammals

A. Postmitotic or Slowly Dividing Cell Populations

In Chapters 2 to 7, we reviewed extensive evidence that mammals have crucial tissues composed of postmitotic or slowly dividing cells, which accumulate DNA damage and age. As reviewed in this chapter, Section II.C, somatic cells of the adult housefly and fruitfly are postmitotic, and these appear to age in a manner similar to postmitotic cells in mammals. As in mammals, aging in these insects seems to be correlated with oxidative damage and decline of neuronal functions.

Mature nematodes are also comprised of a fixed number of nondividing cells. In nematodes, single-strand breaks in DNA accumulate and transcription capacity declines with age. There is also a loss of repair capacity for oxidative DNA damages with age and antioxidants have a beneficial effect on life span. These observations suggest that accumulation of oxidative DNA damage may be the basis of aging in these organisms.

B. Rapidly Dividing Cell Populations

A possible strategy for coping with DNA damage involves rapid replication and replacement of damaged cells by duplication of undamaged ones. This is exemplified by the bacterium *E. coli*. The cambium cells of trees also appear to use this strategy, which permits clonal life spans reaching a thousand or more years. This strategy is used by rapidly dividing cells in mammals, including epithelial cells of the digestive tract and hemopoietic stem cells.

A rapid replication strategy, however, may also lead to aging and death. The protozoan *P. tetraurelia* experiences clonal aging even though undergoing rapid replication. In this case, aging appears to be due to accumulation of damage in the "somatic" DNA of the macronucleus. Presumably this reflects the inability of replication and repair processes to keep pace with the incidence of damage. A high genomic redundancy is acquired after conjugation or autogamy during formation of the macronucleus. This redundancy may initially mask the deleterious effects of DNA damage so that selective weeding out (death) of cells with damaged genomes, as occurs in *E. coli* bacteria, does not occur in paramecium. The information in the multiple genome copies appears to be progressively depleted by DNA damage during successive mitotic cycles and this may account for clonal aging in these organisms.

Filamentous fungi also age during rapid vegetative growth. Clonal aging, here, appears to be triggered by oxidative DNA damage leading to mutation in mitochondria. Selective multiplication of functionally defective forms of mitochondria, and their progressive displacement of the normal mitochondria, appears to be the basis of clonal aging and vegetative death in these organisms.

C. Contributions of Somatic Mutation

In mammals, as in fungi, mutational events triggered by DNA damage may give rise to abnormal replication. However, in mammals, abnormalities of cellular replication, rather than of mitochondrial replication, are often observed. Hartman and Morgan (1985) regarded mutation-induced focal lesions as important factors in mammalian aging. They pointed out that multiple benign focal lesions are very common in humans and that they have an impact on many facets of physiological aging. They presented the hypothesis that effects of such dysfunctioning clusters of cells often are far worse than hypofunctioning or afunctional clusters of cells. They noted that accumulation of multiple small foci could be a factor in limiting life span through disruption of homeostatic mechanisms and by causing "barrier breakdown." They defined barrier breakdown as disruption of tissue-specific cell organizational and product patterns that otherwise would serve to protect stem cells from toxic agents, including mutagens. "Barrier breakdown" leads to localized sensitization to further endogenous and exogenous insults. This disruption, they argued, can result in further somatic mutation and consequences such as ulcers and cancer. Examples of focal lesions described by these authors included a variety of metaplasias (patches of "foreign tissue") in the stomach, intestines, gallbladder, pancreas, liver, and urinary bladder. They also included atherosclerosis, the chief cause of death in the United States and western Europe (Ross and Glomset, 1976).

Atherosclerosis, a disorder of the large arteries, is the primary cause of coronary heart disease and of cerebrovascular disease. Evidence that somatic mutations are involved in atherosclerosis has been reviewed by Bridges (1987). The atherosclerotic lesions are known as plaques. These consist of proliferated smooth muscle cells infiltrated with macrophages and imbedded in a matrix of basement membrane, proteoglycan, and connective tissue. The pathogenesis of the lesions is complex and may involve more than one causative process. Repeated cycles of injury, thrombosis, and lipid accumulation all seem to be involved. The focal proliferation of smooth muscle cells, characteristic of the plaques, appears to be an early step of the process (McGill, 1977). These cells retain their tendency to proliferate faster than other smooth muscle cells when put into tissue culture, indicating that they are not merely responding to an external stimulus but are genetically altered.

Benditt and Benditt (1973) and Benditt (1977) have proposed that the proliferating smooth muscle cells of an atherosclerotic plaque all stem from one mutated cell. By this proposal, the plaque is comparable to a benign tumor of the artery wall. The main evidence for this somatic mutation explanation has been the apparent monoclonal origin of the smooth muscle cells in individual atherosclerotic plaques (Benditt and Benditt, 1973). Hartman and Morgan (1985) have suggested that atherosclerosis can be pictured as a multifocal disorder in which numerous small "benign" tumors (leiomyomas) in the artery walls accumulate and develop with age. Support for these ideas comes from the work of Penn et al. (1986). These workers used DNA from atherosclerotic plaque to transfect NIH 3T3 cells. These cells are standardly used for detecting activated oncogenes such as ras, met, and neu. Upon transfection, they observed an increase in morphologically transformed foci. Using the DNA from these foci, they were able to confer the transformed phenotype in a further round of transfection. Furthermore, the transformed NIH 3T3 cells gave rise to tumors in nude mice. Benditt (1977) suggested that chemical mutagens (e.g., DNA-damaging agents), which can be carried in the blood, may cause atherosclerosis. The results reviewed above suggest that DNA damage in a smooth muscle cell may produce a mutation that activates one or more oncogenes, causing the cell to proliferate into an atherosclerotic plaque.

In this section, we have indicated that somatic mutations, leading to abnormal cellular proliferation, may play a role in human aging. In contrast to the filamentous fungi where abnormal proliferation appears to be a major cause of aging, in humans it appears to be a significant, but secondary, contributor. As discussed in Chapter 5, the major aging phenomena in humans appear to be associated with tissues containing nonreplicating or slowly replicating cells. We discuss this somatic mutation contribution to aging in mammals further in Chapter 15, Section I.H.

D. Contributions of Wear and Tear

In our discussion of trees, above, we noted that although there may be a potential for indefinite clonal proliferation of cambium cells, individual trees age and die due to progressive destruction of their nonliving woody tissue. This destruction of nonliving tissues is caused by external forces such as wind, fire, or disease. Such processes can be referred to as wear and tear. In mammals wear and tear also appears to play some role in aging. For instance, in humans, bacterial decay of teeth and pitting and eroding of articular surfaces where bones contact each other contribute to aging (Strehler, 1977, pp. 130, 185). A more extensive discussion of the role of wear and tear in mammalian aging is given in Chapter 15, Section I.G.

IV. *Aging of Mammals Viewed in a Broad Evolutionary Context*

As reviewed above, several strategies appear to have evolved for responding to DNA damage. Different organisms, or even different tissues within the same organism, may use different strategies. At one extreme, the strategy of continuous replication and replacement of lethally damaged cells may permit a cell population to replicate faster than DNA damage accumulates. The cell population thus avoids aging. This strategy appears to be used by bacteria, some plants, and some subpopulations of mammalian cells. Where it is used in multicellular organisms, the replication strategy is vulnerable to mutationally variant subpopulations that may proliferate abnormally. Examples are senescence plasmids in filamentous fungi and focal lesions (cancer, benign metaplasias, and possibly atherosclerotic plaque) in humans.

Mammals and insects have major cell populations that are nondividing. It seems that the main strategy in these cell populations for coping with DNA damage involves DNA repair. Cell attrition due to accumulating unrepaired DNA damage is partially compensated for by cellular redundancy. This overall strategy avoids the high costs of cell turnover associated with the replication strategy, as well as the problem of mutations that occur during replication, which can cause abnormal proliferation. However, a price is paid in terms of accumulation of DNA damage and aging of the cells in the population.

In each species, the particular strategy or the mix of strategies used for coping with DNA damage and other damage undoubtedly evolved to minimize costs and maximize benefits consistent with the life-style of the organism. Apparently, the specific mix of strategies with which we are most familiar, the one used by humans and other mammals, is but one of several patterns in nature.

References

Aufderheide, K. J. (1984). Clonal aging in *Paramecium tetraurelia*. Absence of evidence for a cytoplasmic factor. *Mech. Ageing Dev.* **28,** 57–66.

Aufderheide, K. J. (1987). Clonal aging in *Paramecium tetraurelia*. II. Evidence of functional changes in the macronucleus with age. *Mech. Ageing Dev.* **37,** 265–279.

Aufderheide, K. J., and Schneller, M. V. (1985). Phenotypes associated with early clonal death in *Paramecium tetraurelia*. *Mech. Ageing Dev.* **32,** 299–309.

Benditt, E. P. (1977). The origin of atherosclerosis. *Sci. Am.* **236(2),** 74–85.

Benditt, E. P., and Benditt, J. M. (1973). Evidence for a monoclonal origin of human atherosclerotic plaques. *Proc. Natl. Acad. Sci. USA* **70**, 1753–1756.

Berger, J. D. (1973). Nuclear differentiation and nucleic acid synthesis in well-fed exconjugants of *Paramecium aurelia*. *Chromosoma* **42**, 247–268.

Bertrand, H. (1983). Aging and senescence in fungi. *In* "Intervention in the Aging Process, Part B: Basic Research and Preclinical Screening" (W. Regelson and F. M. Sinex, eds.), pp. 233–251. Alan R. Liss, New York.

Bridges, B. A. (1987). Are somatic mutations involved in atherosclerosis? *Mutat. Res.* **182**, 301–302.

Cheah, K. S. E., and Osborne, D. J. (1978). DNA lesions occur with loss of viability in embryos of aging rye seed. *Nature (London)* **272**, 593–599.

Cummings, D. J., MacNeil, I. A., Domenico, J., and Matsuura, E. T. (1985). Excision-amplification of mitochondrial DNA during senescence in *Podospora anserina*. DNA sequence analysis of three unique "plasmids." *J. Mol. Biol.* **185**, 659–680.

Cutler, R. G. (1972). Transcription of reiterated DNA sequence classes throughout the lifespan of the mouse. *Adv. Gerontol. Res.* **4**, 219–321.

Cutler, R. G. (1976). Nature of aging and life maintenance processes. *Interdiscipl. Topics Gerontol.* **9**, 83–133.

Epstein, J., and Gershon, D. (1972). Studies on aging in nematodes. IV. The effect of antioxidants on cellular damage and lifespan. *Mech. Ageing Dev.* **1**, 257–264.

Harper, J. L., and White, J. (1974). The demography of plants. *Annu. Rev. Ecol. Syst.* **5**, 419–463.

Hartman, P. E., and Morgan, R. W. (1985). Mutagen-induced focal lesions as key factors in aging: A review. *In* "Molecular Biology of Aging: Gene Stability and Gene Expression" (R. S. Sohal, L. S. Birnbaum, and R. G. Cutter, eds.), pp. 93–136. Raven Press, New York.

Hartman, P. S., Simpson, V. J., Johnson, T., and Mitchell, D. (1988). Radiation sensitivity and DNA repair in *Caenorhabditis elegans* strains with different mean life spans. *Mutat. Res.* **208**, 77–82.

Herman, M. M., Miquel, J., and Johnson, M. (1971). Insect brain as a model for the study of aging. *Acta Neuropath. (Berlin)* **19**, 167–183.

Holmes, G. E., and Holmes, N. R. (1986). Accumulation of DNA damages in aging *Paramecium tetraurelia*. *Mol. Gen. Genet.* **204**, 108–114.

Johnson, T. E. (1987). Aging can be genetically dissected into component processes using long-lived lines of *Caenorhabditis elegans*. *Proc. Natl. Acad. Sci. USA* **84**, 3777–3781.

Johnson, T. E., and Wood, W. (1982). Genetic analysis of lifespan in *Caenorhabditis elegans*. *Proc. Natl. Acad. Sci. USA* **79**, 6603–6607.

Kahn, M., and Enesco, H. E. (1981). Effect of alpha-tocopherol on the lifespan of *Turbatrix aceti*. *Age* **4**, 109–115.

Kahn-Thomas, M., and Enesco, H. E. (1982). Relation between growth rate and life span in alpha-tocopherol cultured *Turbatrix aceti*. *Age* **5**, 46–49.

Klass, M. R., and Smith-Sonneborn, J. (1976). Studies on DNA content, RNA synthesis, and DNA template activity in aging cells of *Paramecium aurelia*. *Exp. Cell Res.* **98**, 63–72.

Klass, M. R., Nguyen, P. N., and DeChavigny, A. (1983). Age-correlated changes in the DNA template in the nematode *Caenorhabditis elegans*. *Mech. Ageing Dev.* **22**, 253–263.

Loehle, C. (1988). Tree life history strategies: The role of defenses. *Can. J. For. Res.* **18**, 209–222.

Martin, R. (1977). A possible genetic mechanism of aging, rejuvenation, and recombination in germinal cells. *ICN-UCLA Symp. Mol. Cell. Biol.* **7**, 355–373.

Mayo, K. A., and Orias, E. (1981). Further evidence for lack of gene expression in the tetrahymena micronucleus. *Genetics* **98**, 747–762.

McGill, H. C. J. (1977). Atherosclerosis: Problems in pathogenesis. *Atheroscler. Rev.* **2**, 27–65.

Miquel, J., and Economos, A. C. (1979). Favorable effects of the antioxidants sodium and magnesium thiazolidine carboxylate on the vitality and lifespan of *Drosophila* and mice. *Exp. Gerontol.* **14**, 279–285.

Miquel, J., Binnard, R., and Howard, W. H. (1973). Effects of DL-alpha-tocopherol on the lifespan of *Drosophila melanogaster*. *The Gerontologist* **13**, 37A.

Miquel, J., Economos, A. C., Bensch, K. G., Atlan, H., and Johnson, J. E. (1979). Review of cell aging in *Drosophila* and mouse. *Age* **2**, 78–88.

Munkres, K. D. (1985). Aging of fungi. *In* "Review of Biological Research in Aging" (M. Rothstein, ed.), pp. 29–43. Alan R. Liss, New York.

Munkres, K. D., and Rana, R. S. (1978). Antioxidants prolong lifespan and inhibit the senescence-dependent accumulation of fluorescent pigment (lipofuscin) in clones of *Podospora anserina* s+. *Mech. Ageing Dev.* **7**, 407–415.

Newton, R. K., Ducore, J. M., and Sohal, R. S. (1989a). Effect of age on endogenous DNA single-strand breakage, strand break induction and repair in the adult housefly, *Musca domestica*. *Mutat. Res.* **219**, 113–120.

Newton, R. K., Ducore, J. M., and Sohal, R. A. (1989b). Relationship between life expectancy and endogenous DNA single-strand breakage, strand break induction and DNA repair capacity in the adult housefly, *Musca domestica*. *Mech. Ageing Dev.* **49**, 259–270.

Osborne, D. J. (1985). Annual plants. *Interdiscipl. Topics Gerontol.* **21**, 247–262.

Osiewacz, H. D., and Esser, K. (1984). The mitochondrial plasmid of *Podospora anserina:* a mobile intron of a mitochondrial gene. *Curr. Genet.* **8**, 299–305.

Panno, J. P., and Nair, K. K. (1984). Chromatin condensation in the aging housefly. *Exp. Gerontol.* **19**, 63–72.

Panno, J. P., and Nair, K. K. (1986). Effects of increased lifespan on chromatin condensation in the adult male housefly. *Mech. Ageing Dev.* **35**, 31–38.

Penn, A., Garte, S. J., Warren, L., Nesta, D., and Mindich, B. (1986). Transforming gene in human atherosclerotic plaque DNA. *Proc. Natl. Acad. Sci. USA* **83**, 7951–7955.

Richter, C., Park, J.-W., and Ames, B. N. (1988). Normal oxidative damage to mitochondrial and nuclear DNA is extensive. *Proc. Natl. Acad. Sci. USA* **85**, 6465–6467.

Ross, R., and Glomset, J. A. (1976). The pathogenesis of atherosclerosis. *N. Engl. J. Med.* **295**, 369–377, 420–425.

Ruddle, D. L., Yengoyan, L. S., Miquel, J., Marcuson, R., and Fleming, J. E. (1988). Propyl gallate delays senescence in *Drosophila melanogaster.* *Age* **11**, 54–58.

Smith-Sonneborn, J. (1979). DNA repair and longevity assurance in *Paramecium tetraurelia.* *Science* **203**, 1115–1117.

Smith-Sonneborn, J. (1987). Aging in protozoa. *In* "Review of Biological Research in Aging," Vol. 3 (M. Rothstein, ed.), pp. 33–40. Alan R. Liss, New York.

Sohal, R. S. (1976). Metabolic rate and lifespan. *Interdiscipl. Topics Gerontol.* **9**, 25–40.

Sohal, R. S. (1981). Metabolic rate, aging and lipofuscin accumulation. *In* "Age Pigments" (R. S. Sohal, ed.), pp. 303–316. Elsevier/North Holland, Amsterdam.

Sohal, R. S., and Buchan, R. B. (1981). Relationship between physical activity and lifespan in the adult housefly, *Musca domestica. Exp. Gerontol.* **16**, 157–162.

Sohal, R. S., Toy, P. L., and Farmer, K. J. (1986). Relationship between life expectancy, endogenous antioxidants and products of oxygen free radical reactions in the housefly, *Musca domestica. Mech. Ageing Dev.* **36**, 71–77.

Strehler, B. L. (1977). "Time, Cells, and Aging." Academic Press, New York.

Targovnik, H. S., Locher, S. E., Hart, T. F., and Hariharan, P. V. (1984). Age-related changes in the excision repair capacity of *Turbatrix aceti. Mech. Ageing Dev.* **27**, 73–81.

Targovnik, H. S., Locher, S. E., and Hariharan, P. V. (1985). Age associated alteration in DNA damage and repair capacity in *Turbatrix aceti* exposed to ionizing radiation. *Int. J. Radiat. Biol.* **47,** 255–260.

Zuckerman, B. M. (1983). The free-living nematode *Caenorhabditis elegans* as a rapid screen for compounds to retard aging. *In* "Intervention in the Aging Process, Part B: Basic Research and Preclinical Screening" (W. Regelson and F. M. Sinex, eds.), pp. 275–285. Alan R. Liss, New York.

Zuckerman, B. M., and Geist, M. A. (1983). Effects of vitamin E on the nematode *Caenorhabditis elegans*. *Age* **6,** 1–4.

Chapter Nine

DNA Repair with Emphasis on Single-Strand Damages

In Chapters 2 and 4, we reviewed evidence that high levels of DNA damage occur in mammalian cells. If, as we assume, genome damage is a problem for all organisms, then DNA repair processes should be widespread in nature. In accord with this expectation, DNA repair processes have been found in a wide range of organisms including viruses (Harm, 1980; Bernstein, 1981), bacteria (Friedberg, 1985), protozoa (Smith-Sonneborn, 1979), fungi (Baker, 1983; Friedberg, 1988), slime molds (Welker and Deering, 1978; Guyer *et al.*, 1986), algae (Daves, 1967), insects (Smith and Dusenberg, 1988), higher plants (e.g., Trosko and Monsour, 1969; Howland, 1975; Jackson and Linskens, 1978; Blaisdell and Warner, 1983; McLennan, 1987), and mammals (Friedberg, 1985).

In this chapter, we first explain that DNA repair processes ordinarily depend on the replacement of damaged information by intact information from a redundant copy. Next, we review the range of DNA repair processes available to overcome single-strand DNA damages (the most common class of DNA damages). We then discuss changes in DNA repair levels that occur with age. Finally, we briefly discuss the less common double-strand DNA damages in relation to aging.

I. *Repair Depends on Redundant Information*

In Chapter 2, Section I, we noted that the basic principles of transmitting accurate information from a source to a destination are independent of the kind of information and the mechanism involved. We illustrated the distinction between mutation and DNA damage by assuming that a string of Latin letters is allowable in the sense of being a proper set of characters for transmitting the English language. In Figure 1 (Chapter 2, Section I), the concept of mutation was shown as an error in transmission of the English word "bubble," which then became "babble." The mutated form, "babble," contains only allowable characters and therefore the error is not recognizable. The concept of damage was illustrated by an error of transmission of the word "bubble," which then became "b#bble." In this case the error can be recognized since the # symbol is not part of the allowed set of symbols. The information lost due to damage ordinarily cannot be recovered unless a redundant copy of the linear sequence with correct information at the point of damage is also transmitted. Redundancy is essential for repair of most types of DNA damage. The exceptions are those damages that can be repaired by direct reversal (see Section II.F in this chapter).

In Figure 10, we illustrate the role of redundancy in error correction. In the upper part of the illustration, part (a), one sees that upon reception of information containing a mutation, redundancy can be used to determine that there is an error. However, there is no way of distinguishing which received string is correct because both contain only allowable symbols. By contrast, in the lower part of the illustration, part (b), one sees that upon reception of information

a) Mutation

Source string: **Bubble**

Received string 1: **Bubble**
Received string 2: **Babble**

b) Damage

Source string: **Bubble**

Received string 1: **Bubble**
Received string 2: **B#bble**

Figure 10   Role of redundancy.

containing a damage, the error can be recognized as a disallowed symbol and so it can be corrected by using information from the undamaged redundant string.

There are limits to the level of redundancy in any informational system since redundancy is costly. Redundancy is costly because it requires an added investment of resources to make the extra copies. In living systems, much of the information encoded in the DNA is critical. We can, therefore, expect to find mechanisms that have evolved to recover lost information through redundancy, but in ways that keep the costs of redundancy to a minimum. These issues will be discussed later in this chapter and also in Chapter 14, Section II.B.

II. *Repair of Single-Strand DNA Damages*

The genomes of most organisms are composed of DNA in the form of a double helix. (The exceptions are viruses with single- or double-stranded RNA genomes or single-stranded DNA genomes. Repair of RNA genomes will be discussed in Chapter 14, Sections II.C and III). Genomes composed of double-stranded DNA have built in informational redundancy because the coding sequences on the two strands are complementary according to the base pairing rules: G pairs with C, and A pairs with T. Therefore, damage localized to one strand, in principal, can be repaired by excising the damage and replacing the lost information by copying from the complementary strand. This general process is referred to as excision repair. A second approach is used for a few particular types of single-strand damages. In these cases, an enzyme can recognize the damage, but instead of removing it, the enzyme can essentially reverse the process by which the damage occurred, thus restoring the original information. We refer to such repair as direct reversal.

In this section (Section II.A–G), we describe excision repair and direct reversal as well as the less well-understood processes of error-prone repair and postreplication recombinational repair. A great deal of research has been published on the various DNA repair pathways and comprehensive reviews are available (see Friedberg [1985] for a detailed review from a biochemical perspective and Wallace [1988] for a recent review of repair processes related to oxidative damage). Because of the vastness of the literature, our discussion in this section will be an overview with emphasis on those aspects of DNA repair that seem most relevant to aging. This overview should also provide a background for Chapter 10 in which we focus on recombinational repair, a specific type of DNA repair process important for understanding the adaptive significance of sexual reproduction.

A. Excision Repair: A Specific Example

One of the most well-understood pathways of excision repair is the pathway of pyrimidine-dimer excision repair in the bacterial virus phage T4 (Fig. 11). Pyrimidine dimers are a type of damage formed in DNA upon exposure to UV light. Presumably, UV is a natural hazard to phage T4, and the pyrimidine-dimer excision repair pathway is retained because of its adaptive benefit. This pathway is thought to have a relatively narrow specificity with respect to the kind of damage it repairs, being limited almost exclusively to pyrimidine dimers. Pathways of excision repair that remove other types of damage are similar to this one, however, so it is instructive to review the steps of this well-studied pathway.

The extensive experimental evidence providing the basis for the pyrimidine-dimer excision repair pathway in phage T4 has been reviewed by Bernstein and Wallace (1983) and Grossman *et al.* (1988). Figure 11 illustrates the repair of one particular type of pyrimidine dimer—the thymine dimer. A thymine dimer is composed of two adjacent, covalently joined pyrimidine bases. As shown in Figure 11, removal of the thymine dimer is initiated by a glycosylase, which catalyzes the uncoupling of one of the two thymine bases from the deoxyribose to which it had been linked. In general, when an excision repair pathway utilizes a glycosylase, the specificity with which the damage is recognized is determined by this enzyme. As will be discussed later, different glycosylases recognize different types of damage. The glycosylase of phage T4 has an additional apyrimidinic endonuclease activity that cleaves the sugar phosphate backbone (also indicated in the first reaction in Fig. 11). In general, apyrimidinic endonucleases are enzymes that cleave the ester linkage between deoxyribose and phosphate at a site in the DNA where the base normally linked to the deoxyribose has been removed. The gene in phage T4 that encodes the enzyme with both glycosylase and apyrimidinic endonuclease activities is referred to as the *denV* gene.

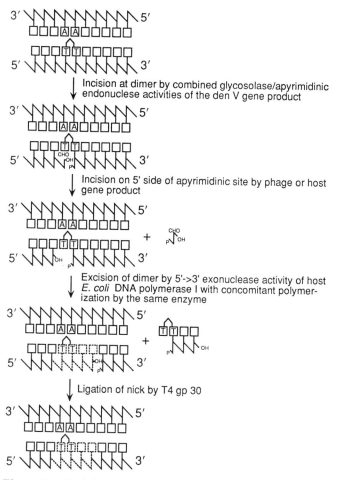

Figure 11   Excision repair in T4-infected cells (adapted from Bernstein and Wallace, 1983).

The two activities of the *denV* gene product are indicated near the uppermost bold arrow in Figure 11.

The next step involves an incision on the 5' side of the apyrimidinic site as indicated by the second arrow in Figure 11. The enzyme that catalyzes this reaction has not yet been identified. The subsequent reaction is the excision of the thymine dimer by the 5' to 3' exonuclease activity of DNA polymerase I. This enzyme is encoded by *E. coli,* the bacterial host of phage T4. The section of polynucleotides containing the thymine dimer that is removed by this exonuclease activity is indicated in Figure 11 by a short separate chain to the right of the main DNA structure. DNA polymerase I also has a polymerizing activity that catalyzes the extension of a polydeoxyribonucleotide chain in a 5′ to 3′ direction by copying the complementary strand of DNA. Thus, as the exonuclease activity of polymerase I removes the damaged section of DNA in a 5′ to 3′ direction, the polymerizing activity inserts correctly paired bases, also in a 5′ to 3′ direction. The patch size is about four nucleotides long (Yarosh et al., 1981). In Figure 11, the correct, newly inserted bases are indicated by dashed lines. The final reaction in the sequence involves the formation of the last phosphodiester bond in the repaired strand. This is catalyzed by an enzyme, polynucleotide ligase, encoded by gene 30 of the phage. The overall consequence of this pathway is the removal of a pyrimidine dimer in one strand of DNA and its replacement by an accurate complementary copy of the undamaged strand. This restores the original sequence.

B. Excision Repair Pathways Based on Glycosylases

In contrast to the glycosylase described above, which has an additional apyrimidinic endonuclease activity, other glycosylases may lack this second activity and rely on a second enzyme for the endonucleolytic reaction. DNA glycosylases, by definition, catalyze the hydrolysis of the N-glycosylic bonds linking bases to the deoxyribose-phosphate backbone. This reaction leaves either an apurinic or apyrimidinic site in the DNA. After removal of the base, these two types of sites are indistinguishable from each other. Therefore, they are conventionally referred to as AP sites. AP sites may arise not only by action of glycosylases, but also by spontaneous hydrolysis of the N-glycosylic bond linking the base to the deoxyribose in DNA. The spontaneous (i.e., heat induced, at body temperature) depurination and depyrimidination of DNA to form AP sites was discussed in Chapter 2, Section II.A, where it was noted that the frequency of such damages may be over 10,000 per mammalian cell per day. Once an AP site is formed, either by action of a glycosylase or by spontaneous hydrolysis, its removal by excision repair requires the action of one or more nucleases. Nucleases that specifically recognize sites of base loss in DNA are called AP endonucleases.

The *denV* gene product indicated in Figure 11 is both a glycosylase and an AP endonuclease. AP endonucleases carry out the incision or nicking of DNA at AP sites by catalyzing the hydrolysis of phosphodiester bonds.

AP endonucleases are common enzymes. A recent review of mammalian AP endonucleases lists two from mouse, two from rat, two from calf, and seven from human (Wallace, 1988). Two classes of AP endonuclease have been described in human. Class I activities nick 3' to the damaged site and class II activities nick 5' to the damaged site. In principle, an AP site can be removed by the sequential action of a 5'-acting and a 3'-acting AP endonuclease (or vice versa) leaving a gap resulting from removal of just the backbone sugar-phosphate molecule from which the purine or pyrimidine was released. In actuality, the excision reaction appears to be more complex, involving the removal of a patch of nucleotides as in Figure 11. After an initial incision by an AP endonuclease, exonucleases can degrade the DNA in the 5' to 3' or 3' to 5' direction at the free ends created by the incisions. These exonucleases apparently are not uniquely involved in excision repair but, rather, may degrade DNA with free ends in other metabolic processes as well.

The DNA glycosylases found in mammalian cells are summarized in Table XIII. The very fact that an enzyme has evolved to remove a particular type of DNA damage implies that the damage reduces fitness in nature and selects for adaptive processes to remove it.

If some fraction of these damages remain unrepaired or decay spontaneously to forms (e.g., single-strand breaks) that remain unrepaired, they may inhibit gene transcription and, therefore, contribute to aging. The repair pathways in which these glycosylases are employed may be a major line of defense against damages that cause aging. As discussed in Chapter 6, Section III.A, the incidence of thymine glycol damage in four mammalian species correlates inversely with species longevity. This supports the concepts that (1) thymine glycol damages are significant in aging and (2) the glycosylase that removes them is a defense against aging.

As discussed in Chapter 6, Section I, some of the damages recognized by these glycosylases are oxidation products of DNA. The damages excised by the last three glycosylases listed in Table XIII are produced in DNA by oxidative reactions. These damaged

Table XIII
Mammalian DNA Glycosylases

Enzyme	Mammalian Sources	References
Uracil–DNA glycosylase	calf thymus	Talpaert-Borle *et al.* (1979, 1982)
	human blast cells	Caradonna and Cheng (1980)
	human KB cells	Anderson and Friedberg (1980)
	rat liver, nuclear and mitochondrial	Colson and Verly (1983), Domena and Mosbaugh (1985)
	human fibroblasts	Kuhnlein *et al.* (1978b)
Hypoxanthine glycosylase	human HeLa cells	Myrnes *et al.* (1982)
	calf thymus and human fibroblasts	Karran and Lindahl (1980)
3-methyladenine glycosylase	rat liver[a]	Cathcart and Goldthwait (1981)
	human lymphoblasts	Brent (1979)
	rat and hamster liver	Margison and Pegg (1981)
Imidazole-ring-opened form of 7-methylguanine	rat and hamster liver	Margison and Pegg (1981)
Urea–DNA glycosylase	calf thymus and human fibroblasts	Breimer (1983)
Hydroxymethyluracil glycosylase	mouse plasmacytoma cells	Hollstein *et al.* (1984)
Thymine-glycol glycosylase[b]	calf thymus	Breimer (1983)
	mouse plasmacytoma cells	Hollstein *et al.* (1984)
	calf thymus	Doetsch *et al.* (1986)
	human	Higgins *et al.* (1987)
	human lymphoblasts	Lee *et al.* (1987)

[a] The 3-methyladenine glycosylase from rat liver also releases 7-methylguanine.
[b] Mammalian thymine glycol glycosylases also have an associated AP endonuclease activity.

structures are urea, hydroxymethyluracil, and thymine glycol. Thymine glycol glycosylase has both a glycosylase and an AP endonuclease activity. Thymine glycol glycosylase is similar in this respect to the *denV* gene product described above, which is also a glycosylase and AP endonuclease. Uracil and hypoxanthine can originate in DNA from the hydrolytic deamination of cytosine and adenine, respectively, as well as by other reactions (Friedberg, 1985: pp. 9–17; Wallace, 1988). Both types of hydrolytic deamination are promoted by heat (Karran and Lindahl, 1980). However, the principal source(s) of these damages remains unknown. The source of 3-methyladenine damages in DNA is also unclear. One hypothesis is that *S*-adenosylmethionine, the major intracellular methyl donor, may contribute the methyl adduct in forming 3-methyladenine from adenine (Rydberg and Lindahl, 1982).

C. AP Endonuclease Defect in Xeroderma Pigmentosum

As discussed in Chapter 7, Section I.B, humans with the inherited condition xeroderma pigmentosum (XP) are defective in excision repair of pyrimidine dimers as well as a variety of other types of DNA damages. There we argued that some of the DNA damages that remain unrepaired in XP may cause features of premature aging, particularly early death of neurons. Evidence was also reviewed that XP cells in complementation groups A and D are defective in an AP endonuclease (Kuhnlein *et al.*, 1976; 1978a; Lambert *et al.*, 1983; Linn et al., 1988). In humans, the excision repair pathway for removing thymine dimers possibly may have similarities to the pathway of thymine-dimer excision repair shown in Figure 11. However, at least one major difference exists between the two pathways in that the initial glycosylase-directed reaction in phage T4 is relatively specific for pyrimidine dimers, whereas the pathway in humans that removes pyrimidine dimers apparently can remove a much wider range of DNA damages. Also, as yet, no evidence indicates that one or more glycosylases are involved in the repair pathway defective in XP individuals. As reviewed in Chapter 7, Section I, numerous studies comparing different mammalian species, including human, have shown that the capacity to repair UV-induced damage correlates with longevity (Table X). Therefore, an excision repair pathway, perhaps one with similarities to that shown in Figure 11, may promote longevity in humans.

D. Preferential Repair of Actively Expressing Genes

Mellon *et al.* (1986) measured the rate of removal of pyrimidine dimers in defined human DNA sequences that were undergoing gene transcription and compared this rate with that for the overall genome. Their results demonstrated a strong preferential rate of removal of dimers from the transcriptionally active gene compared with that in total cellular DNA. The authors suggested that selective removal of pyrimidine dimers from active sequences may be a general characteristic of mammalian DNA repair.

Furthermore, Mellon *et al.* (1987) found selective removal of transcription-blocking DNA damages from the transcribed strand compared with the nontranscribed strand in a mammalian gene. In hamster cells, 80% of UV-induced pyrimidine dimers were removed from the transcribed strand in 4 hr, whereas little repair occurred in the nontranscribed strand of the same gene even after 24 hr. Two alternative models were proposed to explain these observations (Mellon *et al.*, 1987). The chromatin structure of active sequences may be in a more open conformation, which allows the DNA to be more accessible to repair enzymes, or, alternatively, transcription is directly coupled with DNA repair.

E. Excision Repair Pathways not Based on a Glycosylase or AP Endonuclease

Some kinds of base damage in DNA are apparently not recognized by specific DNA glycosylases. Such damages may be excised by a class of endonucleases that make incisions in the DNA backbone near sites of base damage. This class of endonucleases appears to recognize conformational distortions of the DNA structure caused by a wide variety of base damages that are otherwise structurally unrelated to each other.

The most well-studied example of this kind of excision repair involves the uvrABC endonuclease of the bacterium *E. coli* (Grossman *et al.*, 1988). This enzyme can catalyze the formation of a strand break on either side of a pyrimidine dimer. The breaks are separated by about 12 nucleotides. Therefore, the damaged bases may be excised as part of an oligonucleotide structure without requiring exonucleolytic degradation (Rupp *et al.*, 1982; Sancar and Rupp, 1983). Once the damaged bases are removed, the missing nucleotides are replaced by DNA polymerase I, which uses the complementary undamaged strand as a template (Grossman *et al.*, 1988). The final step, the joining of the last newly incorporated nucleotide to the extant parental DNA, is catalyzed by a ligase. These last two steps are similar to the final steps shown in Figure 11.

At present, an excision repair pathway analogous to the above *E. coli* pathway has not been identified in mammalian cells

(Grossman *et al.*, 1988). Because the variety of DNA damages that occur in nature may be very large, specific glycosylases are unlikely to have evolved to handle all damages that might be of significance. Therefore, a repair pathway that could recognize a range of damages by the conformational distortions they cause should be advantageous in mammals. Humans and mouse have homologous excision repair proteins which are both referred to as ERCC-1. Both of these proteins have partial sequence homology with *E. coli* repair proteins uvrA and uvrC (van Duin *et al.*, 1985). Because the uvrA and uvrC proteins are components of the *E. coli* excision repair enzyme uvrABC endonuclease, these results suggest that humans and *E. coli* may have an analogous endonuclease. Friedberg (1985: p. 257) has suggested that some of the genes in human cells that are defective in XP individuals may encode direct-acting nucleases similar to the uvrABC endonuclease of *E. coli*.

F. Direct Reversal Processes for Handling DNA Damages

Two mechanisms for the direct reversal of DNA damage have been extensively studied. These mechanisms are enzymatic photoreactivation of pyrimidine dimers and repair of 0^6-alkylguanine.

1. *Photoreactivation of Pyrimidine Dimers*   As described earlier (this chapter, Section II.A), pyrimidine dimers are formed in DNA upon exposure to UV light. One type of pyrimidine dimer results from covalent linkage between two adjacent pyrimidine bases to form a cyclobutane ring structure. Enzymatic photoreactivation is a light-dependent process involving the enzyme-catalyzed monomerization of the cyclobutyl dimer (Rupert, 1975). The enzyme that catalyzes photoreactivation of dimers is called photoreactivating enzyme, or DNA photolyase. Evidence of DNA photolyase activity has been reported in bacteria as well as in a large number of plants and animals. These include algae, yeast, filamentous fungi, protozoa, molluscs, echinoderms, arthropods, teleost fish, amphibians, reptiles, birds, marsupials, placental mammals, and higher plants (Rupert, 1975). Some of the reports for amphibia, reptiles, birds, and, particularly, mammals involved measurements of DNA photolyase activity in internal organs such as liver, brain, and heart. Because such organs are not naturally exposed to UV-irradiation or visible

light, what the significance of this activity might be is unclear. It would seem that any role that photolyase might have in preventing aging in larger animals should be limited to exposed areas of the skin.

2. *Repair of O^6-alkylguanine* The abnormal DNA bases O^6-methylguanine and O^6-ethylguanine are highly mutagenic if left unrepaired. The mechanism by which they are repaired is best understood in the bacterium *E. coli*. A similar mechanism also seems to exist in mammalian cells although the evidence is somewhat less clear (Friedberg, 1985: pp. 121–125).

In *E. coli,* alkylation damage in DNA activates the synthesis of an enzyme, O^6-methylguanine methyltransferase, whose function is to repair O^6-alkylguanine. The enzyme acts on ethyl groups in the same manner that it acts on methyl groups. The methyl transferase removes the methyl group from O^6-methylguanine in DNA and then transfers it to one of its own cysteine residues, giving rise to S-methylcysteine. Thus, the base damage in DNA is repaired by direct reversal. By transferring the methyl group to itself, the methyltransferase becomes inactivated. As a result, this enzyme has been referred to as a suicide enzyme. Because of this property, the intracellular pool of enzyme may be used up if the incidence of damage is greater than the rate of synthesis of new enzyme.

In mammals, an enzyme has been identified that transfers O^6-alkyl groups from O^6-methylguanine to cysteine, forming S-alkylcysteine in a protein acceptor molecule (e.g., Yarosh *et al.,* 1983; Renard *et al.,* 1983; Harris *et al.,* 1983). As in *E. coli,* the transferase and acceptor activity are apparently produced by the same protein, as judged by the inability to resolve these activities by chromatography (Harris *et al.,* 1983). The biochemical properties of the bacterial and mammalian enzymes are similar (Lindahl *et al.,* 1983), suggesting that the latter also may be suicide enzymes.

The importance of O^6-methylguanine residues in mammalian cells appears to be their tendency to cause mutation (Newbold *et al.,* 1980). O^6-methylguanine residues in DNA presumably cause mutation by mispairing during DNA replication (Loveless, 1969). Clonal proliferation of mutated cells may give rise to focal lesions; therefore, the repair activity of O^6-methyltransferase may be important for avoiding neoplasms in replicating cell populations. In nonreplicating tissues that are prominent in aging, such as muscle and brain,

O^6-methylguanine damages probably have a less significant impact than other more prevalent damages such as strand breaks (see Table I). Therefore, repair of O^6-alkylguanine is probably less important in resisting aging than in resisting cancer.

G. Other, Less Well Defined, Repair Processes

In this section, we briefly review two processes that may play a role in promoting longevity, although currently these processes are not well understood in mammalian cells.

1. *Error-Prone Repair or Damage Tolerance* Evidence for error-prone repair (or tolerance) exists in the bacterium *E. coli,* but evidence for its existence in mammalian cells is only tentative. In *E. coli,* error-prone repair is part of a more general response to DNA damage known as the SOS response (see Friedberg, 1985: pp. 408–445; Peterson *et al.,* 1988, for reviews). DNA damage in *E. coli* initiates a regulatory signal causing the simultaneous derepression (turning on) of a number of genes. The products of some or all of these genes enhance the survival of the cell. The *lex* and *recA* gene products play a key role in the regulation of the SOS response. One aspect of the SOS response is the enhancement in the rate of mutation induced by DNA-damaging agents. When DNA damage is introduced by UV light, most of the mutations that arise depend on the *recA* gene product, which is necessary for turning on the SOS response. These UV-induced mutations are thought to occur through error-prone DNA synthesis that is induced as part of the SOS response. This error-prone synthesis might be associated with a repair process that handles bulky damages such as those caused by UV light. However, the mechanism of error-prone DNA synthesis is not understood in molecular terms. In part, this is due to the fact that the mutational events are infrequent and are not obviously amenable to biochemical analysis. Nevertheless, evidence has been obtained that error-prone DNA synthesis is caused by DNA polymerase III holoenzyme (Bridges *et al.,* 1976), and that the critical component of the holoenzyme may be the product of gene *mutD(dnaQ)* (Echols *et al.,* 1983). The mutD(dnaQ) protein has 3' to 5' exonucleolytic proofreading activity, which can remove mispaired bases during DNA synthesis. By removing base mispairs in the newly forming strand,

incipient mutations are avoided. Relaxation of this proofreading activity may occur during the SOS response and may be involved in error-prone DNA synthesis (Lu *et al.*, 1986). Relaxation of proofreading may allow bases to be added to the nascent strand opposite damages in the template strand as part of an overall repair process. The bases incorporated into the nascent DNA opposite the damaged bases are often incorrect in that they differ from the bases that would have been incorporated had there been no damage. When this nascent strand is completed and undergoes another round of replication, the incorrect bases pair with complementary bases. The end result after this second round of replication is that a new mutant sequence is generated at the site of the original DNA damage.

The inference that error-prone DNA synthesis is part of a DNA repair process is largely based on the fact that error-prone DNA synthesis is induced by DNA damage; however, error-prone DNA synthesis may involve damage tolerance without repair. By this explanation, the accuracy of semiconservative DNA replication is relaxed during the SOS response so that synthesis occurs across template damages without damage removal being part of the process. At present, damage tolerance, rather than repair, seems the more likely explanation for error-prone DNA synthesis. Both error-prone repair and damage tolerance are postulated mechanisms for coping with DNA damage at the cost of generating mutations. One or both of these proposed mechanisms appear to be an infrequent way of dealing with damage, used only as a last resort because of the high cost of deleterious mutation.

The knowledge developed with respect to the *E. coli* SOS response stimulated a search for an analogous response in eukaryotic cells. Some evidence has been obtained in mammalian cells for an enhanced survival and mutation response to UV and other DNA-damaging agents that act as inducing signals in *E. coli* (for review, see Kaufman, 1989). However, whether or not these responses in mammalian cells are the result of derepression of previously repressed genetic functions is uncertain. Furthermore, there currently is no evidence that mammalian cells respond to DNA-damaging agents by expressing a set of genes under control of a common regulatory system analogous to the SOS response in *E. coli*. Evidence has been obtained for UV induction of error-prone DNA synthesis in mammalian cells by DasGupta and Summers (1978), Sarasin and Benoit

(1980), Mezzina *et al.* (1981), Cornelis *et al.* (1982), Gentil *et al.* (1982), and Lytle and Knott (1982). This error-prone synthesis may reflect error-prone repair or damage tolerance involving replication across the damage. If error-prone repair occurs in mammalian cells, whether or not it is significant in promoting longevity is unknown. Error-prone DNA synthesis in mammals might play a role in producing focal lesions that arise by mutation (see Chapter 8, Section III.C) and contribute to aging in this way. Error-prone DNA synthesis in mammalian cells, as in bacteria, appears to reflect a strategy of the cell to repair or bypass potentially lethal damages at the risk of mutation. As noted by Haynes (1988), the existence of this mechanism suggests that maintenance of cellular viability takes precedence over genetic fidelity.

2. *Postreplication Repair* In Chapter 3, Section II, we discussed a model presented by Villani *et al.* (1978) for the possible sequence of events that occurs when a DNA polymerase, during the course of replication, encounters a bulky DNA damage such as a thymine dimer. According to this model, the DNA polymerase "idles" at a damaged site, inserting nucleotides opposite the damage in the new strand but then removing them by its proofreading exonuclease.

The arrest of DNA replication at sites of pyrimidine dimers in template strands is not permanent. There are at least two mechanisms whereby cells could resume DNA synthesis on templates containing replicative blocks. Using these mechanisms improves the probability of cell survival. The first mechanism involves reinitiating DNA synthesis some distance downstream from the blocks, thereby creating gaps or discontinuities in the daughter strands. These gaps are then filled in by some further process. The second mechanism involves DNA replication past the template damage in a continuous manner after an initial arrest. This would involve error-prone DNA synthesis because damaged bases are not copied as accurately as undamaged ones.

In this section, we discuss how gaps may be repaired. One possible way of filling in the gap would be to carry out repair synthesis with relaxed proofreading so that the template strand retaining the damage could be copied. This would be a form of error-prone repair because the gap would be repaired by error-prone synthesis past the thymine dimer, which is not removed. It is not known if such a process actually occurs.

Another way of filling in the gap was proposed by Rupp and Howard-Flanders (1968) and Rupp et al. (1971) for E. coli. The essential features of the model are shown in Figure 12. In the top structure, a thymine dimer formed by UV-irradiation is indicated in one of the two parental strands. Also, newly forming daughter strands are indicated by bold lines. Replicative bypass of the thymine dimer results (second structures from top) in one normal duplex molecule and one containing a gap opposite the dimer. The gap is then filled by a nonreciprocal recombination event. This is labeled "strand transfer" in Figure 12, and the result is shown in the third structures from the top. The recombination event involves physical transfer of a segment of single-stranded DNA from the parental strand of the undamaged duplex to fill in the gap in the damaged duplex. This leaves a temporary gap in the donor strand. This new gap can be filled in by repair synthesis (indicated by a bold line segment in the lowest structure) using the undamaged complementary strand as template. This process is referred to as postreplication recombinational repair (PRRR). It should be noted that the original thymine dimer that initiated the process of PRRR is not removed by the overall process but is merely circumvented. The advantage of PRRR over error-prone replication across the damage or error-prone repair of the gap is that in PRRR the DNA strand inserted opposite the damage contains accurate information. If error-prone repair had been used, the segment opposite the damage would be copied inaccurately and is likely to contain mutations. Thus, PRRR provides an accurate mechanism for tolerating DNA damages, which are in the way of the replicative polymerase.

Experiments with E. coli have yielded results that support the general model shown in Figure 12. The predictions of the model that are supported by evidence are the presence of gaps in newly replicated DNA (Johnson and McNeill, 1978), the specific location of these gaps opposite pyrimidine dimers in template DNA strands (Howard-Flanders et al., 1968), and the rejoining of gapped DNA by a process involving recombination (Smith and Meun, 1970; Rupp et al., 1971).

In mammalian cells, the evidence bearing on the model in Figure 12 is less conclusive than in E. coli. Daughter-strand gaps appear

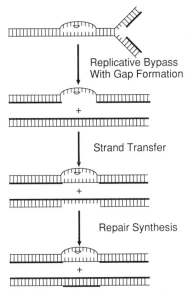

Figure 12   Formation of a postreplicative gap opposite a thymine dimer and filling of the gap by recombinational repair.

to form opposite DNA damages (Kaufmann, 1989), but whether or not the gap filling, as shown, occurs by recombination is unclear. However, evidence indicates that recombinational repair of DNA occurs in mitotic cells of mammals (see Chapter 10, Section III.C, and Chapter 12, Section III.A; also Kaufmann, 1989), and, thus, gap filling may happen by this mechanism in some cases.

III. *Relationship of DNA Repair Levels to Aging*

A. Changes in DNA Repair Capacity with Age

Previously, we reviewed evidence for an early developmental decline in repair capacity as cells differentiate to the postmitotic state in muscle (Chapter 4, Section II.A) and in brain (Chapter 4, Section II.B). In this section, we review a different kind of evidence. We review whether or not repair capacity, after the postmitotic state is reached, declines further in mammalian tissues as the animal ages. The many different studies that have addressed this issue are summarized in Tables XIV and XV. Table XIV lists 25 studies in which no change in repair capacity with age was detected, and Table XV lists 19 studies in which repair capacity declined with age. The methods used to detect repair capacity generally measured some form of excision repair.

On the DNA damage hypothesis of aging, one might expect repair capacity to decline in those tissues in which accumulation of DNA damage leads to a general decline in gene expression. We have previously reviewed evidence for reduced gene expression with age in brain (Chapter 5, Section II.A.1), muscle (Chapter 5, Section II.A.2), liver (Chapter 5, Section II.A.3), lymphocytes (Chapter 5,

Table XIV
Mammalian Tissues in Which No Change in DNA Repair Was Detected with Age

Tissue	Species	Ages Studied	Agent	Assay	Reference
Aortic smooth muscle cells	rat	3–23 mo	X-rays	CS	1[a]
Bone (chondrocytes)	human	23–63 yr	UV	UDS	2
Chondrocytes	rabbit	3 mo & 2 yr	UV	UDS	2
Chondrocytes	rabbit	3 mo & 2 yr	UV	ESS	3
Chondrocytes	rabbit human	3 &39 mo up to 71 yr		O⁶-methylguanine methyl acceptor activity	4
Brain	dog	7 wk to 13 yr	X-rays	ASGS	5
Brain	hamster	4–551 days[b]	UV	UDS	6
Brain	mouse	2 & 22 mo	X-rays	ASGS	7
Kidney	hamster	4–520 days[c]	UV	UDS	8
Lens epithelium	rat	19–172 wk	UV	UDS	9
Liver	mouse	2–22 wk	X-rays	ASGS	7
Lung	hamster	4–520 days[b]	UV	UDS	8
Lymph node	mouse	17–98 wk	UV	UDS	10
Lymphocytes	human	up to 60 yr		O⁶-methylguanine methyl acceptor protein	11
Lymphocytes	human	17–90 yr	UV	CS	12
Lymphocytes	mouse	3–24 mo		uracil–DNA glycosylase; AP endonuclease	13

(continued)

Table XIV (Continued)

Tissue	Species	Ages Studied	Agent	Assay	Reference
Lymphocytes	human	17–74 yr	UV	RS	14
Lymphocytes	human	fetal to 89 yr	X-rays	AE	15
Retina	rat	4–100 wk	4-NQO	UDS	16
Retina	rabbit	6 wk to 7 yr	X-rays	ASGS	17
Skin	human	fetus to 83 yr	UV	CS	18
Skin	human	19–88 yr	UV	UDS	19
Skin	human	young (3 days to 3 yr) and old (84–94 yr)	UV UV, MMS, TPL	CS HCR	20
Skin (keratinocytes)	human	newborn and adults aged 72 & 90 yr	UV	RS	21
Skin (keratinocytes	human	14 & 62 yr	UV	RS	22
Skin	mouse	2 & 18 mo	4-NQO	UDS	23
Skin	rat	6–36 mo	UV	ESS, IA	24
Spermatocyte	human	21–79 yr	UV	UDS	25

Abbreviations: AE, alkaline elution, a very sensitive method for measuring single-strand breaks in DNA; ASGS, alkaline sucrose gradient sedimentation, a method for measuring single-strand breaks; CS, cell survival; DMBA, dimethylbenz(a)anthracene; ESS, thymine dimer-specific endonuclease-sensitive sites; HCR, host cell reactivation; IA, immunoassay based on monoclonal antibody against UV-induced DNA photoproducts; MMS, methyl methanesulfonate; 4-NQO, 4-hydroxyquinoline-1-oxide; RS = repair synthesis; TPL, trimethylpsoralen plus light; UDS, unscheduled DNA synthesis associated with excision repair; UV, ultraviolet light.

[a] References: (1) Rosen et al. (1985); (2) Krystal et al. (1983); (3) Setlow et al. (1983); (4) Lipman et al. (1987); (5) Wheeler and Lett (1974); (6) Gensler (1981b); (7) Ono and Okada (1978); (8) Gensler (1981a); (9) Treton and Courtois (1981); (10) DeSousa et al. (1986); (11) Waldstein et al. (1982); (12) Kutlaca et al. (1982); (13) Barnard et al. (1986); (14) Kovacs et al. (1984); (15) Turner et al. (1982); (16) Ishikawa et al. (1978); (17) Lett (1977); (18) Goldstein (1971); (19) Hennis et al. (1981); (20) Hall et al. (1982); (21) Liu et al. (1982); (22) Liu et al. (1985); (23) Ishikawa and Sakurai (1986); (24) Mullaart et al. (1989); (25) Chandley and Kofman-Alfaro (1971).

[b] A decrease in repair was observed at age 4–57 days, which may be due to developmental regulation.

[c] An increase in repair was observed at age 4–9 days, which may be due to developmental regulation.

Section II.A.4), and kidney (Chapter 5, Section II.A.4). Table XV summarizes evidence for a decline in repair capacity with age in brain (one study), liver (two studies), and lymphocytes (eight studies). In contrast, Table XIV summarizes evidence for lack of change in repair capacity in muscle (one study), brain (three studies), liver (one study), lymphocytes (five studies), and kidney (one study). In skin, a widely investigated tissue, six studies showed no change and four

Table XV

Mammalian Tissues in Which Decreases in DNA Repair Were Detected with Age

Tissue	Species	Ages Studied	Agent	Assay	Reference
Resting chondrocytes	rabbit	3–52 mo	UV	UDS	1[a]
Resting chondrocytes	rabbit	3–36 mo	UV	ESS, AE	2
Brain	mouse	17–98 wk	UV	UDS	3
Colon mucosa	rat	to 13–15 mo	alkylating agents	ASGS	4
Liver	rat	6–32 mo	UV	UDS	5
Liver	rat	3–20 mo	UV bleomycin	UDS	6
Lung (fibroblasts)	mouse	7–132 wk	UV	UDS	7
Lymphocytes	human	13–93 yr	UV	UDS	8
Lymphocytes	human	23–78 yr	X-rays	DSBR	9
Lymphocytes	human	young to 94 yr	X-rays	PSR	10
Lymphocytes	human	18–99 yr	X-rays	CSP	11
Lymphocytes	human	17–90 yr	X-rays	CS	12
Lymphocytes	human	16–104 yr	X-rays[b]	UDS	13
Lymphocytes	human	fetal to 86 yr	bleomycin[c]	PSR	14
Lymphocytes	human	14 mo to 82 yr	UV	IA	15
Skin	rat	28–400 days	electrons	AUND	16
Skin	mouse	2–18 mo	UV	UDS	17
Skin	rat	6–44 mo	UV	UDS	18
Skin	human	17–77 yr	UV	UDS	19

Abbreviations: AUND, alkaline unwinding followed by S1 nuclease digestion; CSP, concanavalin A-stimulated proliferation; DSBR, double-strand break repair as measured by neutral filter elution; ENU, ethyl nitrosourea; PSR, phytohemagglutinin-stimulated replication; SSBR, single-strand break repair. See abbreviations for Table XIV for meaning of AE, ASGS, CS, ESS, IA, UDS, and UV.

[a] References: (1) Lipman and Sokoloff (1985); (2) Lipman and Setlow (1987); (3) DeSousa *et al.* (1986); (4) Kanagalingam and Balis (1975); (5) Plesko and Richardson (1984); (6) Kennah *et al.* (1985); (7) Kempf *et al.* (1984); (8) Lambert *et al.* (1979); (9) Mayer *et al.* (1989); (10) Goodwin (1982); (11) Harris *et al.* (1986); (12)Kutlaca *et al.* (1982); (13) Licastro *et al.* (1982); (14) Seshadri *et al.* (1979); (15) Roth *et al.* (1989); (16) Sargent and Burns (1985); (17) Ishikawa and Sakurai (1986); (18) Vijg *et al.* (1985); (19) Nette *et al.* (1984).

[b] A marked decline in repair with age was found at high X-ray doses, but not at low doses.

[c] Sensitivity to bleomycin, but not mitomycin C was observed to increase with age.

showed a decline in repair capacity with age. The contrasting results of different investigations on any particular tissue may reflect differences in the type of damage measured, species differences, differences in the cell type(s) examined, differences in the ages of "old" animals, and differences in the culturing of cells prior to examination.

Examples of contrasting results for a given tissue are discussed below with respect to brain and liver. We selected these two examples for further discussion because strong evidence indicates a general decline in gene expression in these cases.

DeSousa *et al.* (1986) (Table XV) found a decline with age in excision repair capacity for UV-induced damages in mouse dorsal root ganglia neurons. In one mouse strain, the decline in 97–98-wk-old mice compared with 17–18-wk-old mice was 73% when the dose rate used was 40 J/m² UV light. In another mouse strain, the decline for old mice was 40% at the same dose rate. The three studies on brain in which no decline in repair was detected were performed with dog (Wheeler and Lett, 1974), hamster (Gensler, 1981b), and mouse (Ono and Okada, 1978) (Table XIV). The study on dog was with internal granular layer neurons of the cerebellum, a different type of neuron population than the one studied in mouse by DeSousa *et al.* (1986). Furthermore, the type of damage studied was gamma-ray-induced single-strand breaks rather than UV-induced damage. In the hamster study, UV-induced excision repair was measured as in the study of DeSousa *et al.* (1986). However, whole brain rather than a specific neuron population was examined. In the mouse study by Ono and Okada (1978), gamma-ray-induced strand breaks were measured in the cerebellum, rather than UV-induced damages in dorsal root ganglia. In conclusion, the three studies in which there was a lack of observed decline in repair in the brain used approaches that were different than that of DeSousa *et al.* (1986), thus not invalidating their evidence for a decline in repair capacity in the situation they studied. Niedermuller (1985) and Washington *et al.* (1989), in experiments described below, also reported a decline in repair capacity in the mammalian brain with age.

Mammalian liver repair capacity with age was studied in three laboratories. Plesko and Richardson (1984) found a decrease in UV-induced excision repair in 32-mo-old rat compared with 6-, 14-, and 20-mo-old rat. Kennah *et al.* (1985) also found that excision repair of UV- and bleomycin-induced DNA damages declines in hepatocytes as rats age from 3 mo to 16–20 mo. In contrast, Ono and Okado (1978) found no decline in the capacity to repair gamma-ray-induced single-strand breaks in mouse liver with age. Because this latter study was with a different species and a different type of DNA damage than the first two, the finding of a lack of decline in repair capacity does not invalidate the findings of a decline in liver in the previous two studies.

Most of the 19 studies in which a decline in repair capacity was found with age (Table XV) are not contradicted by the other 25 studies listed in Table XIV, which show no decline in repair capacity; rather, the results indicate that repair declines with age in some instances that are specific with respect to type of repair, tissue or cell type, and species. In other instances, repair does not decline. Any reduction in a repair enzyme may lead to a faster accumulation of DNA damages and, hence, a faster decline in gene expression. Thus, a reduction in repair capacity may accelerate the aging process.

The data from two additional studies on DNA repair levels during aging were not included in Tables XIV and XV because too many separate entries would have been required. The first of these studies was carried out by Niedermuller (1985). He measured four different forms of repair in nine different tissues of the rat with age. The four forms of repair measured were (1) excision repair after treatment with the DNA-damaging agent N-nitrosomethylurea as determined by unscheduled DNA synthesis, (2) single-strand break repair as determined by nucleoid sedimentation after damage by methyl methanesulfonate, (3) double-strand break repair as followed by neutral elution after damage by 4-nitroquinoline-1-oxide, and (4) removal of endonuclease-sensitive sites as measured by velocity sedimentation in alkaline sucrose gradients after damage by gamma-radiation. Repair was measured at the three ages of 9, 18, and 28 mo. The tissues examined were kidney, lung, testis, spleen, skeletal muscle, liver, brain, heart, and duodenum. Excision repair (type 1) was reduced in nearly all organs in the oldest rats (28 mo) compared with both younger age groups. The differences were most significant for liver, spleen, lung, and heart. With regard to single-strand break repair (type 2), only testis and brain lost much of their repair capacity in the 28-mo-old rats. With regard to double-strand break repair (type 3), no significant age-dependent decline was found. With respect to the ability to remove endonuclease-sensitive sites (type 4), a pronounced age-dependent loss of repair capacity was observed in liver, kidney, lung, testis, duodenum, skeletal muscle, and brain. The decline was most significant in brain.

In the second study, Washington et al. (1989) assayed 3-methyladenine–DNA N-glycosylase (MAG) and 0^6-methylguanine DNA methyltransferase (MGMT) activities. These enzymes were measured in the liver, lungs, brain, and ovaries of female mice of two inbred stocks as a function of age. The MGMT levels in adults did not decrease with age from about 2 to 17 mo. A similar result was reported by Woodhead *et al.* (1985) for rats. However, the MAG levels, in general, were significantly lower in the older animals.

The results of the studies by both Niedermuller (1985) and Washington *et al.* (1989) follow a pattern similar to the results summarized in Tables XIV and XV. Measurements of some types of repair capacity in some tissues show a clear decline with age, whereas in other instances decline is not detectable.

B. Repair of Single-Strand Damages in Relation to Aging: An Overview

We now briefly review what is known or can be surmised about the relationship of the repair of single-strand damages to aging. In Chapter 2, Table I, we listed the incidence of various kinds of single-strand damage. The overall rate of damage (mostly single-stranded) was estimated at roughly 72,000 per mammalian cell per day. In Chapter 2, Section II.K, we noted evidence that in some types of cells the rate of repair of some types of damages (i.e., single-strand breaks and 0^6-methylguanine residues) appeared to be adequate to cope with the rate of occurrence of the damage. Nevertheless, we also reviewed evidence that repair levels were low in muscle (Chapter 4, Section II.A) and neuronal tissue (Chapter 4, Section II.B). Furthermore, damages such as single-strand breaks and methylated guanine accumulate with age in muscle (Chapter 4, Section III.A) and neuronal tissue (Chapter 4, Section III.B). Single-strand breaks and/or altered bases also accumulate in liver (Chapter 4, Section IV.B) and other tissues (Chapter 4, Section V). We further reviewed evidence that accumulation of DNA damage provides a reasonable explanation for the principal manifestations of aging (Chapter 5) and that oxidative DNA damage, in particular, may be a major problem (Chapter 6).

Tice and Setlow (1985) noted that cells and organisms are in a delicate balance between the production of damage and its repair, and if the production of damage increases or if the level of repair declines, a new steady-state condition for DNA damage will exist. Even with efficient repair, some level of damage will be present in DNA because the level of damage depends on its relative rates of production and repair. In this chapter, we pointed out that single-strand damages can be repaired by a number of repair processes. Excision repair appears to be a prominent type of repair in somatic cells and may be very important in resisting aging. All forms of excision repair involve the removal of the damaged region from one DNA strand and accurate replacement of the lost information by copying from the undamaged complementary strand. At present, seven different types of specific DNA damage are known to be removed by excision repair in mammalian cells (Table XIII), and this number is likely to increase in the near future due to intensive work in the field. Many studies have been carried out on changes in DNA repair capacity with age in mammals (this chapter, Section III.A). Since gene expression has been observed to decline with age in some tissues (e.g., muscle, brain, liver, lymphocytes, kidney), it would not be surprising if the synthesis of repair enzymes also declined. In a number of specific instances, repair capacity was observed to decline with age (Table XV) but not in other instances (Table XIV). Although we consider the decline in repair capacity to be a secondary effect of DNA damage on gene expression, loss of repair capacity in some tissues in aging individuals may accelerate the aging process.

IV. *Repair of Double-Strand Damages*

Double-strand damages are ones in which both complementary strands of DNA are altered at the same or approximately the same position so that neither strand can be used as an accurate template to repair the other. Excision repair is presumably ineffective in removing these types of damage because this type of repair depends on the intactness of the DNA strand opposite the damaged one. Two particular types of double-strand damages are double-strand breaks and cross-links. In Chapter 2, Section II.H, we estimated that the frequencies of double-strand breaks and cross-links produced by oxidative damage in mammalian cells are 8.8/cell/day and 8.0/cell/day, respectively. We also noted that heat causes double-strand breaks and that these can lead to loss of viability. Thus, two sources of spontaneous DNA damage that are likely to be important, oxidative reactions and heat, produce double-strand damages in significant quantity.

Double-strand damages may also arise indirectly from single-strand damages by the mechanism discussed earlier in this chapter (Section II.G.2), in which the DNA polymerase, upon encountering an unrepaired damage in the template strand, skips over it, leaving a gap opposite the damage. This gap opposite a damage is another kind of double-strand damage.

Evidence suggests that the lethal effect of oxidative reactions in mammalian cells largely reflects double-strand damages. Ward *et al.* (1985) concluded, from experiments on mammalian cells treated with H_2O_2, that DNA single-strand breaks caused by hydroxyl radicals are ineffective in causing cell death and that to produce lethal events "locally multiple damage sites" such as double-strand breaks are required. X-rays induce a spectrum of DNA damages similar to those produced by H_2O_2. Therefore, knowledge concerning these damages may be relevant to natural damages. Blocher and Pohlit (1982) studied the induction and repair of X-ray-induced DNA double-strand breaks in early stationary human cells using neutral sucrose gradient analysis. They showed that cell survival curves could be interpreted on the basis of one unrepaired double-strand break being a lethal event. Van der Schans *et al.* (1982) have also argued on the basis of their results and other data that one double-strand break remaining unrepaired after 2 hr of incubation is lethal for the mammalian cell. Because double-strand breaks, in principle, are less easily repaired than single-strand damages, they may be important contributors to cell impairments leading to aging despite their lower frequency of occurrence (Table I).

As discussed in Section I of this chapter, informational redundancy is essential for the repair of damaged and lost information. In the case of single-strand damages in DNA, the needed redundancy is available in the undamaged complementary strand. In the case of double-strand damages, the source of redundancy must be a second DNA molecule with intact information homologous to that lost in the first molecule. Repair requires the exchange of information between the two DNA molecules. Specifically, repair can occur if the undamaged DNA molecule donates a single-stranded section to the damaged DNA duplex to permit replacement of its lost information. This process is referred to as recombinational repair. Since we indicated that double-strand damages may contribute to aging, repair of double-strand damages may be important in resisting aging. On the other hand, the defect measured after double-strand damage was "lethality" (see above), or the inability to form daughter cells. Such a damaged cell, however, still may be able to carry out other functions. For postmitotic cells, the type of defects produced by double-strand damages may not be any more important than that produced by unrepaired single-strand damages.

In the next chapter, we review evidence that recombinational repair is widespread and is efficient in overcoming a variety of DNA damages, particularly double-strand damages. However, in the rest of the book, we will emphasize the role of recombinational repair of double-strand damages in meiosis, where formation of viable daughter cells is critical, rather than its possible role in aging.

References

Anderson, C. T. M., and Friedberg, E. C. (1980). The presence of nuclear and mitochondrial uracil–DNA glycosylase in extracts of human KB cells. *Nucleic Acids Res.* **8,** 875–888.

Baker, T. I. (1983). Inducible nucleotide excision repair in *Neurospora*. *Mol. Gen. Genet.* **190,** 295–299.

Barnard, J., LaBelle, M., and Linn, S. (1986). Levels of uracil DNA glycosylase and AP endonuclease in murine B- and T-lymphocytes do not change with age. *Exp. Cell Res.* **163,** 500–508.

Bernstein, C. (1981). Deoxyribonucleic acid repair in bacteriophage. *Microbiol. Rev.* **45,** 72–98.

Bernstein, C., and Wallace, S. S. (1983). DNA repair. *In* "Bacteriophage T4" (C. K. Mathews, E. M. Kutter, G. Mosig, and P. B. Berget, eds.), pp. 138–151. American Society for Microbiology, Washington, D.C.

Blaisdell, P., and Warner, H. (1983). Partial purification and characterization of a uracil–DNA glycosylase from wheat germ. *J. Biol. Chem.* **258,** 1603–1609.

Blocher, D., and Pohlit, W. (1982). DNA double strand breaks in Ehrlich ascites tumor cells at low doses of X-rays. II. Can cell death be attributed to double-strand breaks. *Int. J. Radiat. Biol.* **42,** 329–338.

Breimer, L. H. (1983). Urea–DNA glycosylase in mammalian cells. *Biochemistry* **22,** 4192–4197.

Brent, T. P. (1979). Partial purification and characterization of a human 3-methyl-adenine–DNA glycosylase. *Biochemistry* **18,** 911–916.

Bridges, B. A., Mottershead, R. P., and Sedgwick, S. G. (1976). Mutagenic DNA repair in *Escherichia coli*. III. Requirement for a function of DNA polymerase III in ultraviolet mutagenesis. *Mol. Gen. Genet.* **144,** 53–58.

Caradonna, S. J., and Cheng, Y.-C. (1980). Uracil DNA glycosylase. Purification and properties of this enzyme isolated from blast cells of acute myelocytic leukemia patients. *J. Biol. Chem.* **255**, 2293–2300.

Cathcart, R., and Goldthwait, D. A. (1981). Enzymatic excision of 3-methyladenine and 7-methylguanine by a rat liver nuclear fraction. *Biochemistry* **20**, 273–280.

Chandley, A. C. and Kofman-Alfaro, S. (1971). "Unscheduled" DNA synthesis in human germ cells following UV irradiation. *Exp. Cell Res.* **69**, 45–48.

Colson, P., and Verly, W. G. (1983). Intracellular localization of rat-liver uracil–DNA glycosylase. Purification and properties of the chromatin enzyme. *Eur. J. Biochem.* **134**, 415–420.

Cornelis, J. J., Su, Z. Z., and Rommelaere, J. (1982). Direct and indirect effects of ultraviolet light on the mutagenesis of parvovirus H-1 in human cells. *EMBO J.* **1**, 693–699.

DasGupta, U. B., and Summers, W. C. (1978). Ultraviolet reactivation of herpes simplex virus is mutagenic and inducible in mammalian cells. *Proc. Natl. Acad. Sci. USA* **75**, 2378–2381.

Davies, D. R. (1967). UV-sensitive mutants of *Chlamydomonas reinhardi*. *Mutat. Res.* **4**, 765–770.

DeSousa, J., DeBoni, U., and Cinader, B. (1986). Age-related decrease in ultraviolet induced DNA repair in neurons but not in lymph node cells of inbred mice. *Mech. Ageing Dev.* **36**, 1–12.

Doetsch, P. W., Helland, D. E., and Haseltine, W. A. (1986). Mechanism of action of a mammalian DNA repair endonuclease. *Biochemistry* **25**, 2212–2220.

Domena, J. D., and Mosbaugh, D. W. (1985). Purification of nuclear and mitochondrial uracil–DNA glycosylase from rat liver. Identification of two distinct subcellular forms. *Biochemistry* **24**, 7320–7328.

Echols, H., Lu, C., and Burgers, P. M. J. (1983). Mutator strains of *Escherichia coli*, *mutD* and *dnaQ*, with defective exonucleolytic editing by DNA polymerase III holoenzyme. *Proc. Natl. Acad. Sci. USA* **80**, 2189–2192.

Friedberg, E. C. (1985). "DNA Repair." W. H. Freeman and Company, New York.

Friedberg, E. C. (1988). Deoxyribonucleic acid repair in the yeast *Saccharomyces cerevisiae*. *Microbiol. Rev.* **52**, 70–102.

Gensler, H. L. (1981a). The effect of hamster age on UV-induced unscheduled DNA synthesis in freshly isolated lung and kidney cells. *Exp. Gerontol.* **16**, 59–68.

Gensler, H. L. (1981b). Low levels of UV-induced UDS in post-mitotic brain cells in hamsters: Possible relevance to aging. *Exp. Gerontol.* **16**, 199–207.

Gentil, A., Margot, A., and Sarasin, A. (1982). Enhanced reactivation and mutagenesis after transfection of carcinogen-treated monkey kidney cells with UV-irradiated simian virus 40 (SV40) DNA. *Biochemie* **64**, 693–696.

Goldstein, S. (1971). The role of DNA repair in aging of cultured fibroblasts from xeroderma pigmentosum and normals. *Proc. Soc. Exp. Biol. Med.* **137**, 730–734.

Goodwin, J. S. (1982). Changes in lymphocyte sensitivity to prostaglandin E, histamine, hydrocortisone, and X irradiation with age: Studies in a healthy elderly population. *Clin. Immunol. Immunopathol.* **25**, 243–251.

Grossman, L., Caron, P. R., Mazur, S. J., and Oh, E. Y. (1988). Repair of DNA-containing pyrimidine dimers. *FASEB J.* **2**, 2696–2701.

Guyer, R. B., Nonnemaker, J. M., and Deering, R. A. (1986). Uracil–DNA glycosylase activity from *Dictyostelium discoideum*. *Biochim. Biophys. Acta* **868**, 262–264.

Hall, J. D., Almy, R. E., and Scherer, K. L. (1982). DNA repair in cultured human fibroblasts does not decline with donor age. *Exp. Cell Res.* **139**, 351–359.

Harm, W. (1980). "Biological Effects of Ultraviolet Radiation." IUPAB Biophysics Series (F. Hutchinson, W. Fuller, and L. J. Mullins, eds.). Cambridge University Press, London and New York.

Harris, A., Karran, P., and Lindahl, T. (1983). O^6-methylguanine–DNA methyltransferase of human lymphoid cells. *Cancer Res.* **43**, 3247–3252.

Harris, G., Holmes, A., Sabovljev, S. A., Cramp, W. A., Hedges, M., Hornsey, S., Hornsey, J. M., and Bennett, G. C. J. (1986). Sensitivity to X-irradiation of peripheral blood lymphocytes from aging donors. *Int. J. Radiat. Biol.* **50**, 685–694.

Haynes, R. H. (1988). Biological context of DNA repair. *In* "Mechanisms and Consequences of DNA Damage Processing" (E. C. Friedberg and P. C. Hanawalt, eds.), pp. 577–584. Alan R. Liss, Inc., New York.

Hennis, H. L., III, Braid, H. L., and Vincent, R. A., Jr. (1981). Unscheduled DNA synthesis in cells of different shape in fibroblast cultures from donors of various ages. *Mech. Ageing Dev.* **16**, 355–361.

Higgins, S. A., Frankel, K., Cummings, A., and Teebor, G. W. (1987). Definitive characterization of human thymine glycol N-glycosylase activity. *Biochemistry* **26**, 1683–1688.

Hollstein, M. C., Brooks, P., Linn, S., and Ames, B. N. (1984). Hydroxymethyl DNA glycosylase in mammalian cells. *Proc. Natl. Acad. Sci. USA* **81**, 4003–4007.

Howard-Flanders, P., Rupp, W. D., Wilkins, B. M., and Cole, R. S. (1968). DNA replication and recombination after UV-irradiation. *Cold Spring Harbor Symp. Quant. Biol.* **33**, 195–205.

Howland, G. P. (1975). Dark repair of ultraviolet-induced pyrimidine dimers in the DNA of wild carrot protoplasts. *Nature (London)* **254,** 160–161.

Ishikawa, T., and Sakurai, J. (1986). *In vivo* studies on age dependency of DNA repair with age in mouse skin. *Cancer Res.* **46,** 1344–1348.

Ishikawa, T., Takayama, S., and Kitagawa, T. (1978). DNA repair synthesis in rat retinal ganglion cells treated with chemical carcinogens or ultraviolet light *in vitro* with special reference to aging and repair level. *J. Natl. Cancer Inst.* **61,** 1101–1105.

Jackson, J. F., and Linskens, H. F. (1978). Evidence for DNA repair after ultraviolet irradiation of *Petunia hybrida* pollen. *Mol. Gen. Genet.* **161,** 117–120.

Johnson, R. C., and McNeill, W. F. (1978). Electron microscopy of UV-induced post-replication repair of daughter strand gaps. *In* "DNA Repair Mechanisms" (P. C. Hanawalt, E. C. Friedberg, and C. F. Fox, eds.), pp. 95–99. Academic Press, New York.

Kanagalingam, K., and Balis, M. E. (1975). *In vivo* repair of rat intestinal DNA damage by alkylating agents. *Cancer* **36,** 2364–2372.

Karran, P., and Lindahl, T. (1980). Hypoxanthine in DNA: Generation of heat-induced hydrolysis of adenine residues and release in free form by a DNA glycosylase from calf thymus. *Biochemistry* **19,** 6005–6011.

Kaufmann, W. K. (1989). Pathways of human cell post-replication repair. *Carcinogenesis* **10,** 1–11.

Kempf, C., Schmitt, M., Danse, J.-M., and Kempf, J. (1984). Correlation of DNA repair synthesis with aging in mice, evidenced by quantitative autoradiography. *Mech. Ageing Dev.* **26,** 183–194.

Kennah, H. E., Coetzee, M. L., and Ove, P. (1985). A comparison of DNA repair synthesis in primary hepatocytes from young and old rats. *Mech. Ageing Dev.* **29,** 283–298.

Kovacs, E., Weber, W., and Muller, Hj. (1984). Age-related variation in the DNA-repair synthesis after UV-C irradiation in unstimulated lymphocytes of healthy blood donors. *Mutat. Res.* **131,** 231–237.

Krystal, G., Morris, G. M., Lipman, J. M., and Sokoloff, L. (1983). DNA repair in articular chondrocytes. I. Unscheduled DNA synthesis following ultraviolet irradiation in monolayer culture. *Mech. Ageing Dev.* **21,** 83–96.

Kuhnlein, U., Penhoet, E. E., and Linn, S. (1976). An altered apurinic DNA endonuclease activity in group A and group D xeroderma pigmentosum fibroblasts. *Proc. Natl. Acad. Sci. USA* **73,** 1169–1173.

Kuhnlein, U., Lee, B., Penhoet, E. E., and Linn, S. (1978a). Xeroderma pigmentosum fibroblasts of the D group lack an apurinic DNA endonuclease species with a low apparent Km. *Nucleic Acids Res.* **5,** 951–960.

Kuhnlein, U., Lee, B., and Linn, S. (1978b). Human uracil DNA Nglycosidase: Studies in normal and repair defective cultured fibroblasts. *Nucleic Acids Res.* **5,** 112–125.

Kutlaca, R., Seshadri, R., and Mosley, A. A. (1982). Effect of age on sensitivity of human lymphocytes to radiation. A brief note. *Mech. Ageing Dev.* **19,** 97–101.

Lambert, B., Ringborg, U., and Skoog, L. (1979). Age-related decrease of ultraviolet light-induced DNA repair synthesis in human peripheral leukocytes. *Cancer Res.* **39,** 2792–2795.

Lambert, M. W., Lambert, W. C., and Okorodudu, A. O. (1983). Nuclear DNA endonuclease activities on partially apurinic/apyrimidinic DNA in normal human and xeroderma pigmentosum lymphoblastoid and mouse melanoma cells. *Chem. Biol. Interact.* **46,** 109–120.

Lee, K., McCray, W. H. Jr., and Doetsch, P. W. (1987). Thymine glycol–DNA glycosylase/AP endonuclease of CEM-DI lymphoblasts: A human analog of *Escherichia coli* endonuclease III. *Biochem. Biophys. Res. Commun.* **149,** 93–101.

Lett, J. T. (1977). Cellular senescence and the capacity for rejoining DNA strand breaks. *In* "DNA Repair Processes" (W. W. Nichols and D. G. Murphy, eds.), pp. 89–98. Symposia Specialists Inc., Miami, Florida.

Licastro, F., Franceschi, M., Chiricolo, M., Battelli, M. G., Tabacchi, P., Cenci, M., Barboni, F., and Pallenzona, D. (1982). DNA repair after gamma radiation and superoxide dismutase activity in lymphocytes from subjects of far advanced age. *Carcinogenesis* **3,** 45–48.

Lindahl, T., Sedgwick, B., Demple, B., and Karran, P. (1983). Enzymology and regulation of the adaptive response to alkylation agents. *In* "Cellular Responses to DNA Damage" (E. C. Friedberg and B. A. Bridges, eds.), pp. 241–253. Alan R. Liss, New York.

Linn, S., Kim, J., Choi, S.-Y., Kenney, S., Randahl, H., Tomkinson, A. E., Nishida, C., Syvaoja, S., and Krauss, S. W. (1988). The enzymology of mammalian DNA excision repair. *J. Cell Biol.* **107,** 448a.

Lipman, J. M., and Sokoloff, L. (1985). DNA repair by articular chondrocytes. III. Unscheduled DNA synthesis following ultraviolet light irradiation of resting cartilage. *Mech. Ageing Dev.* **32,** 39–55.

Lipman, J. M., and Setlow, R. B. (1987). DNA repair by articular chondrocytes. IV. Measurement of *Micrococcus luteus* endonuclease-sensitive sites by alkaline elution in rabbit articular chondrocytes. *Mech. Ageing Dev.* **40,** 193–204.

Lipman, J. M., Sokoloff, L., and Setlow, R. B. (1987). DNA repair by articular chondrocytes. V. O^{6}methylguanine-acceptor protein activity in resting and cultured rabbit and human chondrocytes. *Mech. Ageing Dev.* **40,** 205–213.

Liu, S. C., Meagher, K., and Hanawalt, P. C. (1985). Role of solar conditioning in DNA repair response and survival of human epidermal keratinocytes following UV irradiation. *J. Invest. Dermatol.* **79**, 330–335.

Liu, S. C., Parsons, C. S., and Hanawalt, P. C. (1982). DNA repair response in human epidermal keratinocytes from donors of different ages. *J. Invest. Dermatol.* **79**, 330–335.

Loveless, A. (1969). Possible relevance of O-6 alkylation of deoxyguanosine to the mutagenicity and carcinogenicity of nitrosamines and nitrosamides. *Nature (London)* **223**, 206–207.

Lu, C., Scheuermann, R. H., and Echols, H. (1986). Capacity of recA protein to bind preferentially to UV lesions and inhibit the editing subunit (epsilon) of DNA polymerase III: A possible mechanism for SOS-inducible targeted mutagenesis. *Proc. Natl. Acad. Sci. USA* **83**, 619–623.

Lytle, C. D., and Knott, D. C. (1982). Enhanced mutagenesis parallels enhanced reactivation of herpes virus in a human cell line. *EMBO J.* **1**, 701–703.

Margison, G. P., and Pegg, A. E. (1981). Enzymatic release of 7-methylguanine from methylated DNA by rodent liver extracts. *Proc. Natl. Acad. Sci. USA* **78**, 861–865.

Mayer, P. J., Lange, C. S., Bradley, M. O., and Nichols, W. W. (1989). Age-dependent decline in rejoining of X-ray-induced DNA double-strand breaks in normal human lymphocytes. *Mutat. Res.* **219**, 95–100.

McLennan, A. G. (1987). The repair of ultraviolet light-induced DNA damage in plants cells. *Mutat. Res.* **181**, 1–7.

Mellon, I., Bohr, V. A., Smith, C. A., and Hanawalt, P. C. (1986). Preferential DNA repair of an active gene in human cells. *Proc. Natl. Acad. Sci. USA* **83**, 8878–8882.

Mellon, I., Spivak, G., and Hanawalt, P. C. (1987). Selective removal of transcription-blocking DNA damage from the transcribed strand of the mammalian DHFR gene. *Cell* **4**, 241–249.

Mezzina, M., Gentil, A., and Sarasin, A. (1981). Simian virus 40 as a probe for studying inducible repair functions in mammalian cells. *J. Supramol. Struct. Cell. Biochem.* **17**, 121–131.

Mullaart, E., Roza, L., Lohman, P. H. M., and Vijg, J. (1989). The removal of UV-induced pyrimidine dimers from DNA of rat skin cells *in vitro* and *in vivo* in relation to aging. *Mech. Ageing Dev.* **47**, 253–264.

Myrnes, B., Guddal, P.-H., and Kroken, H. (1982). Metabolism of dITP in HeLa cell extracts, incorporation into DNA by isolated nuclei and release of hypoxanthine from DNA by a hypoxanthine–DNA glycosylase activity. *Nucleic Acids Res.* **10**, 3693–3701.

Nette, E. G., Xi, Y.-P., Sun, Y.-K., Andrews, A. D., and King, D. W.

(1984). A correlation between aging and DNA repair in human epidermal cells. *Mech. Aging Dev.* **24,** 283–292.

Newbold, R. F., Warren, W., Medcalf, A. S. C., and Amos, J. (1980). Mutagenicity of carcinogenic methylating agents is associated with a specific DNA modification. *Nature (London)* **283,** 596–599.

Niedermuller, H. (1985). DNA repair during aging. *In* "Molecular Biology of Aging: Gene Stability and Gene Expression" (R. S. Sohal, L. S. Birnbaum, and R. G. Cutler, eds.), pp. 173–193. Raven Press, New York.

Ono, T., and Okada, S. (1978). Does the capacity to rejoin radiation-induced DNA breaks decline in senescent mice? *Int. J. Radiat. Biol.* **33,** 403–407.

Peterson, K. R., Ossanna, N., Thliveris, A. T., Ennis, D. G., and Mount, D. W. (1988). Derepression of specific genes promotes DNA repair and mutagenesis in *Escherichia coli*. *J. Bacteriol.* **170,** 1–4.

Plesko, M. M., and Richardson, A. (1984). Age-related changes in unscheduled DNA synthesis by rat hepatocytes. *Biochem. Biophys. Res. Commun.* **118,** 730–735.

Renard, A., Lemaitre, M., and Verly, W. G. (1983). The O^6-alkylguanine transferase activity of rat liver chromatin. *In* "Cellular Responses to DNA Damage" (E. C. Friedberg and B. A. Bridges, eds.), pp. 255–260. Alan R. Liss, New York.

Rosen, E. M., Goldberg, I. D., Myrick, K. V., and Levenson, S. E. (1985). Radiation survival of vascular smooth muscle cells as a function of age. *Int. J. Radiat. Biol.* **48,** 71–79.

Roth, M., Emmons, L. R., Haner, M., Muller, Hj., and Boyle, J. M. (1989). Age-related decrease in an early step of DNA-repair of normal human lymphocytes exposed to ultraviolet-irradiation. *Exp. Cell Res.* **180,** 171–177.

Rupert, C. S. (1975). Enzymatic photoreactivation: Overview. *In* "Molecular Mechanisms for Repair of DNA," part A (P. C. Hanawalt and R. B. Setlow, eds.), pp. 73–87. Plenum, New York.

Rupp, W. D., and Howard-Flanders, P. (1968). Discontinuities in the DNA synthesized in an excision defective strain of *Escherichia coli* following ultraviolet irradiation. *J. Mol. Biol.* **31,** 291–304.

Rupp, W. D., Sancar, A., and Sancar, G. B. (1982). Properties and regulation of the UVR ABC endonuclease. *Biochimie* **64,** 595–598.

Rupp, W. D., Wilde, C. E., III, Reno, D. L., and Howard-Flanders, P. (1971). Exchanges between DNA strands in ultraviolet-irradiated *Escherichia coli*. *J. Mol. Biol.* **61,** 25–44.

Rydberg, B., and Lindahl, T. (1982). Nonenzymatic methylation of DNA by the intracellular methyl group donor S-adenosyl-L-methionine is a potentially mutagenic reaction. *EMBO J.* **1,** 211–216.

Sancar, A., and Rupp, W. D. (1983). A novel repair enzyme: UVR ABC excision nuclease of *Escherichia coli* cuts a DNA strand on both sides of the damaged region. *Cell* **33**, 249–260.

Sarasin, A., and Benoit, A. (1980). Induction of an error-prone mode of DNA repair in UV-irradiated monkey kidney cells. *Mutat. Res.* **70**, 71–81.

Sargent, E. V., and Burns, F. J. (1985). Repair of radiation-induced DNA damage in rat epidermis as a function of age. *Radiat. Res.* **102**, 176–181.

Seshadri, R. S., Morley, A. A., Trainor, K. J., and Sorrell, J. (1979). Sensitivity of human lymphocytes to bleomycin increases with age. *Experientia* **35**, 233–234.

Setlow, R. B., Lipman, J. M., and Sokoloff, L. (1983). DNA repair by articular chondrocytes. II. Direct measurements of repair of ultraviolet and x-ray damage in monolayer cultures. *Mech. Ageing Dev.* **21**, 97–103.

Smith, D. P., and Dusenberg, R. L. (1988). Twelve genes of the alkylation excision repair pathway of *Drosophila melangaster*. *In* "Mechanisms and Consequences of DNA Damage Processing" (E. C. Friedberg and P. C. Hanawalt, eds.), pp. 251–255. Alan R. Liss, New York.

Smith, K. C., and Meun, D. H. C. (1970). Repair of radiation-induced damage in *Escherichia coli*. I. Effect of *rec* mutations on post-replication repair of damage due to ultraviolet radiation. *J. Mol. Biol.* **51**, 459–472.

Smith-Sonneborn, J. (1979). DNA repair and longevity assurance in *Paramecium tetraurelia*. *Science* **203**, 1115–1117.

Talpaert-Borle, M., Clerici, L., and Campagnari, F. (1979). Isolation and characterization of a uracil–DNA glycosylase from calf thymus. *J. Biol. Chem.* **254**, 6387–6391.

Talpaert-Borle, M., Campagnari, F., and Creissen, D. M. (1982). Properties of purified uracil–DNA glycosylase from calf thymus. An in vitro study using synthetic DNA-like substrates. *J. Biol. Chem.* **257**, 1208–1214.

Tice, R. R., and Setlow, R. B. (1985). DNA repair and replication in aging organisms and cells. *In* "Handbook of the Biology of Aging" (C. E. Finch and E. L. Schneider, eds.), pp. 173–224. Van Nostrand Reinhold, New York.

Treton, J. A., and Courtois, Y. (1981). Evolution of the distribution, proliferation and ultraviolet repair capacity of rat lens epithelial cells as a function of maturation and aging. *Mech. Ageing Dev.* **15**, 251–267.

Trosko, J. E., and Mansour, V. H. (1969). Photoreactivation of ultraviolet light-induced pyrimidine dimers in Ginko cells grown *in vitro*. *Mutat. Res.* **7**, 120–121.

Turner, D. R., Griffith, V. C., and Morley, A. A. (1982). Aging in vivo does not alter the kinetics of DNA strand break repair. *Mech. Ageing Dev.* **19**, 325–331.

van der Schans, G. P., Centen, H. B., and Lohman, P. H. M. (1982). DNA lesions induced by ionizing radiation. *Prog. Mutat. Res.* **4**, 285–299.

van Duin, M., van den Tol, J., Warmerdam, P., Odijk, H., Meijer, D., Westerveld, A., Bootsma, D., and Hoeijmakers, J. H. J. (1988). Evolution and mutagenesis of the mammalian excision repair gene *ERCC-1*. *Nuc. Acids Res.* **16**, 5305–5322.

Vijg, J., Mullaart, E., Lohman, P. H. M., and Knook, D. L. (1985). UV-induced unscheduled DNA synthesis in fibroblasts of aging inbred rats. *Mutat. Res.* **146**, 197–204.

Villani, G., Boiteux, S., and Radman, M. (1978). Mechanism of ultraviolet-induced mutagenesis: Extent and fidelity of in vitro DNA synthesis on irradiated templates. *Proc. Natl. Acad. Sci. USA* **75**, 3037–3041.

Waldstein, E. A., Cao, E.-H., Bender, M. A., and Setlow, R. B. (1982). Abilities of extracts of human lymphocytes to remove O^6-methylguanine from DNA. *Mutat. Res.* **95**, 405–416.

Wallace, S. S. (1988). AP endonucleases and DNA glycosylases that recognize oxidative DNA damage. *Environ. Molec. Mutagenesis* **12**, 431–477.

Ward, J. E., Blakely, W. F., and Jones, E. I. (1985). Mammalian cells are not killed by DNA single-strand breaks caused by hydroxyl radicals from hydrogen peroxide. *Radiat. Res.* **103**, 383–392.

Washington, W. J., Foote, R. S., Dunn, W. C., Generoso, W. M., and Mitra, S. (1989). Age-dependent modulation of tissue-specific repair activity for 3-methyladenine and O^6-methylguanine in DNA inbred mice. *Mech. Ageing Dev.* **48**, 43–52.

Welker, D. L., and Deering, R. A. (1978). Genetics of DNA repair in the cellular slime mold, *Dictyostelium discoideum*. In "DNA Repair Mechanisms." ICN-UCLA Symposia on Molecular and Cellular Biology (P. C. Hanawalt, E. C. Friedberg, and C. F. Fox, eds.), pp. 445–448. Academic Press, New York.

Wheeler, K. T., and Lett, J. T. (1974). On the possibility that DNA repair is related to age in non-dividing cells. *Proc. Natl. Acad. Sci. USA* **71**, 1862–1865.

Woodhead, A. D., Merry, B. J., Cao, E.-H., Holehan, A. M., Grist, E., and Carlson, C. (1985). Levels of O^6-methylguanine acceptor protein in tissues of rats and their relationship to carcinogenicity and aging. *J. Natl. Cancer Inst.* **75**, 1141–1145.

Yarosh, D. B., Rosenstein, B. S., and Setlow, R. B. (1981). Excision repair and patch size in UV-irradiated bacteriophage T4. *J. Virol.* **40**, 465–471.

Yarosh, D. B., Rice, M., Ziolkowski, C. H. J., Day, R. S., III, and Scudiero, D. A. (1983). O^6-methylguanine-DNA methyltransferase in human tumor cells. In "Cellular Responses to DNA Damage" (E. C. Friedberg and B. A. Bridges, eds.), pp. 261–270. Alan R. Liss, New York.

Chapter Ten

DNA Repair with Emphasis on Recombinational Repair

Our emphasis in the preceding chapters has been on the DNA damage hypothesis of aging. In the previous chapter, we described DNA repair processes that may be important in resisting aging and promoting longevity. All of the repair processes described used the information available in just one double-stranded DNA molecule. Starting with this chapter, we shift emphasis to the DNA repair (and complementation) hypothesis of sex. A key event in sexual reproduction is the pairing of homologous chromosomes (each equivalent to a double-stranded DNA molecule) and the exchange of DNA between them (Chapter 1, Section I). This process of recombination, we will argue, reflects a form of repair that permits damages in one double-stranded DNA molecule to be replaced by an undamaged DNA sequence from another double-stranded DNA molecule. This form of repair differs from the types discussed previously (Chapter 9) in that it involves the use of the information present in two homologous DNA molecules rather than just one. The type of repair involving exchange of information between two DNA molecules is referred to as recombinational repair.

The following two observations indicate an organism's ability to undergo recombinational repair. First, survival after treatment with a DNA-damaging agent is substantially increased when two or more chromosomes are present in a cell, rather than one. Second, this increase in survival is diminished when a mutation that inactivates a gene necessary for the process of recombination is present.

Recombinational repair has been shown to occur by the above

criteria in several different organisms. For these organisms, recombination between genetic markers (genetic recombination) has also sometimes been measured when *both* DNA damage was introduced and two chromosomes were present. Under these circumstances, it has been consistently found that DNA damage causes an increase in recombination between the genetic markers. Therefore, when DNA damage is observed to increase marker recombination in an otherwise untested organism, this finding, by itself, can be taken as an indication of the occurrence of recombinational repair in the organism. A standard experimental method for increasing DNA damage is to treat an organism with a DNA-damaging agent. However, DNA damage in an organism is also increased when a mutation causing a defect in a DNA repair pathway is present. Thus, when an organism is defective in a repair pathway other than recombinational repair, DNA damage accumulates and genetic recombination would be expected to be higher than in an organism with normal repair. Observation of such an increase would, again, be an indication of the occurrence of recombinational repair. In addition, if an organism is defective in a repair pathway other than recombinational repair, and if a DNA-damaging agent is applied, then genetic recombination should reach even higher levels than would occur if the organism had normal repair. This increase in genetic recombination would occur because the introduced DNA damage, since it cannot be repaired by the defective repair pathway, is repaired by recombinational repair.

In Chapter 11, we will argue that meiosis, a key stage of the sexual cycle, is an adaptation to promote recombinational repair of the DNA passed on to germ cells. In the present chapter, we review the body of evidence showing that recombinational repair of DNA damage is efficient, versatile, and prevalent in nature (for earlier reviews, see also Bernstein *et al.*, 1987; Bernstein, 1983). The evidence on recombinational repair of DNA is derived from experiments with viruses, bacteria, and eukaryotes.

I. *Viruses*

A. Multiplicity Reactivation

Recombinational repair was discovered by Luria (1947) in studies of bacteriophages T2, T4, and T6. When he irradiated these phages with UV, he observed that the ability of these phages to produce progeny was much higher when bacteria were infected by two or more phages than when they were infected by a single phage. The higher survival rate was interpreted as being due to recombination between the damaged parental phage chromosomes (Luria and Dulbecco, 1949). This phenomenon was termed multiplicity reactivation. Multiplicity reactivation is effective against the potentially lethal DNA damages caused by widely different agents including nitrous acid, mitomycin C, ^{32}P, psoralen plus near UV light (PUVA), UV, and X-rays (for reviews, see Bernstein, 1981; Bernstein and Wallace, 1983). Recently, Chen and Bernstein (1987) also showed that potentially lethal oxidative damage introduced by H_2O_2 is subject to efficient multiplicity reactivation. Doses of H_2O_2, UV, mitomycin C, or nitrous acid can be given which permit only 1% survival of progeny-forming ability in single infections (only one phage infecting each cell). These same doses allow, respectively, about 73, 70, 65, and 44% survival when two or more damaged phages were permitted to infect each cell (Chen and Bernstein, 1987; Epstein, 1958; Holmes *et al.*, 1980; Nonn and Bernstein, 1977). This indicates that multiplicity reactivation in phage T4 efficiently overcomes a variety of DNA damages.

Three lines of evidence indicate that multiplicity reactivation is a kind of recombinational repair: (1) by definition, it depends on at least two homologous phage chromosomes infecting the same cell; (2) it requires phage gene functions that are also essential for recombination of phage genetic markers (allelic recombination); and (3) treatment with a DNA-damaging agent increases recombination of genetic markers. Since we have already discussed point (1) as the defining characteristic of multiplicity reactivation, we will now discuss points (2) and (3).

With respect to point (2), seven gene products are necessary both for multiplicity reactivation of UV-irradiated phage T4 and for allelic recombination. These gene products (gps) are gp uvsX, which promotes homologous pairing and strand exchange between DNA molecules and is analogous to the *recA* protein of *E. coli* (Griffith and Formosa, 1985; Yonesaki *et al.*, 1985); gp32, which is a helix-destabilizing protein that enhances the pairing reaction; gps 46 and 47, which have an exonuclease activity that produces single-strand gaps from nicks and may promote the joining of two homologous DNA molecules; and gps 59, uvsW, and uvsY, whose functions are not yet understood (for reviews, see Bernstein, 1981; Bernstein and Wallace, 1983). The genes required for multiplicity reaction of phage damaged by UV are similar to those required for repair of damages caused by other agents. This has been shown for phage damaged by nitrous acid (Nonn and Bernstein, 1977), mitomycin C (Holmes *et al.*, 1980), and H_2O_2 (Chen and Bernstein, 1987).

Phage lambda, another type of bacteriophage, also undergoes multiplicity reactivation after UV treatment (Kellenberger and Weigle, 1958; Baker and Haynes, 1967; Huskey, 1969). This multiplicity reactivation depends either on the host recombination protein *recA* or on a phage recombination function, red. The absence of both gene products results in loss of phage multiplicity reaction (Huskey, 1969).

With respect to point (3) above, allelic recombination in phage T4 has been shown to be increased by UV (Epstein, 1958), nitrous acid (Fry, 1979), mitomycin C (Holmes *et al.*, 1980), PUVA (Miskimins *et al.*, 1982), X-rays (Harm, 1958), and ^{32}P (Symonds and Ritchie, 1961). In phage lambda, allelic recombination has been shown to be increased by UV-irradiation (e.g., Baker and Haynes, 1967; Lin and Howard-Flanders, 1976).

Chan (1975) reported that recombination in phage T4 is stimulated by UV-irradiation and this stimulation is enhanced in the excision repair-defective mutant $denV^-$. Similarly, in phage lambda, recombination of UV-irradiated phage was stimulated by a defect in excision repair (Lin and Howard-Flanders, 1976). These findings imply that UV-induced DNA damages, which are ordinarily repaired by excision repair, are handled by recombinational repair when excision repair is deficient, and this added recombinational repair is reflected in additional allelic recombination.

Multiplicity reactivation of UV-irradiated phage T4 occurs without introduction of new mutations (Yarosh *et al.*, 1978). Multiplicity reactivation of phage phiX174 treated with proflavin plus light also was found to be nonmutagenic (Piette *et al.*, 1978). These results imply that recombinational repair is an error-free form of repair. If recombinational repair were error-prone, the benefit of removing damages would be diminished by the increased chance of introducing deleterious mutations.

Luria (1947) showed that the recombination events that occur during multiplicity reactivation are not random, because phage chromosomes carrying 20–50 lethal hits do not inactivate a co-infecting undamaged chromosome. Sturtevant (as quoted by Luria and Dulbecco, 1949) originally proposed that multiplicity reactivation takes place by stimulated recombination acts at the sites of damages. Nonn and Bernstein (1977) suggested that this stimulated recombination may involve transfer of a short length of undamaged DNA strand from one chromosome to replace a damaged strand in another chromosome. Further data in support of this model of multiplicity reactivation are discussed by Bernstein and Wallace (1983).

Phages other than T4 and lambda also have been shown to undergo multiplicity reactivation. These are the *E. coli* phages T1 (Tessman and Ozaki, 1957), T2, T5, and T6 (Luria, 1947, Luria and Dulbecco, 1949) and phiX174 (Piette *et al.*, 1978), and the *Salmonella* phage V_1 (Bernstein, 1957). In addition to the bacteriophages, a number of animal viruses undergo multiplicity reactivation. These include herpes virus (Selsky *et al.*, 1979; Hall and Scherer, 1981; Hall *et al.*, 1980), adenovirus type 12 and simian virus 40 (Yamamoto and Shimijo, 1971), influenza virus (Barry, 1961), reovirus (McClain and Spendlove, 1966), and vaccinia virus (Abel, 1962). Coffin (1979) has also presented a model for recombinational repair in the retroviruses. This process was postulated to occur when DNA is copied from damaged RNA templates.

The presence of multiplicity reactivation in a virus indicates the occurrence of recombinational repair. The finding of multiplicity reactivation in so many viruses implies that recombinational repair is widespread among viruses. This contributes to our point that recombinational repair is prevalent in nature.

B. Postreplication Recombinational Repair

In addition to multiplicity reactivation, evidence indicates that phage T4 undergoes postreplication recombinational repair (PRRR) (for review, see Bernstein and Wallace, 1983, and for a description of the process see Chapter 9, Section II.G.2). The existence of PRRR was originally proposed for phage T4 by Harm (1964). It involves many of the same gene functions that are used in multiplicity reactivation; however, PRRR, unlike multiplicity reactivation, can take place in single infections, because recombination is between the daughter chromosomes formed after replication of the initially infecting single chromosome. Thus, this is a kind of recombinational repair that only requires a single chromosome that can replicate.

C. Prophage Reactivation

Some phages can undergo another type of recombinational repair called prophage reactivation (for reviews, see Devoret *et al.*, 1975; Bernstein, 1981). In this case, a damaged infecting phage chromosome is repaired by recombination with a homologous phage genome integrated into the bacterial chromosome and existing in a prophage state. Prophage reactivation of UV-irradiated phage lambda depends on the *E. coli recA* and phage lambda *red* gene products (Blanco and Devoret, 1973), both of which are also employed in phage lambda allelic recombinational and multiplicity reactivation.

D. Summary of Virus Recombinational Repair

In summary, recombinational repair is a common form of repair among viruses, occurring in a range of bacterial and animal viruses. Recombinational repair is manifested as multiplicity reactivation, PRRR, and prophage reactivation. Multiplicity reactivation, in particular, appears to be effective against the damages caused by a wide variety of DNA-damaging agents.

II. *Bacteria*

In this section, we review evidence that double-strand damages, as well as other types of damage, are overcome by recombinational repair. In some cases, this process has been shown to be very efficient.

A. Cross-Link Repair

Treatment of *E. coli* with PUVA produces interstrand cross-links in the cells' DNA. Interstrand cross-links are a type of double-strand DNA damage. Cole *et al.* (1976; 1978) have found that these cross-links can be repaired by a process in which there is first an uncoupling of the cross-link from one strand, leaving a gap in that strand. This uncoupling, they suggested, is catalyzed by enzymes that include the endonuclease encoded by genes *uvrA* and *uvrB* (see Chapter 9, Section II.E). As noted by Sancar and Rupp (1983) in discussing the activity of this endonuclease, the two-cut activity of the enzyme may be particularly useful in cases where both strands of the DNA are damaged, such as occurs with a cross-link. They suggested that a molecule with cuts in one strand on both sides of a cross-link would be an excellent substrate for initiating recombinational repair. The information lost in the single-strand gap then appears to be obtained from another homologous chromosome by recombinational exchange. The recombination step requires the RecA protein. *Escherichia coli* cells under good growth conditions have more than one homologous chromosome present in each cell.

The RecA protein of *E. coli* plays an essential role in genetic recombination and in promoting cell survival following exposure to agents that damage chromosomal DNA. Studies *in vitro* on the action of RecA protein on DNA substrates suggest that the enzyme performs key steps in recombinational repair. RecA protein promotes homologous pairing and strand exchange between duplex DNA molecules if one is partially single-stranded (Hahn *et al.*, 1988).

Even though a single unrepaired cross-link is sufficient to inactivate a cell, a wild-type cell can repair and therefore recover from 65 psoralen cross-links (Cole, 1971). This indicates that recombinational repair is very efficient in overcoming this type of damage.

B. Double-Strand Breaks

When *E. coli* are irradiated with X-rays, double-strand breaks are introduced in the cells' DNA. Double-strand breaks are another type of double-strand damage. Krasin and Hutchinson (1977) observed that when four or five chromosomes were in a cell, 50–70% of the double-strand breaks were removed during a period of postirradiation incubation. However, repair of double-strand breaks was not detected when an average of only 1.3 genomes were present per cell, even though, under the same conditions, repair of single-strand breaks was very efficient. Krasin and Hutchinson (1977, 1981) found that double-strand break repair requires the RecA protein as well as other gene products inducible by DNA damage. When cells were defective in the RecA protein, a single unrepaired double-strand break was apparently lethal. Picksley *et al.* (1984) reported that inducible double-strand break repair requires the *recN* gene, which is also necessary for recombination proficiency. The involvement of the RecA and RecN gene products and the need for more than one chromosome imply that double-strand breaks in *E. coli* are overcome by recombinational repair.

C. Gaps Opposite Thymine Dimers

In Chapter 9, Section II.G.2, we discussed postreplication recombinational repair in *E. coli*. This process is initiated when gaps are formed in newly replicated DNA strands opposite pyrimidine dimers. A gap opposite a pyrimidine dimer can be regarded as a double-strand damage because information is lost in both strands of the DNA in the same region. We have reviewed evidence (Chapter 9, Section II.G.2) that the postreplicative gaps are filled by recombinational transfer of undamaged single-strand segments from the homologous sister chromosome. This permits the dimer to be bypassed during replication. About 50 pyrimidine dimers are required to inactivate a cell when excision repair is absent and PRRR is, apparently, the only significant repair process present (Howard-Flanders *et al.*, 1968). This suggests that PRRR permits efficient bypass of about 98% of pyrimidine dimers. In common with other recombinational repair processes in *E. coli*, PRRR requires the RecA protein (Smith and Meun, 1970).

D. Recombinational Repair during Conjugation

Conjugation in *E. coli* is a process by which male Hfr cells (Hfr stands for high frequency of recombination) transfer DNA via a cytoplasmic bridge to recipient female F⁻ cells. When a donor Hfr cell is irradiated with UV light prior to conjugal transfer of DNA, the replication normally associated with transfer still occurs, but gaps are formed in the new DNA strand opposite the pyrimidine dimers in the parental strand (Howard-Flanders *et al.*, 1968). The gaps are then filled in by *recA*-dependent recombination with the F⁻ DNA. The frequency of allelic recombination is increased by UV-irradiation of the donor. This reflects recombinational repair following the conjugative transfer of DNA.

E. Recombinational Repair is Effective against Many Types of Damage

DeFlora *et al.* (1984) presented evidence that recombinational repair is versatile in overcoming lethal DNA damages. They reported the results of testing a wide variety of compounds for genotoxicity (DNA-damaging ability) in three *E. coli* strains. These were (1) wild-type, (2) an excision repair-defective strain, and (3) a strain defective in both excision repair and recombinational repair. After testing 135 compounds, the strain defective in both excision and recombinational repair was found to be considerably more sensitive to the lethal effects of genotoxic agents than the strain defective only in excision repair. This increased sensitivity was likely due to the inability of the double mutant to carry out recombinational repair. Therefore, these results indicate that recombinational repair is used for protection against a wide variety of DNA damages in *E. coli*.

F. Recombination Is Increased by Mutations in Mismatch Repair Genes

Escherichia coli has a mismatch repair system that recognizes noncomplementary base pairs in DNA and corrects them (Radman and Wagner, 1984). This process appears to involve localized excision and repair synthesis. In regions of DNA where the specific sequence GATC is methylated at the adenine residue, mismatch repair does not occur. However, in the DNA regions immediately behind the replicative fork, GATC sequences are transiently under-methylated. This allows mismatch repair to operate only on newly synthesized strands and thus to correct replication errors. This repair system requires the following genes: *mutH*, *mutL*, *mutS*, *mutU*, and *dam*, where the *dam* gene codes for adenine methylase. Mutants defective in each of these genes were found to have abnormally high levels of intragenic recombination (Feinstein and Low, 1986) or recombination between chromosomal duplications (Konrad, 1977). When mismatch repair is faulty, the result likely is damage to the DNA. The stimulation of recombination by such presumed damages may reflect recombinational repair of the damages.

G. Recombinational Repair in Bacteria Other than *E. coli*

As discussed earlier in this chapter (Section II.A–D), the *recA* gene is necessary for recombinational repair in *E. coli*. Genes similar to *recA* have been identified in other bacteria primarily by their ability to functionally substitute for the *recA* gene and/or by the immunological cross-reaction of their product with the RecA protein. These bacteria are *Proteus mirabilis* (Eitner *et al.*, 1982; West *et al.*, 1983), *Salmonella typhimurium* (Pierre and Paoletti, 1983), *Rhizobium meliloti* (Better and Helinski, 1983), *Bacillus subtilis* (deVos *et al.*, 1983; Lovett and Roberts, 1985), *Proteus vulgaris, Erwinia carotovora*, and *Shigella flexneri* (Keener *et al.*, 1984), *Vibrio cholerae* (Goldberg and McKalanos, 1986), the cyanobacterium *Anabaena variabilis* (Owttrim and Coleman 1987) and *Haemophilus influenzae* (Setlow *et al.*, 1988). These findings suggest that recombinational repair is common in bacteria.

Dodson and Hadden (1980) have presented evidence that *B. subtilis* undergoes a PRRR process that is similar to the more well-studied example in *E. coli* (this chapter, Section II.C, and Chapter 9, Section II.G.2). This process requires the *B. subtilis recE* gene. The *recE* gene is similar to the *E. coli recA* gene (deVos *et al.*, 1983; Lovett and Roberts, 1985). Other evidence for the occurrence of recombinational repair in bacteria comes from studies of the photosynthetic cyanobacterium *Synechococcus RZ* (Kolowsky and Szalay, 1986). These authors showed that in this organism double-strand gaps in DNA are repaired by a recombinational process.

H. Transformation in *B. subtilis*

Although *E. coli* has efficient recombinational repair, it seems to use this mechanism primarily during asexual reproduction rather than during a sexual process. Selander and Levin (1980) have presented evidence that sexual interactions in *E. coli* are infrequent in nature. The two chromosomes participating in recombinational repair in an *E. coli* cell generally are daughter chromosomes that have not yet segregated into separate cells. In contrast to *E. coli*, in nature *B. subtilis* seems to frequently undergo transformation, a sexual process (Graham and Istock, 1979). In natural *B. subtilis* transformation, the cells take in naked DNA from the surrounding medium. This DNA may be provided by lysed *B. subtilis* cells or actively exported by healthy cells, because *B. subtilis* releases DNA into the environment during growth (Dubnau, 1982; Stewart and Carlson, 1986). The bacteria that take up transforming DNA may then incorporate this DNA into their genome by recombination. Natural transformation in *B. subtilis* appears to be a highly evolved trait. DNA entry results from a complex, energy-requiring, developmental process. That is, natural transformation in *B. subtilis* is not due to passive entry of DNA into the cell or artificial manipulation of the bacterial population. For a *B. subtilis* cell to bind, take up, and recombine exogenous DNA into its genome, it must enter a special physiological state referred to as competence. Genetic control of transformation includes genes involved in both recombination and DNA repair. Although much work has been done on the mechanisms involved in transformation, studies on the evolutionary function of natural transformation have only been reported recently.

Michod *et al.* (1988) carried out experiments to determine whether or not the function of transformation is to provide a DNA template for the recipient bacterial cell to use in recombinational repair. They treated a *B. subtilis* population with UV and measured survival of transformed and total cells as a function of UV dose in two kinds of experimental treatments. Cells were either transformed before treatment with UV (DNA-UV) or after treatment with UV (UV-DNA).

The results showed a qualitative difference in the relationship between the survival of transformed cells (sexual cells) and total cells (primarily asexual cells) in the two experiments. In the UV-DNA experiments, transformed cells had a greater average survival than total cells. Such a result is expected if DNA added after damaging treatment is used for repair. However, in the DNA-UV experiments, this relationship was reversed. When DNA was added first, transformation should have already taken place before exposure to UV. Thus, presumably, the donor DNA was no longer available to contribute to recombinational repair.

Transformation results in the introduction of DNA fragments in several pieces, which together comprise about 1% of the total genome (Fornili and Fox, 1977). Therefore, this DNA has only about a 1% chance of being the DNA needed for repair at each of the damaged sites induced by UV in the *B. subtilis* recipient cell. Thus, to explain the enhanced survival of transformed cells in the UV-DNA experiments, Michod *et al.* (1988) pointed out that there may be targeted uptake of DNA homologous to damaged sites.

These results generally support the hypothesis that the adaptive function of bacterial transformation is to provide template for recombinational repair of DNA damage. However, *B. subtilis* can carry out other repair processes besides recombinational repair. To sustain the recombinational repair hypothesis, it is necessary to rule out a major role for these other processes in the enhanced survival of transformed cells in the UV-DNA experiments. Wojciechowski *et al.* (1989) used different DNA-repair-deficient mutants to study the change in transformation rate (ratio of transformants to total cells) in UV-DNA experiments compared with DNA-UV experiments. They found that the greater survival of transformed cells relative to total cells in UV-DNA experiments than in DNA-UV experiments did not depend on excision repair or inducible SOS-like repair (these two repair processes are discussed in Chapter 9, Sections II.A and II.G.1). This result further supports the hypothesis that the adaptive function of competence is to bring DNA into the cell for use as a template in the recombinational repair of DNA damage.

I. Summary of Bacterial Recombinational Repair

Evidence was obtained in *E. coli* for efficient recombinational repair of three kinds of DNA double-strand damage: cross-links, double-strand breaks, and gaps opposite pyrimidine dimers. Recombinational repair was also implicated to some extent in repairing damages from 135 genotoxic treatments as well as damages left in the absence of other repair processes. This suggests that recombinational repair in *E. coli*, like multiplicity reactivation in viruses, is an effective mechanism for overcoming different types of damages. This repair process in *E. coli* depends on the RecA protein as well as the presence of more than one copy of the bacterial chromosome. Furthermore, judging from the prevalence of *recA*-like genes in other bacteria, recombinational repair is likely prevalent among bacterial species. Experiments on the sexual process of transformation in *B. subtilis* indicate that it functions in providing template for recombinational repair of DNA damages.

III. *Eukaryotes*

In Sections I and II of this chapter, we reviewed evidence indicating that recombinational repair is common in viruses and bacteria and that it efficiently overcomes a variety of damages, particularly double-strand damages. In this section, we review evidence that permits us to extend these conclusions to eukaryotes. We consider simple fungi first and then discuss multicellular eukaryotes.

A. Fungi

Most of the experiments on recombinational repair in fungi were done with the yeast *Saccharomyces cerevisiae*.

1. *Cross-Link Repair*   Experiments by Magana-Schwencke *et al.* (1982) on recombinational repair of DNA cross-links in the yeast *S. cerevisiae* indicate that the mechanism is similar to that of *E. coli*. In the yeast studies, as in those of *E. coli*, cross-links were introduced by treatment with PUVA. The repair process was initiated by an incision, presumably leading to the unlinking of the cross-link from one DNA strand and the creation of a gap in that strand. The filling in of the gap depends on the rad-51 gene product, which is necessary for recombination. When this repair process is deficient, the induction of only a single cross-link is lethal, whereas in repair-proficient cells it takes about 120 psoralen cross-links per genome to induce one lethal hit. This indicates that recombinational repair of cross-links is very efficient.

2. *Double-Strand Break Repair*   Ionizing radiation has been used to introduce double-strand breaks in *S. cerevisiae*. Repair of these double-strand breaks depends on two chromosomes being present, either sister chromatids or homologous chromosomes (Resnick and Martin, 1976). *Gene 52*, which is required for allelic recombination (either in the absence or presence of a DNA-damaging agent, [for review, see Kunz and Haynes, 1981]), is also required for repair of double-strand breaks (Ho, 1975; Resnick and Martin, 1976). These observations indicate that in yeast, as in *E. coli*, the repair of double-strand breaks is a recombinational process. Repair of double-strand breaks appears to be an efficient process in yeast. In wild-type yeast, an average of about 35 double-strand breaks is required to produce one lethal hit, whereas in a *rad-52* mutant, lacking recombinational repair, only 2.2 double-strand breaks were required to produce a lethal hit (Resnick and Martin, 1976).

Double-strand breaks introduced by methyl methanesulfonate also seem to be subject to recombinational repair in yeast, because two homologous chromosomes are required for their repair (Chlebowicz and Jachymczyk, 1979). However, genes required for recombination were not tested in this instance.

3. *Repair of Other Types of Damage* Many different DNA-damaging agents have been shown to increase allelic recombination in yeast. For instance, mitotic recombination is stimulated by UV (Wilkie and Lewis, 1963), X-rays (Manny and Mortimer, 1964), methyl methanesulfonate (Snow and Korch, 1970), N-methyl-N'-nitro-N-nitrosoguanidine (Ryttman and Zetterberg, 1976), nitrous acid and ethyl methanesulfonate (Davies *et al.,* 1975) and mitomycin C (Holliday, 1964). Simmon (1979) tested 101 heterogeneous chemicals including carcinogens and cocarcinogens for stimulation of mitotic recombination in *S. cerevisiae.* Of these, 44 proved to be recombinogenic. If stimulation of recombination by a DNA-damaging agent is a valid indication of recombinational repair in yeast, then these results indicate that recombinational repair is used against a wide variety of types of DNA damage.

4. *Excision Repair Mutants Have Increased UV-Induced Recombination Saccharomyces cerevisiae* mutants defective in the *rad-3* gene have reduced excision repair of UV-induced pyrimidine dimers. These mutants also have an increased frequency of UV-induced recombination at low doses (Kunz and Haynes, 1981), suggesting that pyrimidine dimers not removed by excision repair are handled by recombinational repair.

5. *Recombinational Repair in the Fission Yeast* Schizosaccharomyces Pombe Phipps *et al.* (1985), in a review of DNA repair in *S. pombe,* characterized recombinational repair as a major form of repair in this organism. Gentner *et al.* (1978) described recombinational repair of UV-induced damages in *S. pombe* as depending on (1) the presence of two genome copies, either sister chromatids or homologous chromosomes, and (2) on the rad-1 gene product, which is also necessary for allelic recombination.

6. *Recombinational Repair in the Smut Fungus* Ustilago maydis Kmiec and Holloman (1982, 1984) showed that the rec-1 gene product of the smut fungus *U. maydis* has properties similar to the

RecA protein of *E. coli* in that it promotes pairing of homologous chromosomes. A *rec-1* mutant of *U. maydis* undergoes abnormal meiosis, and it also has increased sensitivity to UV, ionizing radiation, and nitrosoguanidine. These properties suggest that the *rec-1* gene is needed for recombinational repair (Holliday *et al.*, 1976). The finding that UV-irradiation and mitomycin C treatment increase mitotic recombination in *U. maydis* (Holliday, 1961, 1964) is also suggestive of the occurrence of recombinational repair in this organism.

7. Summary of Fungal Recombinational Repair Good evidence indicates the presence of recombinational repair in fungi, similar to that in viruses and bacteria. In *S. cerevisiae*, repair of double-strand breaks and cross-links depends on a gene required for genetic recombination. Also, repair of double-strand breaks requires the presence of two homologous chromosomes. Recombinational repair in *S. cerevisiae* is efficient in overcoming different types of damages, as it is in phage T4 and in *E. coli*. In wild-type *S. cerevisiae*, about 35 X-ray-induced double-strand breaks or 122 psoralen cross-links are required to produce one lethal hit, implying efficient repair of these rather different types of damage. Recombination in *S. cerevisiae* is stimulated by a wide variety of DNA-damaging agents, and the UV-stimulation of recombination is enhanced when excision repair is defective. This suggests that recombinational repair is versatile with respect to the types of damage it can repair. Two other fungi, *S. pombe* and *U. maydis*, also appear to undergo recombinational repair, suggesting that the process is common among fungi.

B. *Drosophila*

Drosophila melanogaster is a fruit fly used extensively in genetic research. Recombinational repair appears to be present in *Drosophila*, as evidenced by the finding that mutants defective in allelic recombination are also sensitive to several DNA-damaging agents. Mutants defective in genes *mei-9* and *mei-41* have reduced spontaneous meiotic recombination and increased sensitivity to X-rays, UV light, methyl methanesulfonate, nitrogen mustard, and 2-acetylaminofluorene (Baker *et al.*, 1976). Gatti *et al.* (1980) presented evidence that the normal products of genes *mei-9* and *mei-41* function to remove X-ray-induced damages that, if unrepaired, can lead to the formation of chromosome aberrations. Gene *mei-41*, but not gene *mei-9*, is required for postreplication repair of UV-induced damages (Boyd and Setlow, 1976). Postreplication repair in *Drosophila* is measured by the formation of high-molecular weight DNA strands in the presence of damaged templates and may involve recombination in *Drosophila*, as it does in *E. coli* (Chapter 9, Section II.G.2). The *mei-9* mutant is also defective in excision repair of UV-induced pyrimidine dimers (Boyd *et al.*, 1976). However, whether this excision repair function is essential for the overall meiotic recombination process or meiotic recombination and excision repair are separate functions under common control by the mei-9 gene product is not known.

Mutants defective in another gene, *mus101*, also had reduced meiotic recombination and increased sensitivity to methyl methanesulfonate, X-rays, and nitrogen mustard (Boyd, 1978). Furthermore, postreplication repair in the *mus101* mutant was defective (Boyd and Setlow, 1976). The *mei-41* and *mus-101* mutants in *Drosophila* are similar to the recombinational repair mutants in phage T4, *E. coli*, and yeast, which have concomitant deficiencies in genetic recombination, and in postreplication repair as well as reduced resistance to DNA-damaging agents. Thus, genes *mei-41, mus-101*, and possibly *mei-9* may be required for recombinational repair. The presence of recombinational repair in *Drosophila* is also suggested by the finding that recombination of alleles is stimulated by a variety of DNA-damaging agents, including UV light (Prudhommeau and Proust, 1973; Martensen and Green, 1976), X-rays (Abbadessa and Burdick, 1963), and mitomycin C (Shewe *et al.*, 1971).

In summary, evidence in *Drosophila* indicates that at least two genes function in recombinational repair. Mutations in these genes lead to increased sensitivity to a variety of DNA-damaging agents, defective postreplication repair of UV-induced damages, and reduced meiotic recombination. The finding of such genes in *Drosophila* implies that recombinational repair is a significant process in multicellular organisms.

C. Mammalian Cells

Hamster ovary cells can carry out repair of double-strand breaks in DNA introduced by ionizing radiation or bleomycin; mutant cells that are sensitive to X-rays or bleomycin are defective in this repair (Kemp *et al.*, 1984; Weibezahn *et al.*, 1985; Robson *et al.*,

1989). Human fibroblast cells are also able to repair DNA double-strand breaks caused by ionizing radiation and bleomycin (Coquerelle *et al.*, 1987). Furthermore, Song *et al.* (1985) have shown that double-strand breaks enhance homologous recombination in mammalian cells and extracts and that the breaks may be the initiation sites for the recombination event. Because double-strand breaks are overcome by recombinational repair in *E. coli* and yeast, a similar repair pathway might be used in mammalian cells. Experiments by Resnick and Moore (1979), however, suggest that such a mechanism is, at best, inefficient in hamster ovary cells. On the other hand, investigations by Brenner *et al.* (1986) indicate that recombinational repair of double-strand breaks and gaps in mouse L cells between homologous plasmids is efficient.

Coogan and Rosenblum (1988) measured repair of double-strand breaks following X-irradiation in rat spermatogenic cells during different stages of germ cell formation. These stages were spermatagonia and preleptotene spermatocytes, pachytene spermatocytes, and spermatid spermatocytes. Pachytene spermatocytes demonstrated the greatest repair capability. Because pachytene is the stage of meiosis in which recombination occurs, these findings may reflect meiotic recombinational repair of the double-strand breaks.

Repair of DNA cross-links in human lymphoblastoid cells was analyzed by Matsumoto *et al.* (1989). Cross-links were introduced by treatment with mitomycin C, and their repair was analyzed by dena-turation–renaturation gel electrophoresis. Cells from persons with the inherited condition Fanconi anemia (FA) were compared with cells from normal individuals. FA cells were found to be deficient in the ability to remove DNA cross-links. In various FA cell lines, the deficiency in removal correlated with sensitivity to mitomycin C as measured by cell survival. Earlier in this chapter (Sections II.A and III.A.1), we described evidence that cross-link removal in *E. coli* and yeast occurs by a recombinational repair process. Considering this evidence, the finding of cross-link removal in humans and the identi-fication of mutant individuals defective in this process suggests that recombinational repair of cross-links may occur in humans as well.

Hollstein and McCann (1979) compiled 18 reports in the litera-ture on various chemical carcinogens and mutagens (DNA-damaging agents) that stimulate sister-chromatid exchange (SCE) in mammalian cells *in vitro*. Such stimulation indicates recombinogenic activity. Because recombinational repair in lower organisms is generally accompanied by other manifestations of recombination (e.g., exchange of genetic markers), the observed stimulation of SCEs in mammalian cells may also reflect recombinational repair.

D. Plants

Small supernumerary chromosomes, called B chromosomes, are common in plants. These are not homologous to the standard autosomal chromosomes and may not segregate regularly during nuclear division. In many plant species, including maize, the B chromosomes increase chiasma (recombination) frequency (see Bell, 1982: p. 411, for references). Staub (1984) irradiated maize germinal tissue prior to meiosis and assayed resistance to X-rays by pollen viability. He found that B chromosomes in maize cause an increase in resistance to X-irradiation-induced DNA damage. These findings suggest that B chromosomes may be maintained in natural populations, at least in part, because they enhance recombinational repair.

Repair of bleomycin-induced DNA double-strand breaks has been reported in the broad bean *Vicia faba* (Angelis *et al.,* 1989). Again, we can argue by analogy with the use of recombinational repair to remove double-strand breaks in *E. coli* and yeast (this chapter, Sections II.B and III.A.2) that recombinational repair also may be used to handle these damages in the broad bean plant.

The results of nine studies, on the effect of DNA-damaging treatments given at meiosis on recombination, was summarized by Westerman (1967). These studies were carried out on plants (*Lilium, Tradescantia,* and *Chlamydomonas*), as well as insects (*Drosophila* and *Schistocerca*). When treatments were given during the period subsequent to premeiotic DNA replication and prior to chiasma formation (i.e., during zygotene or early pachytene) recombination generally increased. Again, if the stimulation of recombination by a DNA-damaging agent is a valid indication of recombinational repair, then these results indicate that recombinational repair occurs in these species.

Schvartzman (1987), in a review of the literature on the induction of SCEs in higher plants by DNA-damaging agents, suggested that the most important mechanism giving rise to SCEs is replication of damaged DNA templates. He also compiled evidence from three plant species (*V. faba, Allium cepa,* and *Secale cereale*) that SCEs can be increased by a variety of DNA-damaging agents. This evidence suggests that higher plant cells may use recombinational repair to overcome DNA damages.

IV. *Overview*

In Chapter 9, Section IV, we argued that only recombinational repair can overcome double-strand damages and that such damages are common in nature. In this chapter, we reviewed evidence that recombinational repair is prevalent among viruses, bacteria, and fungi and that it occurs in *Drosophila,* mammals, and plants. Recombinational repair is effective against the damages introduced by a wide variety of agents and, in particular, is very efficient in overcoming double-strand damages in both *E. coli* and *S. cerevisiae.* These findings provide the basis for the first part of the repair (and complementation) hypothesis of sex, that recombination is an adaptation for repairing DNA damage (Chapter 1, Section II). In Chapter 13, we deal with the second part of this hypothesis: that outcrossing is selected by the advantage of complementation.

A number of different molecular models for recombinational repair have been proposed. For example, Cole *et al.* (1978) suggested a model for recombinational repair of DNA cross-links and Resnick (1976) has proposed a model for recombinational repair of double-strand breaks. In Chapter 11, we will discuss current knowledge of the mechanism of recombinational repair of double-strand breaks.

The systems used for studying recombinational repair reviewed here were chosen for experimental convenience, in most cases, and thus involve asexually, rather than sexually, reproducing cells (e.g., mitotic recombination in yeast and sister chromosome recombination in *E. coli* and mammalian cells). If efficient recombinational repair can occur in vegetative cells, we presume that recombinational repair in the germ line should be at least as efficient, because rates of recombination in meiosis are substantially higher than in mitosis (see Chapter 12, Section III.B). Furthermore, as we argue in the next chapter, meiosis appears to be an adaptation for promoting recombinational repair of DNA.

References

Abbadessa, R., and Burdick, A. N. (1963). The effect of x-irradiation on somatic crossing over in *Drosophila melanogaster*. *Genetics* **48**, 1345–1356.

Abel, P. (1962). Multiplicity reactivation and marker rescue with vaccinia virus. *Virology* **17**, 511–519.

Angelis, K. J., Veleminsky, J., Rieger, R., and Schubert, I. (1989). Repair of bleomycin-induced DNA double-strand breaks in *Vicia faba*. *Mutat. Res.* **212**, 155–157.

Baker, B. S., Boyd, J. B., Carpenter, A. T. C., Green, M. M., Nguyen, T. D., Ripoll, P., and Smith, P. D. (1976). Genetic controls of meiotic recombination and somatic DNA metabolism in *Drosophila melanogaster*. *Proc. Natl. Acad. Sci. USA* **73**, 4140–4144.

Baker, R. M., and Haynes, R. H. (1967). UV-induced enhancement of recombination among lambda bacteriophages in UV-sensitive host bacteria. *Mol. Gen. Genet.* **100**, 166–167.

Barry, R. D. (1961). The multiplication of influenza virus. II. Multiplicity reactivation of ultraviolet irradiated virus. *Virology* **14**, 398–405.

Bell, G. (1982). "The Masterpiece of Nature: The Evolution and Genetics of Sexuality." University of California Press, Berkeley.

Bernstein, A. (1957). Multiplicity reactivation of ultraviolet-irradiated Vi-phage II of *Salmonella typhi*. *Virology* **3**, 286–298.

Bernstein, C. (1981). Deoxyribonucleic acid repair in bacteriophage. *Microbiol. Rev.* **45**, 72–98.

Bernstein, C., and Wallace, S. S. (1983). DNA repair. *In* "Bacteriophage T4" (C. K. Mathews *et al.*, eds.), pp. 138–151. American Society for Microbiology, Washington, D.C.

Bernstein, H. (1983). Recombinational repair may be an important function of sexual reproduction. *BioScience* **33**, 326–331.

Bernstein, H., Hopf, F. A., and Michod, R. E. (1987). The molecular basis of the evolution of sex. *Adv. Genet.* **24**, 323–370.

Better, M., and Helinski, D. R. (1983). Isolation and characterization of the *RecA* gene of *Rhizobium meliloti*. *J. Bacteriol.* **155**, 311–316.

Blanco, M., and Devoret, R. (1973). Repair mechanisms involved in prophage reactivation and UV reactivation of UV-irradiated phage. *Mutat. Res.* **17**, 293–305.

Boyd, J. B., (1978). DNA repair in *Drosophila*. *In* "DNA Repair Mechanisms" (P. C. Hanawalt, E. C. Friedberg, and C. F. Fox, eds.), pp. 449–452. Academic Press, New York.

Boyd, J. B., and Setlow, R. B. (1976). Characterization of post-replication repair in mutagen-sensitive strains of *Drosophila melanogaster*. *Genetics* **84**, 507–526.

Boyd, J. B., Golino, M. D., and Setlow, R. B. (1976). The mei 9[a] mutant of *Drosophila melanogaster* increases mutagen sensitivity and decreases excision repair. *Genetics* **84**, 527–544.

Brenner, D. A., Smigocki, A. C., and Camerini-Otero, R. D. (1986). Double-strand gap repair results in homologous recombination in mouse L cells. *Proc. Natl. Acad. Sci. USA* **83**, 1762–1766.

Chan, V. L. (1975). On the role of V gene ultraviolet-induced enhancement of recombination among T4 phages. *Virology* **65**, 266–267.

Chen, D., and Bernstein, C. (1987). Recombinational repair of hydrogen peroxide-induced damages in DNA of phage T4. *Mutat. Res.* **184**, 87–98.

Chlebowicz, E., and Jachymczyk, W. J. (1979). Repair of MMS-induced DNA double-strand breaks in haploid cells of *Saccharomyces cerevisiae,* which requires the presence of a duplicate genome. *Mol. Gen. Genet.* **167**, 279–286.

Coffin, J. M. (1979). Structure, replication, and recombination of retrovirus genomes: Some unifying hypotheses. *J. Gen. Virol.* **42**, 1–26.

Cole, R. S. (1971). Inactivation of *Escherichia coli,* F′ episomes at transfer and bacteriophage lambda by psoralen plus 360-nm light: Significance of deoxyribonucleic acid crosslinks. *J. Bacteriol.* **107**, 846–852.

Cole, R. S., Levitan, D., and Sinden, R. R. (1976). Removal of psoralen interstrand crosslinks from DNA of *Escherichia coli:* Mechanism and genetic control. *J. Mol. Biol.* **103**, 39–59.

Cole, R. S., Sinden, R. R., Yoakum, G. H., and Broyles, S. (1978). On the mechanism for repair of crosslinked DNA in *E. coli* treated with psoralen and light. *In* "DNA Repair Mechanisms," ICN-UCLA Symposia on Molecular and Cellular Biology (P. C. Hanawalt, E. C. Friedberg, and F. C. Fox, eds.), pp. 287–290. Academic Press, New York.

Coogan, T. P., and Rosenblum, I. Y. (1988). DNA double-strand damage and repair following gamma-irradiation in isolated spermatogenic cells. *Mutat. Res.* **194**, 183–191.

Coquerelle, T. M., Weibezahn, K. F., and Lucke-Huhle, C. (1987). Rejoining of double strand breaks in normal human and ataxia-telangiectasia fibroblasts after exposure to ^{60}Co gamma-rays, [241]Am alpha-particles or bleomycin. *Int. J. Radiat. Biol.* **51**, 209–218.

Davies, P. J., Evans, W. E., and Parry, J. M. (1975). Mitotic recombination induced by chemical and physical agents in the yeast *Saccharomyces cerevisiae. Mutat. Res.* **29**, 301–314.

DeFlora, S., Zanacchi, P., Camoirano, A., Bennicelli, C., and Badolati, B. S. (1984). Genotoxic activity and potency of 135 compounds in the Ames reversion test and in a bacterial DNA repair test. *Mutat. Res.* **133**, 161–168.

Devoret, R., Blanco, M., George, J., and Radman, M. (1975). Recovery of phage lambda from ultraviolet damage. *In* "Molecular Mechanisms for

Repair of DNA," part A (P. C. Hanawalt and R. B. Setlow, eds.), pp. 155-171. Plenum, New York.

deVos, W. M., deVries, S., and Venema, G. (1983). Cloning and expression of the *E. coli recA* gene in *Bacillus subtilis*. *Gene* **25**, 301-308.

Dodson, L. A., and Hadden, C. T. (1980). Capacity for post-replication repair correlated with transducibility in *rec⁻* mutants of *Bacillus subtilis*. *J. Bacteriol.* **144**, 608-615.

Dubnau, D. (1982). Genetic transformation in *Bacillus subtilis*. *In* "The Molecular Biology of the Bacilli," Vol. 1 (D. A. Dubnau, ed.), pp. 148-175. Academic Press, New York.

Eitner, G., Adler, B., Lanzov, V. A., and Hofemeister, J. (1982). Interspecies *recA* protein substitution in *Escherichia coli* and *Proteus mirabilis*. *Mol. Gen. Genet.* **185**, 481-486.

Epstein, R. H. (1958). A study of multiplicity reactivation in bacteriophage T4-K12(lambda) complexes. *Virology* **6**, 382-404.

Feinstein, S. I., and Low, K. B. (1986). Hyper-recombining recipient strains in bacterial conjugation. *Genetics* **113**, 13-33.

Fornili, S. L., and Fox, M. S. (1977). Electron microscope visualization of the products of *Bacillus subtilis* transformation. *J. Mol. Biol.* **113**, 181-191.

Fry, S. E. (1979). Stimulation of recombination in phage T4 by nitrous acid-induced lesions. *J. Gen. Virol.* **43**, 719-722.

Gatti, M., Pimpinelli, S., and Baker, B. S. (1980). Relationships among chromatid interchanges, sister chromatid exchanges, and meiotic recombination in *Drosophila melanogaster*. *Proc. Natl. Acad. Sci. USA* **77**, 1575-1579.

Gentner, N. E., Werner, M. M., Hannan, M.A., and Nasim, A. (1978). Contribution of a caffeine-sensitive recombinational repair pathway to survival and mutagenesis in UV-irradiated *Schizosaccharomyces pombe*. *Mol. Gen. Genet.* **167**, 43-49.

Goldberg, I., and McKalanos, J. J. (1986). Cloning of the *Vibrio cholerae recA* gene and construction of a *Vibrio cholerae recA* mutant. *J. Bacteriol.* **165**, 715-722.

Graham, J. B., and Istock, C. A. (1979). Gene exchange and natural selection cause *Bacillus subtilis* to evolve in soil culture. *Science* **204**, 637-639.

Griffith, J., and Formosa, T. (1985). The uvsX protein of bacteriophage T4 arranges single-stranded and double-stranded DNA into similar helical nucleoprotein filaments. *Biol. Chem.* **260**, 4484-4491.

Hahn, T.-R., West, S., and Howard-Flanders, P. (1988). RecA-mediated strand exchange reactions between duplex DNA molecules containing damaged bases, deletions and insertions. *J. Biol. Chem.* **263**, 7431-7436.

Hall, J. D., and Scherer, K. (1981). Repair of psoralen-treated DNA by genetic recombination in human cells infected with herpes simplex virus. *Cancer Res.* **441,** 5033–5038.

Hall, J. D., Featherston, J. D., and Almy, R. E. (1980). Evidence for repair of ultraviolet light-damaged herpes virus in human fibroblasts by a recombination mechanism. *Virology* **105,** 490–500.

Harm, W. (1958). Multiplicity reactivation, marker rescue and genetic recombination in phage T4 following x-ray inactivation. *Virology* **5,** 337–361.

Harm, W. (1964). On the control of UV-sensitivity of phage T4 by the gene *x*. *Mutat. Res.* **1,** 344–354.

Ho, K. S. Y. (1975). Induction of DNA double-strand breaks by X-rays in a radiosensitive strain of the yeast *Saccharomyces cerevisiae. Mutat. Res.* **30,** 327–334.

Holliday, R. (1961). Induced mitotic crossing-over in *Ustilago maydis. Genet. Res.* **2,** 231–248.

Holliday, R. (1964). The induction of mitotic recombination by mitomycin C in *Ustilago* and *Saccharomyces. Genetics* **50,** 323–335.

Holliday, R., Halliwell, R. E., Evans, M. W., and Rowell, V. (1976). Genetic characterization of *rec-1,* a mutant of *Ustilago maydis* defective in repair and recombination. *Genet. Res.* **27,** 413–453.

Hollstein, M., and McCann, J. (1979). Short-term tests for carcinogens and mutagens. *Mutat. Res.* **65,** 133–226.

Holmes, G. E., Schneider, S., Bernstein, C., and Bernstein, H. (1980). Recombinational repair of mitomycin C lesions in phage T4. *Virology* **103,** 299–310.

Howard-Flanders, P., Rupp, W. D., Wilkins, B., and Cole, R. S. (1968). DNA replication and recombination after ultraviolet-irradiation. *Cold Spring Harbor Symp. Quant. Biol.* **33,** 195–207.

Huskey, R. J. (1969). Multiplicity reactivation as a test for recombination function. *Science* **164,** 319–320.

Keener, S. L., McNamee, K. P., and McEntee, K. (1984). Cloning and characterization of *recA* genes from *Proteus vulgaris, Erwinia caratovora, Shigella flexneri,* and *Escherichia coli B/r. J. Bacteriol.* **160,** 153–160.

Kellenberger, G., and Weigle, J. (1958). Étude au moyen des rayons ultraviolets de l'interaction entre bactériophage tempéré et bactérie hôte. *Biochem. Biophys. Acta* **30,** 112–124.

Kemp, L. M., Sedgwick, S. G., and Jeppo, P. A. (1984). X-ray sensitive mutants of Chinese hamster ovary cells defective in double-strand break rejoining. *Mutat. Res. DNA Repair Rep.* **132,** 189–196.

Kmiec, E. B., and Holloman, W. K. (1982). Homologous pairing of DNA molecules promoted by a protein from *Ustilago. Cell* **29,** 367–374.

Kmiec, E. B., and Holloman, W. K. (1984). Synapsis promoted by *Ustilago rec-1* protein. *Cell* **36,** 593–598.

Kolowsky, K. S., and Szalay, A. A. (1986). Double-strand gap repair in the photosynthetic procaryote *Synechococcus R2*. *Proc. Natl. Acad. Sci. USA* **83**, 5578–5582.

Konrad, E. B. (1977). Method for isolation of *Escherichia coli* mutants with enhanced recombination between chromosomal duplications. *J. Bacteriol.* **130**, 167–172.

Krasin, F., and Hutchinson, F. (1977). Repair of DNA double-strand breaks in *Escherichia coli* which requires *recA* function and the presence of a duplicate genome. *J. Mol. Biol.* **116**, 81–89.

Krasin, F., and Hutchinson, F. (1981). Repair of DNA double-strand breaks in *Escherichia coli* cells requires synthesis of proteins that can be induced by UV light. *Proc. Natl. Acad. Sci. USA* **78**, 3450–3453.

Kunz, B. A., and Haynes, R. H. (1981). Phenomenology and genetic control of mitotic recombination in yeast. *Annu. Rev. Genet.* **15**, 57–89.

Lin, P.-F., and Howard-Flanders, P. (1976). Genetic exchanges caused by ultraviolet photoproducts in phage lambda DNA molecules: The role of DNA replication. *Molec. Gen. Genet.* **146**, 107–155.

Lovett, C. M., Jr., and Roberts, J. (1985). Purification of a *recA* protein analogue from *Bacillus subtilis*. *J. Biol. Chem.* **260**, 3305–3313.

Luria, S. E. (1947). Reactivation of irradiated bacteriophage by transfer of self-producing units. *Proc. Natl. Acad. Sci. USA* **33**, 253–264.

Luria, S. E., and Dulbecco, R. (1949). Genetic recombinations leading to production of active bacteriophage from ultraviolet inactivated bacteriophage particles. *Genetics* **34**, 93–125.

Magana-Schwencke, N., Henrique, J.-A. P., Chanet, R., and Moustacchi, E. (1982). The fate of 8-methoxypsoralen photo-induced crosslinks in nuclear and mitochondrial yeast DNA: Comparison of wild-type and repair deficient strains. *Proc. Natl. Acad. Sci. USA* **79**, 1722–1726.

Manny, T. R., and Mortimer, R. K. (1964). Allelic mapping in yeast by X-ray-induced mitotic reversion. *Science* **143**, 581–583.

Martensen, D. V., and Green, M. M. (1976). UV-induced mitotic recombination in somatic cells of *Drosophila melanogaster*. *Mutat. Res.* **36**, 391–396.

Matsumoto, A., Vos, J.-M. H., and Hanawalt, P. C. (1989). Repair analysis of mitomycin C-induced DNA crosslinking in ribosomal RNA genes in lymphoblastoid cells from Fanconi's anemia patients. *Mutat. Res.* **217**, 185–192.

McClain, M. E., and Spendlove, R. S. (1966). Multiplicity reactivation of reovirus particles after exposure to ultraviolet light. *J. Bacteriol.* **92**, 1422–1429.

Michod, R. E., Wojciechowski, M. F., and Hoelzer, M. A. (1988). DNA repair and the evolution of transformation in the bacterium *Bacillus subtilis*. *Genetics* **118**, 31–39.

Miskimins, R., Schneider, S., Johns, V., and Bernstein, H. (1982). Topoisomerase involvement in multiplicity reactivation of phage T4. *Genetics* **101**, 157–177.

Nonn, E., and Bernstein, C. (1977). Multiplicity reactivation and repair of nitrous acid-induced lesions in bacteriophage T4. *J. Mol. Biol.* **116**, 31–47.

Owttrim, G. W., and Coleman, J. R. (1987). Molecular cloning of a recA-like gene from the cyanobacterium *Anabaena variabilis. J. Bacteriol.* **169**, 1824–1829.

Phipps, J., Nasim, A., and Miller, D. R. (1985). Recovery, repair and mutagenesis in *Schizosaccharomyces pombe. Adv. Genet.* **23**, 1–72.

Picksley, S. M., Attfield, P. V., and Lloyd, R. G. (1984). Repair of DNA double-strand breaks in *E. coli* K-12 requires a functional *recN* gene product. *Mol. Gen. Genet.* **195**, 267–274.

Pierre, A., and Paoletti, C. (1983). Purification and characterization of *recA* protein from *Salmonella typhimurium. J. Biol. Chem.* **258**, 2870–2874.

Piette, J., Calberg-Bacq, C. M., and Van deVorst, A. (1978). Photodynamic effect of proflavine on phiX174 bacteriophage, its DNA replicative form and its isolated single-stranded DNA: Inactivation, mutagenesis and repair. *Mol. Gen. Genet.* **167**, 95–103.

Prudhommeau, C., and Proust, J. (1973). UV irradiation of polar cells of *Drosophila melanogaster* embryos. V. A study of the meiotic recombination in females with chromosomes of a different structure. *Mutat. Res.* **23**, 63–66.

Radman, R., and Wagner, R. (1984). Effects of DNA methylation on mismatch repair, mutagenesis, and recombination in *Escherichia coli. Curr. Top. Microbiol. Immunol.* **108**, 23–28.

Resnick, M. A. (1976). The repair of double-strand breaks on DNA: A model involving recombination. *J. Theor. Biol.* **59**, 97–106.

Resnick, M. A., and Martin, P. (1976). The repair of double-strand breaks in the nuclear DNA of *Saccharomyces cerevisiae* and its genetic control. *Mol. Gen. Genet.* **143**, 119–129.

Resnick, M. A., and Moore, P. D. (1979). Molecular recombination and the repair of DNA double-strand breaks in CHO cells. *Nucleic Acids Res.* **6**, 3145–3160.

Robson, C. N., Harris, A. L., and Hickson, I. D. (1989). Defective repair of DNA single- and double-strand breaks in the bleomycin- and X-ray-sensitive Chinese hamster ovary cell mutant, BLM-2. *Mutat. Res.* **217**, 93–100.

Ryttman, H., and Zetterberg, G. (1976). Induction of mitotic recombination with N-methyl-N'-nitro-N-nitrosoguanidine (MNNG) in *Saccharomyces cerevisiae*. A comparison between treatment in vitro and in the host-mediated assay. *Mutat. Res.* **34**, 201–216.

Sancar, A., and Rupp, W. D. (1983). A novel repair enzyme: uvrABC excision nuclease of *Escherichia coli* cuts a DNA strand on both sides of the damaged region. *Cell* **33**, 249–260.

Schvartzman, J. B. (1987). Sister-chromatid exchanges in higher plant cells: Past and perspectives. *Mutat. Res.* **181**, 127–145.

Selander, R. K., and Levin, B. R. (1980). Genetic diversity and structure in *Escherichia coli* populations. *Science* **31**, 545–547.

Selsky, C. A., Henson, P., Weichselbaum, R. R., and Little, J. B. (1979). Defective reactivation of ultraviolet light-irradiated herpes virus by a Bloom's syndrome fibroblast strain. *Cancer Res.* **39**, 3392–3396.

Setlow, J. K., Spikes, D., and Griffin, K. (1988). Characterization of the rec-1 gene of *Haemophilus influenzae* and behavior of the gene in *Escherichia coli*. *J. Bacteriol.* **170**, 3876–3881.

Shewe, M. J., Suzuki, D. T., and Erasmus, U. (1971). The genetic effects of mitomycin C in *Drosophila melanogaster*. II. Induced meiotic recombination. *Mutat. Res.* **12**, 269–279.

Simmon, V. F. (1979). *In vitro* assays for recombinogenic activity of chemical carcinogens and related compounds with *Saccharomyces cerevisiae* D3. *J. Natl. Cancer Inst.* **62**, 901–909.

Smith, K. C., and Meun, D. H. C. (1970). Repair of radiation-induced damage in *Escherichia coli*. I. Effect of rec mutations on post-replication repair of damage due to ultraviolet radiation. *J. Mol. Biol.* **51**, 459–472.

Snow, R., and Korch, C. T. (1970). Alkylation induced gene conversion in yeast: Use in fine structure mapping. *Mol. Gen. Genet.* **107**, 201–208.

Song, K.-Y., Chekuri, L., Rauth, S., Ehrlich, S., and Kucherlapati, R. (1985). Effect of double-strand breaks on homologous recombination in mammalian cells and extracts. *Mol. Cell. Biol.* **5**, 3331–3336.

Staub, R. W. (1984). The influence of B chromosomes on the susceptibility of maize to gamma irradiation induced DNA damage. Ph.D. Thesis, University of Arizona, Tucson.

Stewart, G. J., and Carlson, C. A. (1986). The biology of natural transformation. *Annu. Rev. Microbiol.* **40**, 211–235.

Symonds, N., and Ritchie, D. A. (1961). Multiplicity reactivation after decay of incorporated radioactive phosphorus in phage T4. *J. Mol. Biol.* **3**, 61–70.

Tessman, I., and Ozaki, T. (1957). Multiplicity reactivation of bacteriophage T1. *Virology* **4**, 315–327.

Weibezahn, K. F., Lohrer, H., and Herrlich, P. (1985). Double-strand break repair and G2 block in Chinese hamster ovary cells and their radiosensitive mutants. *Mutat. Res.* **145**, 177–183.

West, S. C., Countryman, J. K., and Howard-Flanders, P. (1983). Purification and properties of the *recA* proteins from *Proteus mirabilis*. *J. Biol. Chem.* **258,** 4648–4654.

Westerman, M. (1967). The effect of X-irradiation on male meiosis in *Schistocerca gregaria* (Forskal). I. Chiasma frequency response. *Chromosoma* **22,** 401–416.

Wilkie, D., and Lewis, D. (1963). The effect of ultraviolet light on recombination in yeast. *Genetics* **48,** 1701–1716.

Wojciechowski, M. F., Hoelzer, M. A., and Michod, R. E. (1989). DNA repair and the evolution of transformation in *Bacillus subtilis*. II. Role of inducible repair. *Genetics* **121,** 411–422.

Yamamoto, H., and Shimijo, H. (1971). Multiplicity reactivation of human adenovirus type 12 and simian virus 40 irradiated by ultraviolet light. *Virology* **45,** 529–531.

Yarosh, D. B. (1978). UV-induced mutation in bacteriophage T4. *J. Virol.* **26,** 265–271.

Yonesaki, T., Rao, Y., Minagawa, T., and Takahashi, H. (1985). Purification and some of the functions of the products of bacteriophage T4 recombination genes, *uvsX* and *uvsY*. *Eur. J. Biochem.* **148,** 127–134.

Chapter Eleven

Meiosis and
Meiotic Recombination

Meiosis is the key stage of the sexual cycle in eukaryotes (Chapter 13, Section II). Here we review the stages of meiosis and the mechanism of meiotic recombination. Next we summarize the evidence bearing on the current double-strand break repair model of meiotic recombination. We show that this model is consistent with meiotic recombination being an adaptation for repair of DNA. This point of view was presented previously by Bernstein *et al.* (1987, 1988).

I. *Events of Meiosis and of Meiotic Recombination*

A. Meiosis and Mitosis Compared

Meiosis is often described in comparison with mitosis; therefore, we summarize the essential features of both processes here. In organisms with true nuclei (eukaryotes), the production of germ cells occurs by meiosis. All other eukaryotic cell divisions occur by mitosis. In a multicellular species such as humans, all cell divisions of the body (somatic cell divisions) occur by mitosis, whereas formation of sperm or egg cells occurs by meoisis. Even in the germ line, the cell divisions prior to germ cell formation are by mitosis.

In humans, mitosis occurs only in cells with two sets of chromosomes (diploid cells). In other organisms, such as some fungi, mitosis occurs only in cells with a single set of chromosomes (haploid cells) and in many species, especially plants, mitosis occurs in both diploid and haploid cells. We will not discuss the cytological features of meiosis and mitosis as is conventionally done in biology texts. Rather, we will describe the two processes at the level of DNA and genetic information.

The process of mitosis is rather simple. In terms of genetic information, each chromosome in a cell is equivalent to a double-stranded DNA molecule. Replication of each DNA molecule occurs by the separation of the two strands, each of which then acts as a template for the synthesis of a new complementary strand. When each chromosome (DNA molecule) has been replicated, the cell divides, and the two copies of each chromosome segregate in such a way that each of the two daughter cells gets one copy. As a general rule, the two daughter cells contain the same genetic information as each other and as their progenitor cell. However, genetic variation can arise through infrequent mutation and, in a diploid cell, by rare recombination between nonsister homologous chromosomes. In general, mitosis is a process of replication and division to produce two, usually identical, cells from one cell.

Compared with mitosis, which has two basic steps, DNA replication and cell division, meiosis has four basic steps. These are (1) DNA replication, (2) lining up and intimate pairing of homologous chromosomes leading to recombination, (3) cell division, and (4) another cell division. Through meiosis, a diploid cell produces four haploid daughter cells.

In a diploid cell, two equivalent sets of chromosomes are present. Each of the two sets contains several different (i.e., nonhomologous) chromosomes. One set of chromosomes traces its information back to a haploid germ cell from one sexual parent (e.g., the egg of the mother), and the other traces back to a haploid germ cell of the other parent (e.g., the sperm of the father). As in mitosis, meiosis is initiated by the replication of each of the chromosomes. After replication, each formerly diploid cell now has four complete sets of chromosomes, two sets tracing their lineage back to each original parent. The chromosomes at this stage are conventionally called chromatids. The next step is a lining up and intimate pairing of each set of four homologous chromatids. Recombination, involving exchange of DNA segments, can occur between any two of the four homologous chromatids. This systematic alignment of homologous chromatids and subsequent recombination is the central feature distinguishing meiosis from mitosis. In Section B, below, we will discuss the process of meiotic recombination in more detail.

After recombination has occurred, there are two successive cell divisions resulting in four cells. The four sets of chromotids present at meiosis segregate in such a way during these two divisions that each of the resulting four cells has only one set of chromosomes. These haploid cells may then differentiate further into germ cells. Although the events preceding and following meiosis differ from species to species, the events of meiosis itself are consistent with few exceptions (for discussion of one exceptional case, see Chapter 12, Section IV.B). The diploid line is regenerated when two haploid germ cells fuse by the process of syngamy or fertilization.

B. Double-Strand Break Repair Model of Meiotic Recombination

The recombination events occurring at meiosis have long been regarded as central to the function of meiosis. The problem of how recombination occurs is one of the classical problems in genetic research and a great deal of detailed information has been accumulated on this process. The goal of this work has been to achieve a model of recombination that explains, as simply as possible, the accumulated experimental data. Early work on the mechanism of meiotic recombination, starting in the 1950s, involved primarily genetic analysis. This was later supplemented by biochemical and molecular approaches. The large body of evidence in this area has been reviewed by Kushev (1974), Stahl (1979), and Whitehouse (1982). A molecular model of recombination that embodied the evidence available up to 1975 was proposed by Meselson and Radding (1975). This model dominated thinking concerning the mechanism of recombination for much of the decade subsequent to its presentation.

In 1983, however, Szostak et al. proposed a model for meiotic recombination, which they refer to as the double-strand break repair model. This model, like its earlier counterpart, explained the extensive data obtained from fine structure genetic analyses as well as knowledge of the physical characteristics of DNA and the enzymatic reactions involved in its processing (as reviewed by Orr-Weaver and Szostak, 1985). However, the double-strand break repair model accommodated more recent experimental evidence not readily ac-

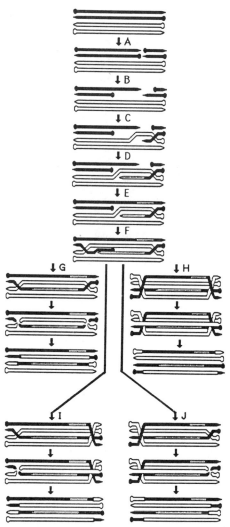

Figure 13 Model of meiotic recombination adapted from Szostak *et al.* (1983). Each recombination event at meiosis involves only two of the four chromatids present. These two meiotic chromatids are depicted, whereas the other two are left out for clarity. Each of the two involved chromatids consist of double-stranded DNA with two antiparallel strands. These are drawn without their helical twists, and the 5' end of the segment shown is indicated by a circle and the 3' end by a point. The model shows details of recombination within a region one or a few genes in length, with most of the chromatids extending to the left and right of the diagrammed segments.

During step (A), a double-strand break or gap is formed in one DNA duplex. In step (B), an exonuclease acts on each strand in a 5' to 3' direction (i.e., from the circle end to the pointed end in the figure) so that the initial counted for by the older Meselson-Radding model. It is worth noting that the design of the model was not based on any assumptions about the adaptive significance of recombination, i.e., whether the primary function of recombination is to promote DNA repair, allelic varia- tion, or any other possible function. The evidence supporting this model is discussed below in Section II of this chapter.

Currently, the double-strand break repair model is the most authoritative one for general recombination, although a few addi- tional changes have been suggested since it was first proposed. In a

---

break becomes flanked by single-stranded DNA with 3' ends. These single strands with 3' ends are indicated by the lines with pointed ends shown extending into the gap. During step (C), one of the 3' ends invades another homologous chromatid. Invasion involves the pairing of the 3'-ended strand with one strand of the undamaged chromatid and the displacement of the undamaged strand's former partner. This former partner strand forms a displacement loop (D-loop) as shown after step (C). In this process, a Holli- day junction (a junction where two single strands cross) is created. In step (D), the D-loop is enlarged by elongation through DNA synthesis primed by the invading strand. This elongation proceeds until the other 3'-ended single strand can pair with the D-loop from the other side. The newly synthesized DNA is represented by a stippled 3' segment. During step (E), the gap in the upper (black) strand is filled by repair synthesis (represented as a stippled segment) using part of the D-loop as template. In step (F), the single strand that initially invaded the D-loop continues to elongate by further DNA synthesis. This results in enlargement of the D-loop by branch migration and strand displacement. The displaced strand migrates to the lower chromatid forming a heteroduplex region (one black, one white strand). In going from step (F) to (G–J), degradation of extra lengths of single strands may be required prior to ligation (i.e., strand joining). In steps (G–J), each of the two Holliday junctions can exist in two equivalent forms (Meselson and Radding, 1975) and can thus be cut and resolved in two alternative ways. Therefore, at this stage there are four possible configurations. When the two Holliday junctions are resolved in the same sense—i.e., cutting the inner strands at both junctions (G) or the outer strands at both junctions (H)—two chroma- tids are formed that have not exchanged flanking alleles. That is, the flanking portions of each of the final recombinant chromatids have the same color, either both black or both white. When the two Holliday junctions are re- solved in the opposite sense, with one inner pair cut and one outer pair cut (I and J), two chromatids are formed whose flanking alleles have been ex- changed.

recent review, Thaler and Stahl (1988) suggested some modifications of the original model on the basis of current data, but the fundamental concept does not seem to be seriously affected. In addition, Nickoloff *et al.* (1989), on the basis of experiments in yeast, have suggested that double-strand breaks can stimulate several alternative mechanisms of recombinational repair. However, at the present time, the double-strand break repair model embodies our best understanding of how meiotic recombination occurs at the molecular level.

In the original double-strand break repair model of Szostak *et al.* (1983), an initial double-strand break in a chromatid is converted at the next step into a double-strand gap. This chromatid then receives information to fill the gap from the homologous chromatid with which it is intimately paired. The succeeding process of exchange is explained in detail by the model, as described above. However, the most significant feature of the model from the perspective of the repair hypothesis is the first step. The formation of a double-strand break or a double-strand gap is what one would expect if meiotic recombination were an adaptation to remove any double-strand damages in a simple direct fashion; i.e., an obvious way of removing a double-strand damage in DNA is simply to chop it out, which would leave a double-strand break or gap.

The pathway of meiotic recombination shown in Figure 13 is essentially the same as the model of Szostak *et al.* (1983; see their fig. 8), except that, for clarity, we have filled in some of the intermediate steps implied, but not included, in their original version.

II. *Evidence Bearing on the Double-Strand Break Repair Model of Meiotic Recombination*

We describe experiments, below, that either directly support the double-strand break repair model of recombination, or else are consistent with it. We also discuss how each type of evidence fits with the hypothesis that repair is the adaptive function of recombination.

A. Double-Strand Breaks and Gaps Stimulate Recombination

The initial steps in the double-strand break repair model of meiotic recombination, reviewed above, are the formation of a double-strand break and then the extension of the break into a double-strand gap. These key features of the model are based on several lines of evidence reviewed by Orr-Weaver and Szostak (1985). One of the most important experiments was the demonstration that when a DNA segment is removed from a yeast chromosome to form a double-strand gap, the lost information can be precisely restored by recombination with a homologous chromosome. These gaps were found to stimulate recombination by as much as 1,000-fold, and the process required the product of gene *RAD52*, which is essential for spontaneous allelic recombination (Orr-Weaver *et al.*, 1981). Direct experimental demonstrations that a double-strand break can initiate meiotic recombination have been achieved in the yeasts *S. cerevisiae* (Kolodkin *et al.*, 1986; Ray *et al.*, 1988) and *S. pombe* (Klar and Miglio, 1986).

In principal, any double-strand damage can be removed enzymatically, leaving a double-strand break. The double-strand break repair model of meiotic recombination provides a mechanism for coupling this removal with accurate restoration of the lost information from another homologous DNA molecule. Although this model was designed to explain spontaneous meiotic and mitotic recombination, it also suggests a general mechanism for recombinational repair. Thus, this model is consistent with the hypothesis that repair is the primary function of recombination.

B. Gene Conversion and Exchange of Flanking Regions

The double-strand break repair model of Szostak *et al.* (1983) was also designed to explain observations from fine-structure genetic analyses. As described earlier, a diploid cell gives rise by meiosis to four haploid cells. The diploid cell contains two sets of homologous chromosomes, one from each parent. After the chromosomes replicate, four copies of each chromosome are present, two from each original parent. If the two parents differ in a particular gene, we can represent the two forms (called alleles) as *A* and *a*. These two forms, for instance, might represent a difference of a single nucleotide among the hundreds encoding a particular gene. Among the four cells produced by each meiosis from an *Aa* diploid, there will ordinarily be exactly two cells with the *A* allele and two with the *a* allele. This can be represented as 2*A*:2*a*.

Infrequently, 3*A*:1*a* or 1*A*:3*a* products arise from a single meiosis. This phenomenon is referred to as "gene conversion" because one of the four alleles appears to be converted into the alternative type. In the late 1950s, geneticists came to realize that these infrequent gene conversion events represented instances where the *A* and *a* alleles happened to be at the exact location of a recombination event. During the following decades, gene conversion was studied intensively in a number of model systems to generate information on the mechanism of recombination. Of particular interest, from the point of view that recombination is an adaptation for DNA repair, are the studies in yeast showing that gene conversion is increased by UV (Ito and Kobayashi, 1975; Hannan *et al.*, 1980) and ionizing radiation (Raju *et al.*, 1971). These results would be expected if gene conversions are a consequence of recombinational repair of DNA damage.

One of the generalizations emerging from studies of gene conversion was that, when a gene conversion event occurred, the chromosomal regions flanking the event were often exchanged. The frequency of exchange varies depending on the particular pair of alleles studied. In most studies, the flanking regions were exchanged less than 50% of the time (Whitehouse, 1982: p. 321). In eight studies involving four species of fungi, the frequency of exchange of flanking regions was on average 36% (for summaries of these data see Whitehouse, 1982: p. 320). In addition to the evidence from fungi, data from half-tetrad analyses using compound autosomes in *D. melanogaster* showed that one-third of recombination events in the *rosy* gene resulted in recombination of flanking regions (Chovnick *et al.*, 1970).

By examining Figure 13 we can see both how gene conversion and how exchange of flanking regions are explained by the double-strand break repair model. Recombination occurs when four chromosomes (i.e., four DNA copies) are present in the cell. Assume the four chromosomes have the genetic alleles *A, A, a* and *a,* respectively, prior to the recombination event. Any individual recombination event involves only two of the four chromosomes. Furthermore, assume that a recombination event occurs between a chromosome carrying the allele *A* and a chromosome carrying the allele *a,* and that the event itself is in the region marked by alleles *A* and *a.* Because only two of the four DNA copies in the cell are involved in the event, from the two noninvolved copies one product will be *A* and one will be *a.* According to the double-strand break repair model, one of two DNA molecules involved in recombination will lose information in the region of the double-strand gap and this information is restored by copying from the other DNA molecule of the pair. Assume that the deleted region contained the *A* allele, and the copied region contained the *a* allele. Then the DNA molecule from which this information was removed, when restored, would have undergone "gene conversion" to become an *a* allele. Of course, gene conversion can occur in either direction, so that *a* may be converted into *A* instead.

Gene conversion can also arise in regions of the DNA duplex where the two strands of a DNA duplex have received genetic information from different parents. These are indicated in Figure 13 by the chromatids in the bottom row with a black region in one DNA strand opposite a white region in the other strand. Such regions are referred to as heteroduplex regions. In such a region, one strand may carry the genetic information for the *A* allele and the other strand may carry the information for the *a* allele. If these alleles are encoded by a different DNA base pair (e.g., AT for the *A* allele, GC for the *a* allele) then in the heteroduplex region a mismatch would exist (e.g., AC). Enzymes are thought to correct such mispairs. Thus, an AC mispair would be corrected to either AT (allele *A*) or GC (allele *a*). This process of mispair correction in the heteroduplex regions can also give rise to gene conversion. In the double-strand break repair model, as shown in Figure 13, heteroduplex regions are generated on either side of the initial break, so this manner of gene conversion is also consistent with an overall recombinational repair function.

The model in Figure 13 also explains why gene conversion is frequently associated with exchange of flanking regions. In two pathways, G and H in the figure, flanking regions are not exchanged. In the other two pathways, I and J, they are exchanged. Whether flanking region exchange occurs or does not occur depends on how the two Holliday junctions on either side of the gene conversion event are resolved. If the two Holliday junctions are resolved in the same sense, as in G and H, there is no exchange of outside regions. If they are resolved in the opposite sense, as in I and J, then flanking regions will be exchanged. In principle, each Holliday junction should be resolved in either of its two possible configurations with equal likelihood (Meselson and Radding, 1975). Therefore, the four end products of pathways G, H, I, and J should occur with equal frequency. If this were the general rule, gene conversion events should be associated with exchange of flanking regions 50% of the time. Instead, flanking regions are observed to be exchanged 36% of the time. This implies that pathways G and H are somewhat more common than I and J.

It should be noted that most of the genetic variation generated by recombination arises from pathways I and J, because the flanking regions to the left and right of the recombination event are generally very long, containing many different genes. Relatively little genetic variation is generated by pathways G and H because the region involved in information transfer is relatively small, being about the size of one or a few genes.

In Chapter 1, Section III.B, we mentioned that the variation hypothesis of sex, as distinct from the repair hypothesis, postulates that the main adaptive function of sexual reproduction is the generation of genetic diversity. From this perspective, pathways I and J in Figure 13 should be favored over G and H, in contrast to what is actually observed. This bias away from the generation of variation is evidence against the variation hypothesis and, by implication, supports the repair hyphothesis. We will return to this issue in Chapter 15, Section II.B.3, when we discuss the variation hypothesis of sex.

C. Polarity of Gene Conversion

A gradient in gene conversion frequency from one end of a gene to the other, or over an extended region of the gene, is a common finding in fungi (Whitehouse, 1982: pp. 287–298). To explain this polarity, all recent models of recombination invoke preferential initiation of the recombination event at specific sites. Genetic evidence suggests that polarity of gene conversion results from the preferential initiation of recombination at fixed positions outside the gene. Polarity of gene conversion is consistent with the double-strand break repair model. If initiation occurs by a break at a fixed site, then polarity may reflect the variable size of the region of gap repair and heteroduplex DNA. Thus, a site at an increasing distance from the initiation site has a decreasing likelihood of falling within a gap or within heteroduplex DNA. Nicolas et al. (1989) identified a specific initiation site for meiotic gene conversion in the promoter region of the ARG4 gene of the yeast S. cerevisiae. They also found that the chromosome on which initiation occurs is the recipient during gene conversion, as would be expected by the double-strand break repair model. Sun *et al.* (1989) showed further that a double-strand break appears at the *ARG4* recombination initiation site at the time of meiotic recombination and that the broken DNA molecules end in single-stranded tails several hundred nucleotides long. These data provide strong circumstantial support for the double-strand break repair model of recombination.

Sun *et al.* (1989) mapped break sites in or near three different promoters, which may be recombination initiation sites. In addition, Voelkel-Meimen *et al.* (1987) showed that a sequence that stimulates mitotic recombination, *HOT1*, is identical to the rDNA promoter. Because promoters are DNA sequences at which gene transcription begins, these results suggest a relationship between the start of transcription and the occurrence of double-strand breaks leading to recombination. Voelkel-Meimen *et al.* (1987) concluded, on the basis of their observations, that transcription by RNA polymerase I stimulates genetic recombination.

When transcription is started from a promoter, the DNA becomes locally unwound and an RNA polymerase begins messenger RNA synthesis using one of the DNA strands as a template. These mechanical activities and the local single-strandedness of the DNA may make the promoter region particularly vulnerable to the occurrence of DNA-damaging events, which then initiate recombinational repair. This speculation is supported by the evidence cited in Chapter 3, Section I, that a number of DNA-damaging agents bind preferentially to transcriptionally active regions.

In recent experiments with *D. melanogaster*, Clark *et al.* (1988) found that recombination can be initiated at a large number of sites within a particular gene. If, as we hypothesize, recombination is a response to DNA damage, then this result implies that while damages may occur preferentially in vulnerable regions of the DNA, such as promoters, they occur elsewhere as well.

III. *Implications*

In Chapter 10 we reviewed evidence that recombinational repair in a range of species is effective in overcoming a variety of DNA damages, especially double-strand damages. In this chapter, we explained that the large amount of evidence on the mechanism of meiotic recombination is currently interpreted by authorities in this field in terms of a double-strand break repair model. We outlined some of the experimental evidence supporting this interpretation. We suggest that this model provides a general mechanism for repairing damages, particularly double-strand damages, at meiosis.

References

Bernstein, H., Hopf, F. A., and Michod, R. E. (1987). The molecular basis of the evolution of sex. *Adv. Genet.* **24**, 323–370.

Bernstein, H., Hopf, F. A., and Michod, R. E. (1988). Is meiotic recombination an adaptation for repairing DNA, producing genetic variation, or both? *In* "The Evolution of Sex: An Examination of Current Ideas" (B. Levin and R. Michod, eds.), pp. 139–160. Sinauer, New York.

Chovnick, A., Ballantyne, G. H., Baillie, D. L., and Holm, D. G. (1970). Gene conversion in higher organisms: Half-tetrad analysis of recombination within the *rosy* cistron of *Drosophila melanogaster*. *Genetics* **66**, 315–329.

Clark, S. H., Hilliker, A. J., and Chovnick, A. (1988). Recombination can initiate and terminate at a large number of sites within the *rosy* locus of *Drosophila melanogaster*. *Genetics* **118**, 261–266.

Hannan, M. A., Calkins, J., and Lassiwell, W. L. (1980). Recombinagenic and mutagenic effects of sunlamp (UV-B) irradiation in *Saccharomyces cerevisiae*. *Mol. Gen. Genet.* **177**, 577–580.

Ito, T., and Kobayashi, K. (1975). Studies on induction of mitotic gene conversion by ultraviolet irradiation II. Action spectra. *Mutat. Res.* **30**, 43–54.

Klar, A. J. S., and Miglio, L. M. (1986). Initiation of meiotic recombination by double chain breaks in *S. pombe*. *Cell* **46**, 725–731.

Kolodkin, A. L., Klar, A. J. S., and Stahl, F. W. (1986). Double-strand breaks can initiate meiotic recombination in *S. cerevisiae*. *Cell* **46**, 733–740.

Kushev, V. V. (1974). "Mechanisms of Genetic Recombination." Plenum, New York.

Meselson, M. S., and Radding, C. M. (1975). A general model for genetic recombination. *Proc. Natl. Acad. Sci. USA* **72**, 358–361.

Nickoloff, J. A., Singer, J. D., Hoekstra, M. F., and Heffron, F. (1989). Double-strand breaks stimulate alternative mechanisms of recombinational repair. *J. Mol. Biol.* **207**, 527–541.

Nicolas, A., Treco, D., Schultes, N. P., and Szostak, J. W. (1989). An initiation site for meiotic gene conversion in the yeast *Saccharomyces cerevisiae*. *Nature* **338**, 35–39.

Orr-Weaver, T. L., and Szostak, J. W. (1985). Fungal recombination. *Microbiol. Rev.* **49**, 33–58.

Orr-Weaver, T., Szostak, J., and Rothstein, R. (1981). Yeast transformation: A model system for the study of recombination. *Proc. Natl. Acad. Sci. USA* **78**, 6354–6358.

Raju, M. R., Gnanapurani, M., Stackler, B., Martins, B. I., Madhvanath, U., Howard, J., Lyman, J. T., and Mortimer, R. K. (1971). Induction of heteroallelic reversions and lethality in *Saccharomyces cerevisiae* exposed to radiations of various LET (^{60}Co gamma-rays, heavy ions and pi$^-$ mesons) in air and nitrogen atmosphere. *Radiat. Res.* **47**, 635–643.

Ray, A., Siddigi, I., Kolodkin, A. L., and Stahl, F. W. (1988). Intrachromosomal gene conversion induced by a DNA double-strand break in *Saccharomyces cerevisiae*. *J. Mol. Biol.* **201**, 247–260.

Stahl, F. W. (1979). "Genetic Recombination: Thinking about It in Phage and Fungi." Freeman, San Francisco.

Sun, H., Treco, D., Schultes, N. P., and Szostak, J. W. (1989). Double-strand breaks at an initiation site for meiotic gene conversion. *Nature* **338**, 87–90.

Szostak, J. W., Orr-Weaver, T. L., Rothstein, R. J., and Stahl, F. W. (1983). The double-strand break repair model for recombination. *Cell* **33**, 25–35.

Thaler, D. S., and Stahl, F. W. (1988). DNA double-chain breaks in recombination of phage lambda and of yeast. *Annu. Rev. Genet.* **22**, 169–197.

Voelkel-Meimen, K., Keil, R. L., and Roeder, G. S. (1987). Recombination-stimulating sequences in yeast ribosomal DNA correspond to sequences regulating transcription by RNA polymerase I. *Cell* **48**, 1071–1079.

Whitehouse, H. L. K. (1982). "Genetic Recombination." Wiley, New York.

Chapter Twelve

Meiosis Viewed as an Adaptation for DNA Repair

In Chapter 11, we pointed out that the data on the mechanism of meiotic recombination are consistent with this process being an adaptation for DNA repair. In this chapter, we review other kinds of evidence bearing on the function of meiosis. Although this evidence is heterogeneous, we show that it also is consistent with the view that meiosis is an adaptation to promote DNA repair. Some of the views presented in this chapter were also presented by Bernstein *et al.* (1987, 1988).

I. *DNA Repair during Meiosis*

A central feature of meiosis is the intimate pairing of homologous chromatids during which recombination takes place. In this section, we review available evidence bearing on the supposition that meiotic recombination reflects recombinational repair. During recombinational repair, presumably by the double-strand break repair model, a type of DNA synthesis occurs. This is referred to as DNA repair synthesis. This type of DNA synthesis should, by our supposition, be experimentally detectable during the stage of meiosis when DNA pairing and recombination occur. The zygotene–pachytene stage of meiotic prophase is when pairing and recombination occur during spermatogenesis (Peacock, 1970). Therefore, DNA repair synthesis should be evident in spermatogenesis during the zygotene–pachytene stage if meiotic recombination reflects DNA repair.

A. Spermatogenesis

Kofman-Alfaro and Chandley (1971) measured repair synthesis in spermatogenic cells of the mouse following irradiation with UV or X-rays. Among unirradiated spermatocytes they found a small amount of DNA synthesis at the zygotene–pachytene stage. After UV- or X-irradiation, DNA repair synthesis was maximal in late zygotene–early pachytene. Similar results were obtained by Coogan and Rosenblum (1988), as discussed in Chapter 10, Section III.C. The authors concluded that the large amount of radiation-induced synthesis at zygotene–pachytene may reflect a recombinational repair process that is normally operative at a much reduced level in unirradiated "control" cells at this time. They also noted that the low amount of DNA repair synthesis occurring at the other meiotic stages may be more comparable with that observed in somatic cells. Chandley and Kofman-Alfaro (1971) performed experiments using human spermatogenic cells, which were similar to those described earlier using the mouse. The results obtained were consistent with, and therefore confirmed, the basic findings using the mouse.

Hotta and Stern (1971) analyzed the DNA synthesis occurring during meiotic prophase in microsporocytes (pollen mother cells) of *Lilium*. They concluded that the DNA synthesized during pachytene has the characteristics of DNA repair replication, and that chiasma formation (recombination) is temporally associated with DNA repair synthesis in *Lilium*. Similarly, Stubbs and Stern (1986) presented evidence that in unirradiated mouse spermatocytes there is recombination-associated repair during pachytene.

Howell and Stern (1971) reported the appearance of three enzymatic activities during the zygotene–pachytene stages of meiosis in lily. These enzymes were a DNA endonuclease, a polynucleotide kinase, and a polynucleotide ligase. All of these enzymes may reasonably be expected to be involved in the breakage and repair steps of recombination, and the authors proposed that these activities compose a breakage and reunion mechanism responsible for genetic recombination at meiosis.

Hotta *et al.* (1985) identified two proteins, found in both mouse and lily, with enzymatic properties similar to the RecA protein of *E. coli* (see Chapter 10, Section II.A, for a discussion of the role of the *E. coli* RecA protein in recombinational repair). These RecA-like proteins were purified from somatic and meiotic cells and were therefore designated "s-rec" and "m-rec" to indicate their respective tissues of origin. The m-rec protein was isolated from both mouse spermatocytes and lily microsporocytes. Hotta *et al.* (1985) showed that the m-rec protein is most active during the late zygotene and pachytene stages in lily and during the less precisely defined prophase stages in mouse spermatocytes. The authors suggested that if the activity of the m-rec protein is as necessary for recombination in eukaryotes as the RecA protein is for recombination in prokaryotes, the meiocytes may be poised for recombinogenic activity during the prophase interval. Since the RecA protein of *E. coli* has been shown to be necessary for recombinational repair (Chapter 10, Sections II.A–D), the amplification of the m-rec protein during meiosis is consistent with our hypothesis that meiosis is an adaptation for repair.

Orlando *et al.* (1988) presented evidence that DNA synthesis in mouse spermatogenesis involves DNA polymerase-beta. Mammalian DNA polymerase beta is regarded as a DNA polymerase specific for repair (Kornberg, 1982: p. S46). DNA polymerase-alpha, the enzyme responsible for chromosomal DNA replication, is present in premeiotic cells but is shut off in meiotic and postmeiotic mouse spermatocytes. These results indicate that DNA polymerase-beta activity, alone, is involved in DNA synthesis in pachytene spermatocytes and, thus, the type of DNA synthesis occurring during meiosis is DNA repair synthesis.

In summary, the evidence discussed in this section supports the hypothesis that meiotic recombination during spermatogenesis is a DNA repair process. This evidence includes the following findings: (1) In mouse and human, DNA repair synthesis subsequent to DNA-damaging treatment is maximal during the zygotene–pachytene stage of meiosis when recombination occurs. (2) In lily, during zygotene–pachytene, recombination is associated with DNA replication that has the characteristics of repair synthesis. (3) A protein similar to the RecA protein of *E. coli,* which has a key role in recombinational repair in *E. coli,* is found in mouse and lily. It is most active during the zygotene–pachytene period of meiosis when recombination occurs. (4) The mammalian DNA polymerase identified with DNA repair synthesis, rather than the one identified with chromosomal replication, is the only one active during meiotic recombination in the mouse.

B. Oogenesis

As summarized by Swanson *et al.* (1967: p. 53), by the fifth month of prenatal life human oocytes have formed and have reached diplonema. Diplonema is a stage of meiosis following pachytene in which the four chromatids are evident and chromatid pairing is apparent. Following diplonema, chromatids become much more diffuse (less compactly coiled), the state they will remain in for 12–50 yr. This long period is referred to as the dictyotene, or dictyate, stage. The dictyate stage is succeeded by the final stages of meiosis and ovulation. This sequence of chromatid behavior also occurs in oocytes of other vertebrates, including mice, amphibians, and some fishes, as well as the spermatocytes of some insects. In amphibians, the diffusion of the chromatids is associated with the production of RNA and protein and an increase in the mass of the cytoplasm. These features are generally observed in eggs that go through a long period of development to store reserve material in the form of yolk.

Mature mouse oocytes at the dictyate stage have a high capacity for repairing the DNA damage induced by UV and other agents (Masui and Pederson, 1975; Brazill and Masui, 1978). Immature oocytes in the dictyate stage, from 2- to 3-day-old mice, can undergo repair of UV-induced damage as well (Pederson and Mangia, 1978). Guli and Smyth (1988) also reported that UV-irradiated dictyate cells of the mouse had increased DNA repair. They were unable to detect repair of UV-induced damages in the preceding stages of meiosis: leptotene, zygotene, or pachytene. Guli and Smyth concluded that, in the mouse, the pattern of DNA repair capacity during oogenesis is very different from that during spermatogenesis, suggesting that the sexes have evolved different ways of ensuring the integrity of their genetic DNA. They proposed that, in the male, a rejuvenatory repair step during recombination at pachytene may well be a significant factor in correcting accumulated damage. However, they noted that their study suggests that this step is lacking in the female and the same outcome may be achieved by the high capacity for repair exhibited by the long-lived dictyate oocyte. Bernstein (1979) has suggested that DNA damages that arise during this long period may be, in part, removed by mechanisms dependent on chromosome pairing. We now speculate that the apparent delay in recombinational repair from the zygotene–pachytene period to the dictyate stage may be an adaptation to allow the oocyte to repair DNA damages occurring during the long period in which reserve material is prepared and stored.

In the fruitfly *Drosophila*, meiotic recombination is limited to oogenesis and does not occur in spermatogenesis. Baker *et al.* (1976) presented evidence that at least two genes of *Drosophila* have a role in both meiotic recombination and DNA repair (see also Chapter 10, Section III.B). For instance, mutants defective in one of the more well studied of these genes, *mei-41*, has a decreased frequency of meiotic recombination as well as increased sensitivity to a variety of DNA-damaging agents. These *Drosophila* mutants are also defective in a form of postreplication repair (Boyd and Setlow, 1976) that, by analogy with bacteria (Rupp *et al.*, 1971), may involve recombination. Although the defect in meiotic recombination was only expressed in females, the DNA repair defects occurred in the somatic cells of both males and females. These results link meiotic recombination and DNA repair as joint functions of individual genes and thus support the hypothesis that meiotic recombination is a repair process.

In summary, the evidence discussed in this section, although not extensive, is consistent with the idea that meiosis during oogenesis is an adaptation for DNA repair. This idea is based on these findings: (1) Vertebrate oocytes have a high capacity for DNA repair during the dictyate stage of meiosis, a stage when recombination may also occur. (2) In the fruitfly *Drosophila*, mutants defective in meiotic recombination during oogenesis are also defective in DNA repair.

II. *Premeiotic Replication*

The DNA replication that precedes the prophase stage of meiosis is conventionally referred to as "premeiotic replication." Despite this convention, this DNA replication can be thought of as the initial step of the meiotic process, rather than as a step preceding it.

Because premeiotic replication seems to be a universal feature of meiosis, it is reasonable to ask what adaptive function it serves. Bernstein *et al.* (1988) have argued that premeiotic replication could generate a gap opposite a damage, and this may aid in the recognition and recombinational repair of single-strand damages. This viewpoint is discussed in Section A below. Bernstein *et al.* have also contended that if, as argued by others (see Chapter 15, Section II.B), the function of meiosis were to generate genetic diversity, then premeiotic replication would be wasteful. This viewpoint is discussed in Section B below.

A. Recognition of Single-Strand Damages

As discussed in Chapter 10, Section II.C, evidence in *E. coli* indicates that replication of DNA containing pyrimidine dimers leads to gaps in the new strands opposite the damages (Rupp and Howard-Flanders, 1968). These gaps promote recombinational repair (West *et al.*, 1982). Movement of the DNA polymerase along the template strand is likely to be blocked by any unrepaired bulky DNA damage. Therefore, gaps may be produced opposite other types of bulky damages as well as pyrimidine dimers. Such gaps have a distinctive and constant molecular structure, which is independent of the structure of the original damage and therefore may serve as a universal initiator of recombinational repair. The next step in initiation of recombinational repair might involve excision of the damage opposite the single-strand gap (by an enzyme such as the nuclease controlled by the *RAD52* gene in yeast [Resnick *et al.*, 1984]). This would produce a double-strand gap. Such gaps could start the mechanism of double-strand break repair described in Chapter 11, Section I.B.

Structural variation among DNA damages may limit the efficiency of excision repair mechanisms. Because naturally occurring DNA damages are a mixture of many types, a large proportion of damage types are probably present in low frequency. A large number of infrequent damage types makes it improbable that excision repair enzymes might have evolved that recognize all specific types of DNA damage, especially those types represented in low frequency. Another difficulty with evolving enzymes specific for each type of damage arises from the fact that DNA normally undergoes transient configurational variations associated with gene expression and these variations may also be of many types. Such variations, again, may impede the recognition of specific DNA damages. Recombinational repair mechanisms, however, may be able to distinguish DNA damage from transient configurational variations in DNA structure by the following device. As the DNA polymerase moves along the DNA template during replication and the duplex unwinds, such transient structural variations should be released and damages should become readily distinguishable as impediments to polymerase movement. The various single-strand damages that block polymerase movement could be translated into a "common currency" when the polymerase produces a gap opposite the damage in the new strand. This gap could then initiate recombinational repair via the double-strand break repair model.

The mechanism described above for converting single-strand damages into a "common currency" signaling the start of double-strand break repair would avoid the alternative of evolving excision repair enzymes to distinguish each of the wide variety of DNA damages from the normal variations in DNA structure. Our above assumption that excision repair is not highly efficient in removing single-strand damages is supported by evidence in bacteriophages. One of the most well-studied excision repair systems, pyrimidine-dimer excision repair in phage T4, has been shown by Pawl et al. (1976) to be only 45–60% efficient. Furthermore, evidence reviewed by Bernstein (1983) shows that recombinational repair is able to overcome a wide variety of lethal damages to DNA, as is expected by the view that recombinational repair is triggered by a general signal (which could be a gap opposite any damage). In conclusion, premeiotic replication may serve the useful function of generating single-strand gaps opposite damages, thus converting difficult-to-recognize single-strand damages into a standard signal for the start of recombinational repair. Premeiotic replication may thus put DNA into a form where it could easily be repaired by recombinational repair when chromosomes are brought into juxtaposition at meiosis.

B. Premeiotic Replication Makes Sister-Chromatid Exchanges Possible

Premeiotic replication generates two copies of each homologous chromosome in the cell. The two daughter chromatids that are formed upon replication can also be referred to as sister chromatids. Recombination between two daughter–sister chromatids is referred to as sister-chromatid exchange (SCE). Because of premeiotic replication, SCE becomes possible during meiosis. That premeiotic replication allows formation of SCEs has implications for the adaptive function of meiosis. We first ask if there is a bias against SCEs during the four-chromatid stage of meiosis when exchange occurs both between nonsister chromatids and between sister chromatids. By comparing, in males, the levels of SCE for the X chromosome and the autosomes, it is possible to determine if there is a preference for non-SCE when both types of recombination are possible (the X chromosome has only a sister homologue, whereas autosomes have both sister and nonsister homologues). As reported by Peacock (1970), there is no bias against SCEs when non-SCEs are possible, because SCEs occur in the X chromosome, where they are as frequent per unit chromosome length as in the autosomes.

SCEs cannot generate genetic variation because the sister chromatids are replicas of each other. It has been proposed by others (Chapter 1, Section III.B, and Chapter 15, Section II.B) that the main adaptive advantage of sex is the production of genetic variation. If meiotic recombination is an adaptation to promote variation then premeiotic replication should not be necessary and, indeed, would be counterproductive because premeiotic replication promotes SCE. There is no adaptive advantage for SCE by the variation hypothesis. On the other hand, premeiotic replication may promote the nonspecific recognition of DNA damages and therefore is advantageous if meiotic recombination is a mechanism of DNA repair. Thus, the generality of premeiotic replication argues against the variation hypothesis (for further discussion, see Chapter 15, Section II.B).

III. Role of Mitotic Recombination Compared With Meiotic Recombination

In this section, we present evidence that mitotic recombination may serve a similar repair function to meiotic recombination, but that in meiosis this function is carried out more efficiently than in mitosis.

A. Mitotic Recombination May Be a Repair Process

As pointed out in the introduction to Chapter 10, two observations (1 and 2, below) when present together indicate an ability to undergo recombinational repair, and two further observations (3 and 4, below) suggest the presence of recombinational repair. (1) Resistance to a DNA-damaging agent is considerably greater when two chromosomes are present in a cell than when just one is present. (2) This resistance is diminished by mutations in genes required for recombination. (3) DNA-damaging agents increase recombination. (4) Mutants defective in a repair pathway other than recombinational repair have increased levels of recombination. Each of these observations have been made with mitotic cells. Taken together, the observations suggest that mitotic recombination is a recombinational repair process. Each type of observation is summarized briefly below.

First, we consider whether or not resistance to a DNA-damaging agent is greater when two sets of chromosomes are present in a cell (diploid), rather than when only one set of chromosomes is present (haploid). One type of mitotic cell that can be studied in both haploid form and diploid form is a yeast cell. As summarized by Game (1983), numerous workers have shown that in the yeast *S. cerevisiae,* diploid cells are more resistant to UV light than haploid cells. This resistance is generally thought to arise from repair systems that operate only when two homologous copies of DNA are present.

Second, we consider whether or not mutations in genes required for recombination also cause sensitivity to DNA-damaging agents in mitotic cells. This consideration has two parts. First, there must be genes that are known to be required for mitotic recombination. Second, mutants defective in these genes must be sensitive to DNA-damaging agents. Observations along these two lines have been made both in yeast and in mammalian cells.

A recent review of DNA repair in the yeast *S. cerevisiae* (Friedberg, 1988) describes the genes considered to be involved in mitotic recombinational repair. This review indicates that genes *RAD51, RAD52,* and *RAD54* are essential for repair of ionizing radiation-induced DNA damage and are essential for meiotic and mitotic recombination. In addition, *RAD50, RAD53, RAD55, RAD56,* and *RAD57* are required for repair of ionizing radiation damage as well as recombination, but a mutation in any one of these latter genes does not cause as large a defect in repair or recombination as a mutation in a gene of the first group. A mutation in *RAD9* also almost completely eliminates spontaneous or UV-induced intergenic and intragenic mitotic recombination, while causing sensitivity to UV-radiation (Kowalski and Laskowski, 1975).

A mutant line of mammalian Chinese hamster ovary (CHO) cells, EM9, has been extensively studied. As summarized by Hoy *et*

al. (1987) and Thacker (1989), mutant EM9 cells have been characterized as having enhanced sensitivity to some, but not all, alkylating
agents, significant sensitivity to X-irradiation, and hampered ability
to rejoin DNA strand breaks. Both Hoy *et al.* (1987) and Thacker
(1989) showed that EM9 was only about 37% as efficient in carrying
out homologous recombination as wild-type CHO cells. Thus, mutant EM9 meets our criteria for being both sensitive to DNA damaging agents and deficient in recombination.

Third, we consider whether or not DNA-damaging agents increase recombination in mitotic cells. In Chapter 10, we reviewed
work indicating that many DNA-damaging agents increase recombination in mitotic cells of yeast and *Ustilago maydis.* In particular,
UV light, X-rays, and methyl methanesulfonate stimulate mitotic
recombination in yeast (Chapter 10, Section III.A.3), and UV and
mitomycin C stimulate mitotic recombination in *U. maydis* (Chapter
10, Section III.A.6). Furthermore, mitomycin C has been shown to
increase mitotic recombination as measured by SCEs in hamster
cells (Perry and Evans, 1975) and mouse spermatagonia (Allen and
Latt, 1976), and mitotic crossing-over in the fungi *U. maydis* and *S.
cerevisiae* (Holliday, 1964).

DNA-damaging agents were shown to stimulate recombination
in mitotic cells of the fruit fly *Drosophila.* Clements and Vogel (1988)
showed that alkylating agents (methyl methanesulfonate and ethyl
nitrosourea) and mutagenic antibiotics (adriamycin hydrochloride
and daunomycin hydrochloride) all increased twin-spots, a measure
of somatic recombination, in *Drosophila.* X-rays have also been
shown to stimulate mitotic recombination in *Drosophila* (Abbadessa
and Burdick, 1963).

Finally, we consider whether or not mitotic cells that are defective in a DNA repair pathway other than recombinational repair have
increased levels of recombination. Mutations in the *RAD6* and
RAD18 genes (involved in undefined DNA repair pathway[s]; see
Burtscher *et al.* [1988] and Friedberg [1988] for reviews) result in
increased levels of mitotic recombination (Boram and Roman, 1976;
Kern and Zimmermann, 1978). Similarly, a human xeroderma pigmentosum cell line, defective in excision repair, showed a fourfold
higher rate of spontaneous recombination than cells that were wild-
type for excision repair (Bhattacharyya *et al.*, 1990).

Overall, the results summarized in this section are consistent with the idea that mitotic recombination, like meiotic recombination, is a recombinational repair process. Indeed, Whitehouse (1982: p. 341) has concluded that with present evidence it is reasonable to suppose a common mechanism for recombination between sister chromatids and between homologous chromosomes, both at mitosis and meiosis.

B. Frequencies of Recombination Events in Meiosis and Mitosis

We will next show that (1) recombination events are more frequent in a single meiosis than in a single mitosis, as would be expected on the hypothesis that meiosis is specifically an adaptation for recombinational repair, and (2) when recombination events are summed for all somatic cells, the total amount of recombination is far higher in mitotic than meiotic cells, a conclusion compatible with the repair hypothesis but difficult to explain on the variation hypothesis (Chapter 1, Section II, and Chapter 15, Section II.B).

We first estimate the average number of recombination events occurring during a single meiosis or mitosis as well as the total number of such events occurring during all meioses and mitoses in a single human lifetime. We have chosen to carry out these calculations for women because an estimate of the total lifetime production of oocytes is available (Baker, 1963).

The number of recombination events during meiosis can be approximated by the number of chiasmata (cytologically visible exchanges presumed to reflect recombination events; see Swanson *et al.*, 1967: p. 67) per oocyte during meiosis. The average number of chiasmata per bivalent is 1.89 (Jagiello *et al.*, 1976) and there are 23 bivalents in humans. Therefore, the number of recombination events per meiosis is about 23×1.89, or 43. The number of oocytes present in a female at birth is approximately 2×10^6 (Baker, 1963). Therefore, the total number of recombination events in a woman's oocyte population is approximately $43 \times 2 \times 10^6$, or 8.6×10^7.

In human somatic cells, the estimated frequency of SCEs is 0.12 per chromatid at each cell cycle (Brewen and Peacock, 1969). We consider only SCEs because the frequency of nonsister chromatid exchanges is very small in somatic cells. Because humans have 46 chromosomes, the frequency of recombination per somatic cell duplication cycle is approximately 0.12×46, or about 5.5. To estimate the total number of recombination events in somatic cells, we need to know the average number of somatic cells in a woman at full development, the average number of times a somatic cell turns over, and the average number of recombination events per cell at each generation. We estimate the number of cells in the body to be 6×10^{12}. The average number of times a cell turns over in a lifetime is difficult to estimate because some cells do not turn over at all, and some turn over very rapidly. The average number of turnovers per cell in a human lifetime is certainly greater than one. The total number of recombination events in somatic cells during a woman's lifetime is therefore greater than $5.5 \times 6 \times 10^{12}$, or 3.3×10^{13}. Thus, the proportion of total recombination events that occur during meiosis in a woman's lifetime compared with those that occur during mitosis is less than $8.7 \times 10^{7}/3.3 \times 10^{13}$, or 2.6×10^{-6}.

In this one case, in which a rough calculation is possible, meiotic recombination events clearly comprise an exceedingly small fraction of the total recombination events (both meiotic and mitotic) that occur in all cells in one human female throughout a lifetime. On the other hand, each meiotic cell experiences more recombination events on average than a somatic cell (approximately 43 vs. 5.5, or a ratio of about 8 : 1). If the function of recombination is repair, these calculations suggest that recombinational repair is significant for cells generally, but that it is especially important in the production of germ cells. As pointed out by Bernstein *et al.* (1988), if the function of recombination were to promote genetic variation, then the vast majority of recombination events, those occurring in somatic cells, would be wasteful because they produce no genetic variation. Thus, allelic variation is unlikely to be the general adaptive function of recombination (for further discussion, see Chapter 15, Section II.B).

C. Meiosis Occurs at a Stage When Unrepaired Damages are Extremely Costly

Why is meiosis limited to the cell cycle preceding gamete formation? We consider that unrepaired DNA damages in gametes have a unique cost that justifies the investment in producing them through meiosis rather than mitosis. A parent contributes only a single gamete to each progeny, and this cell must be as free as possible of

DNA damage. No other type of cell in a multicellular organism so directly determines reproductive success as the gamete. In general, we think, this simple fact explains the position in the sexual cycle of meiosis, a uniquely powerful repair process.

Another factor that may determine the position of meiosis prior to gamete formation is the common strategy, among multicellular organisms, of provisioning their eggs with metabolic products to be used by the developing embryo. Such provisioning involves considerable metabolic activity, which produces oxidative free radicals as by-products. These, in turn, would cause DNA damage. Thus, the common strategy of provisioning the egg may increase DNA damage and as a result increase the need for DNA repair. Consequently, there is a special adaptive value of meiotic DNA repair during oogenesis.

The recombinational repair occurring during meiosis, as well as other repair and protective processes, may be just barely able to cope with DNA damage. Austin (1972) noted that, in humans, about half of all fertilized eggs fail to produce surviving embryos, as is typical of mammals generally. Using a statistical model, Roberts and Lowe (1975) postulated an even greater loss of all human conceptions, 78%, most of them occurring before the first missed period. Recently, Wilcox et al. (1988) concluded that, in humans, 31% of embryos miscarry after implantation, often before the woman realizes she is pregnant. In plants, the failure of fertilized ovules to mature as seeds is common (Bawa and Webb, 1984). Although the causes of these failures in mammals and plants have not been determined, evidence indicates that DNA damage is a significant factor.

Recently Boue et al. (1985) concluded that cytogenetic studies on spontaneous and induced abortions and on perinatal deaths have shown that a great frequency and variety of chromosome abnormalities are present in these conceptions. We consider that these chromosome abnormalities are likely due to DNA damages, probably present in the egg or sperm, although some damages may have occurred in later embryonic cell divisions.

D. Recombination between Nonsister Homologues is Rare in Mitosis but Frequent in Meiosis

In the somatic cells of higher eukaryotes such as humans, nonsister chromosomes do not ordinarily pair and, therefore, do not ordinarily recombine. Occasional recombination between nonsister homologous chromosomes in somatic cells is inferred from the appearance of twin spots. Twin spots are adjacent patches of somatic tissue with genetic constitutions different from each other and from the parental tissue, that can be explained by a single recombination event in a somatic cell. Although conspicuous under certain circumstances (e.g., in individuals with Bloom syndrome; see Festa *et al.*, 1979), twin spots are ordinarily rare in humans. On the other hand, pairing of nonsister homologous chromosomes (chromatids) and recombination between them is a central feature of meiosis. This process, in principle, can allow repair of double-strand damages that were present in a chromosome prior to meiotic replication, because information lost in one chromosome can be provided by the other through recombination. Thus, a key feature of meiosis, pairing of nonsister homologous chromatids, may allow the recombinational repair of a class of potentially lethal damages that are not repairable in any other way.

E. Coping with DNA Damage during the Mitotic Divisions of the Germ Line Prior to Meiosis, and the Need for Special Repair during Meiosis

Double-strand damages occurring in a cell population of a multicellular organism can be eliminated in one of two ways (Bernstein *et al.*, 1985a, 1985b). The damaged cells can undergo recombinational repair or else cellular selection can occur. Selection against damaged cells occurs when cells try to replicate. Then cells without lethal damages usually succeed in replicating and cells with unrepaired lethal damages die or, at least, do not replicate. Cellular selection can only occur when undamaged cells that can replicate are present. The presence of undamaged cells is more likely when damage rates are low and/or when the pool of cells is large. Furthermore, cellular selection is more effective when cell division rates are high. For example, in the case of the frequently dividing spermatogonia in animals, cells with unrepaired lethal DNA damages may simply die and be effectively replaced by the duplication of undamaged ones.

We assume that mitotically dividing germline cells in multicellular organisms are able to undergo the same kinds of repair processes as somatic cells. In particular, double-strand damages arising in mitotic cells after DNA replication but before cell division may be dealt with by recombinational repair between sister chromatids. Given these mechanisms for coping in mitotic cells, why is there the need for meiotic recombination during the final division of the germ line when there was no need prior to it?

We consider that the answer to this question lies in the fact that a parent contributes only a single haploid gamete to each of its offspring. This gamete must be free of all double-strand damages, because even one double-strand damage contributed by either gamete will kill the fertilized egg (zygote) when it attempts to divide. How can a parent ensure that the one cell it uses to make an offspring is free of all double-strand damages? This is an especially critical problem in light of the metabolic activity and, hence, likely high rate of damage, that occurs during the production of gametes, especially eggs. During the mitotic division of cells in the germ line, the ability to successfully replicate would certify that a cell was free of double-strand damages. However, if such a cell had unrepaired single-strand damages, DNA replication would tend to leave gaps opposite the damage. Thus, new double-strand damages would arise in the daughter cells at positions where only single-strand damages had existed in the parent. As discussed in Section II.A of this chapter, there are reasons to expect that excision repair may not be very efficient at repairing all common single-strand damages of a given type nor various less common single-strand damages. Therefore, double-strand damages arising as postreplicative gaps opposite single-strand damages might be a serious problem. Hence, although successful replication guarantees that the parent cell was free of double-strand damages, it does not ensure that the daughter cells are free of them. The only way of ensuring this is to produce daughter cells by a process that can efficiently repair all damages. This process, we think, is meiotic recombinational repair.

IV. Organisms Not Using Meiosis or Having Reduced Meiotic Recombination in Males

A. Vegetative Reproduction

Some organisms apparently are able to cope with DNA damage without meiosis. In the previous section, we mentioned cellular selection as a factor in coping with lethal DNA damages. A population of cells could survive so long as there are resources to support division of undamaged cells and the average number of double-strand damages per genome per replication cycle (m) is sufficiently small so that most cells will be without double-strand damage. When m is small, cellular selection can allow effective vegetative reproduction. This occurs in many plants where, during vegetative reproduction, many cells are passed on to an offspring. Some common examples of vegetative reproduction are reproduction of bermuda grass or strawberries by runners.

B. Parthenogenesis by Endomitosis

White (1973) estimated that less than a thousand parthenogenic "species" have been described in the animal kingdom, compared with over a million sexual species; i.e., there are over a thousand sexual genetic systems among animals to every thelytokous one. Thelytoky is the production of females from unfertilized eggs. The relatively low frequency of parthenogenesis in animals suggests that it is not a generally successful strategy.

A significant number of the most successful and conspicuous parthenogens in animals undergo two sequential premeiotic chromosome replications, followed by an apparently normal meiosis to produce diploid eggs (Schultz, 1971; White, 1973: chap. 19). The extra duplication is referred to as an "endomitotic" doubling. Pairing at meiosis is presumably between sister homologues. Species that use this mode of parthenogenesis include a common lizard in the southwestern United States (*Cnemidophorus*), reptiles (*Ambystoma*), fish (*Poeciliopsis*), grasshopper (*Moraba*), earthworms, planarians, and spider beetles (*Ptinus*). At the four-chromosome (chromatid) stage of meiosis, pairing is between chromatids derived from only one initial chromosome (Cuellar, 1971; White, 1973; Cole, 1984). Because there is no recombination between nonsister homologues, the maternal genome should be passed on intact to daughters. If recombination between nonsister homologues were allowed, homozygosity would increase and accumulated deleterious recessive alleles would express, leading to a decline in fitness among daughters. Because recombination between nonsister homologues presumably does not occur in these species, parthenogenesis by endomitosis might seem to be an ideal strategy; i.e., it reaps the benefit of meiotic repair while avoiding the expression of deleterious recessive alleles.

The question then arises: Why does this method of reproduction not predominate over sexual reproduction? The answer may be that in this form of parthenogenesis double-strand damages occurring before the first premeiotic replication cannot be repaired, as all pairing partners are derived from the same chromosome and there is no intact template corresponding to the damaged site. This problem does not arise in conventional meiosis because recombination can occur between nonsister chromatids (Chapter 11, Section I). Therefore, if double-strand damages are common before premeiotic replication, parthenogenesis by endomitosis is an unsatisfactory option.

C. Differences in Recombination between Females and Males

In general, in a sexual species, both the females and males undergo meiotic recombination, although often at different rates. That is, females tend to have higher rates than males (Bell, 1982: chap. 5; Trivers, 1988). *Drosophila* males, which rarely undergo allelic recombination or gene conversion (Chovnick *et al.*, 1970), are an extreme example of this trend. However, meiotic recombination in *Drosophila* males can be induced by X-rays (Hannah-Alava, 1965), suggesting that recombinational repair occurs when the level of DNA damage is high. The general tendency of females to have higher spontaneous frequencies of recombination can be explained on the basis of the higher metabolic activity, and hence higher levels of DNA-damaging oxidative-free radicals produced during oogenesis, as well as the longer time of exposure of the DNA in the oocyte. However, in contrast to the general tendency, females of some species such as silkworm moths (Sturtevant, 1915; Tazima, 1964: p. 47) do not undergo meiotic recombination, whereas males from these species do. We cannot explain these exceptional cases.

White and Lalouel (1988) have noted in humans that the ratio of recombination in women to that in men varies from chromosome to chromosome and from region to region within the chromosome. On chromosome 12, for example, the overall rate of recombination seen when the chromosome was passed on by a woman was about 1.7-fold higher than when it was passed on by a man. Also, on chromosome 13, recombination frequencies are several times higher in women than in men. However, on chromosome 11 the opposite was true in one interval, and in an adjacent interval the two sexes showed similar recombination frequencies. The explanation for the local differences is unknown but might reflect differential gene expression in males and females prior to meiotic recombination. If genes undergoing expression are more vulnerable to DNA damage than nonexpressing genes (see Chapter 3, Section I), the distribution of damages, and therefore of recombinational repair events, would be expected to be different in males and females.

V. Immortality of the Germ Line: Aging of the Somatic Line

Weismann (1889) first proposed the theory that germ cells are potentially immortal, compared with an aging soma. He noted (p. 33) that all higher forms possess an undying succession of reproductive cells, although this may be but poor consolation to the conscious individual that perishes. In some organisms, such as insects and animals, germs line cells are set aside early in embryonic development, whereas in plants, discrimination occurs late in development. Potential immortality of the germ line appears to be a consequence of the rejuvenation that occurs at each generation of meiotically produced gametes. This suggests that the explanation for potential immortality lies in features of gametogenesis that are distinct from events in somatic cells. Medvedev (1981) has reviewed the range of genetic and biochemical factors relevant to the potential immortality of the germ line. According to Medvedev, the key events include the elimination or reversal of some DNA changes in germ cells through recombination. On the repair hypothesis of sex, meiotic recombinational repair during gamete formation is likely a major factor contributing to the potential immortality of the germ line.

VI. Summary

In Section I, we reviewed evidence that during the course of meiosis in spermatogenesis, DNA repair synthesis is maximal at the stages when recombination occurs. In contrast to spermatogenesis, oogenesis, in some species, has a characteristic very long delay during the dictyate stage of meiosis. During this period when the cells are vulnerable to the occurrence of DNA damage, repair synthesis associated with recombination occurs. Also, in *Drosophila* and yeast, genes that have a joint role in both meiotic recombination and repair of DNA damage have been identified. These findings support the hypothesis that meiotic recombination is a repair process.

In Section II, we considered the adaptive function of premeiotic replication, a universal feature of meiosis. We suggested that premeiotic replication may aid in the repair of single-strand damages by generating gaps opposite such damages, which could then initiate efficient recombinational repair. On the other hand, if we start with the hypothesis that meiosis is an adaptation for promoting variation rather than DNA repair, premeiotic replication is difficult to explain because it allows SCEs, a type of recombination that generates no variation.

In Section III, we reviewed evidence suggesting that mitotic recombination is a recombinational repair process. Next, we pointed out that recombination is more frequent in individual cells undergoing meiosis than in those undergoing mitosis, consistent with the hypothesis that meiosis is adapted to promote recombinational repair compared with mitosis. However, even in mitotically dividing cells recombinational repair between sister chromatids seems to be a significant process. We calculated that, in women, the total number of recombination events occurring in all somatic cells by SCEs is vastly greater than the number of meiotic recombination events occurring in all egg cells. Because the somatic recombination events (i.e., SCEs) do not produce genetic variation, such variation is unlikely to be the general adaptive function of recombination. In other words, why would a complex process such as recombination take place in somatic cells if it bestows no selective advantage?

We also noted that recombination between nonsister homologues should allow repair of double-strand damages that had occurred prior to DNA replication. Recombination of nonsister homologues is characteristic of meiosis, but not of mitosis. Again, this suggests that meiosis is a special adaptation to promote more effective DNA repair. Thus, we proposed that the reason meiosis occurs prior to germ cell formation is to ensure that these cells, which are critically important for reproductive success, are free of lethal DNA damages.

In Section IV, we noted that vegetative reproduction and parthenogenesis occur in some species, particularly among plants. These organisms may cope with DNA damage through cellular selection without the necessity of meiosis. Parthenogenesis is relatively rare in animals, suggesting that it is not a generally successful reproductive strategy in animals. One well-studied example of parthenogenesis involves an extra premeiotic replication, leading to formation of a diploid egg. We argued that this is not a predominant form of reproduction because it is unable to repair double-strand damages present prior to premeiotic replication. We also noted that recombination frequencies in the female of a species are generally higher than in the male. This may be due to the more active metabolism in oogenesis and consequent higher level of oxidative free radical production as well as the longer time of DNA exposure during oogenesis compared with spermatogenesis. This active metabolism occurs during a period in the meiotic cycle when recombinational repair can remove damages caused by this activity.

In Section V, we suggested that the potential immortality of the germ line compared with the aging of the soma is due, at least in part, to repair of DNA damages during meiosis.

In Chapter 10, we presented evidence that recombinational repair occurs in a wide range of species, is versatile in overcoming many types of damage, and, in particular, is efficient in removing double-strand damages. In Chapter 11, we explained that current understanding of the mechanism of meiotic recombination supports the hypothesis that it is an adaptation for DNA repair. In the present chapter, we reviewed additional features of meiosis that bear on its function and have concluded that, in general, these features either support, or are consistent with, the hypothesis that meiosis is an adaptation for DNA repair.

References

Abbadessa, R., and Burdick, A. B. (1963). The effect of X-irradiation on somatic crossing over in *Drosophila melanogaster*. *Genetics* **48,** 1345–1356.

Allen, J. W., and Latt, S. A. (1976). Analysis of sister chromatid exchange formation in vivo in mouse spermatogonia as a new test system for environmental mutagens. *Nature (London)* **260,** 449–451.

Austin, C. R. (1972). Pregnancy losses and birth defects. *In* "Reproduction in Mammals," Chapter 5 (C. R. Austin and R. V. Short, eds.). Cambridge University Press, London.

Baker, B. S., Boyd, J. B., Carpenter, A. T. C., Green, M. M., Nguyen, T. D., Ripoll, P., and Smith, D. P. (1976). Genetic controls of meiotic recombination and somatic DNA metabolism in *Drosophila melanogaster*. *Proc. Natl. Acad. Sci. USA* **73**, 4140–4144.

Baker, T. G. (1963). A quantitative and cytological study of germ cells in human ovaries. *Proc. Roy. Soc. London B* **158**, 417–433.

Bawa, K. S., and Webb, C. J. (1984). Flower, fruit and seed abortion in tropical forest trees. Implications for the evolution of paternal and maternal reproductive patterns. *Am. J. Bot.* **71**, 736–751.

Bell, G. (1982). "The Masterpiece of Nature: The Evolution and Genetics of Sexuality." University of California Press, Berkeley.

Bernstein, C. (1979). Why are babies young? Meiosis may prevent aging of the germ line. *Persp. Biol. Med.* **Summer 1979**, 539–544.

Bernstein, H. (1983). Recombinational repair may be an important function of sexual reproduction. *BioScience* **33**, 326–331.

Bernstein, H., Byerly, H. C., Hopf, F. A., and Michod, R. E. (1985a). The evolutionary role of recombinational repair and sex. *Int. Rev. Cytol.* **96**, 1–28.

Bernstein, H., Byerly, H. C., Hopf, F. A., and Michod, R. E. (1985b). DNA repair and complementation: The major factors in the origin and maintenance of sex. *In* "Origin and Evolution of Sex" (H. O. Halvorsen, ed.), pp. 29–45. Alan R. Liss, New York.

Bernstein, H., Hopf, F. A., and Michod, R. E. (1987). The molecular basis of the evolution of sex. *Adv. Genet.* **24**, 323–370.

Bernstein, H., Hopf, F. A., and Michod, R. E. (1988). Is meiotic recombination an adaptation for repairing DNA, producing genetic variation, or both? *In* "The Evolution of Sex: An Examination of Current Ideas" (B. Levin and R. Michod, eds.), pp. 139–160. Sinauer, New York.

Bhattacharyya, N. P., Maher, V. M., and McCormick, J. J. (1990). Intrachromosomal homologous recombination in human cells which differ in nucleotide excision repair capacity. *Mutat. Res.* **234**, 31–41.

Boram, W. R., and Roman, H. (1976). Recombination in *Saccharomyces cerevisiae:* A DNA repair mutation associated with elevated mitotic gene conversion. *Proc. Natl. Acad. Sci. USA* **73**, 2828–2832.

Boue, A., Boule, J., and Gropp, A. (1985). Cytogenetics of pregnancy wastage. *Adv. Hum. Gen.* **14**, 1–57.

Boyd, J. B., and Setlow, R. B. (1976). Characterization of postreplication repair in mutagen-sensitive strains of *Drosophila melanogaster*. *Genetics* **84**, 507–526.

Brazill, J. L., and Masui, Y. (1978). Changing levels of UV light and carcinogen-induced unscheduled DNA synthesis in mouse oocytes during meiotic maturation. *Exp. Cell Res.* **112**, 121–125.

Brewen, J. G., and Peacock, W. J. (1969). The effect of tritiated thymidine on sister-chromatid exchange in a ring chromosome. *Mutat. Res.* **7**, 433–440.

Burtscher, H. J., Cooper, A. J., and Couto, L. B. (1988). Cellular responses to DNA damage in the yeast *Saccharamyces cerevisiae*. *Mutat. Res.* **194**, 1–8.

Chandley, A. C., and Kofman-Alfaro, S. (1971). "Unscheduled" DNA synthesis in human germ cells following UV irradiation. *Exp. Cell Res.* **69**, 45–48.

Chovnick, A., Ballantyne, G. H., Baille, D. L., and Holm, D. G. (1970). Gene conversion in higher organisms: Half-tetrad analysis of recombination within the rosy cistron of *Drosophila melanogaster*. *Genetics* **66**, 315–329.

Clements, J., and Vogel, E. W. (1988). Somatic recombination and mutation assays in *Drosophila:* A comparison of the response of two different strains to four mutagens. *Mutat. Res.* **209**, 1–5.

Cole, C. J. (1984). Unisexual lizards. *Sci. Am.* **250**, 94–100.

Coogan, T. P., and Rosenblum, I. Y. (1988). DNA double-strand damage and repair following gamma-irradiation in isolated spermatogenic cells. *Mutat. Res.* **194**, 183–191.

Cuellar, O. (1971). Reproduction and the mechanism of meiotic restitution in the parthenogenetic lizard *Cnemidophorus uniparens*. *J. Morphol.* **133**, 139–165.

Festa, R. S., Meadows, A. T., and Boshes, R. A. (1979). Leukemia in a black child with Bloom's syndrome. *Cancer* **44**, 1507–1510.

Friedberg, E. C. (1988). Deoxyribonucleic acid repair in the yeast *Saccharomyces cerevisiae*. *Microbiol. Rev.* **52**, 70–102.

Game, J. C. (1983). Radiation-sensitive mutants and repair in yeast. *In* "Yeast Genetics: Fundamental and Applied Aspects," Chapter 4 (J. F. T. Spencer, D. M. Spencer, and A. R. W. Smith, eds.). Springer-Verlag, New York.

Guli, C. L., and Smyth, D. R. (1988). UV-induced DNA repair is not detectable in pre-dictyate oocytes of the mouse. *Mutat. Res.* **208**, 115–119.

Hannah-Alava, A. (1965). The premeiotic stages of spermatogenesis. *Adv. Genet.* **13**, 157–226.

Holliday, R. (1964). The induction of mitotic recombination by mitomycin C in *Ustilago* and *Saccharomyces*. *Genetics* **50**, 323–335.

Hotta, Y., and Stern, H. (1971). Analysis of DNA synthesis during meiotic prophase in *Lilium*. *J. Mol. Biol.* **55**, 337–355.

Hotta, Y., Tabata, S., Bouchard, R. A., Pinon, R., and Stern, H. (1985). General recombination mechanisms in extracts of meiotic cells. *Chromosoma* **93**, 140–151.

Howell, S. H., and Stern, H. (1971). The appearance of DNA breakage and repair activities in the synchronous meiotic cycle of *Lilium*. *J. Mol. Biol.* **55,** 357–378.

Hoy, C. A., Fuscoe, J. C., and Thompson, L. H. (1987). Recombination and ligation of transfected DNA in CHO mutant EM9, which has high levels of sister-chromatid exchange. *Mol. Cell. Biol.* **7,** 2007–2011.

Jagiello, G., Ducayen, M., Fang, J. S., and Graffeo, J. (1976). Cytological observations in mammalian oocytes. *Chrom. Today* **5,** 43–64.

Kern, R., and Zimmermann, F. K. (1978). The influence of defects in excision and error prone repair on spontaneous and induced mitotic recombination and mutation in *Saccharomyces cerevisiae*. *Mol. Gen. Genet.* **161,** 81–88.

Kofman-Alfaro, S., and Chandley, A. C. (1971). Radiation-initiated DNA synthesis in spermatogenic cells of the mouse. *Exp. Cell Res.* **69,** 33–44.

Kornberg, A. (1982). "DNA Replication (Supplement)." W. H. Freeman and Co., San Francisco.

Kowalski, S., and Laskowski, W. (1975). The effect of three *rad* genes on survival, inter- and intragenic mitotic recombination in *Saccharomyces*. *Molec. Gen. Genet.* **136,** 75–86.

Masui, Y., and Pederson, R. A. (1975). Ultraviolet light-induced unscheduled DNA synthesis in mouse oocytes during meiotic maturation. *Nature (London)* **257,** 705–706.

Medvedev, Z. A. (1981). On the immortality of the germ line: Genetic and biochemical mechanisms. A review. *Mech. Aging Dev.* **17,** 331–359.

Orlando, P., Geremia, R., Frusciante, C., Tedeschi, B., and Grippo, P. (1988). DNA repair synthesis in mouse spermatogenesis involves DNA polymerase beta activity. *Cell Differ.* **23,** 221–230.

Pawl, G., Taylor, R., Minton, K., and Friedberg, E. C. (1976). Enzymes involved in thymine dimer excision in bacteriophage T4-infected *Escherichia coli*. *J. Mol. Biol.* **108,** 99–109.

Peacock, W. J. (1970). Replication, recombination and chiasmata in *Goniaea australasiae* (Orthopetara:Acrididae). *Genetics* **65,** 593–617.

Pederson, R. A., and Mangia, F. (1978). Ultraviolet light-induced unscheduled DNA synthesis by resting and growing mouse oocytes. *Mutat. Res.* **49,** 425–429.

Perry, P., and Evans, H. J. (1975). Cytological detection of mutagencarcinogen exposure by sister chromatid exchange. *Nature (London)* **258,** 121–125.

Resnick, M. A., Chow, T., Nitiss, J., and Game, J. (1984). Changes in the chromosomal DNA of yeast during meiosis in repair mutants and the possible role of deoxyribonuclease. *Cold Spring Harbor Symp. Quant. Biol.* **49,** 639–649.

Roberts, C. J., and Lowe, C. R. (1975). Where have all the conceptions gone? *Lancet* **i,** 498–499.

Rupp, W. D., and Howard-Flanders, P. (1968). Discontinuities in DNA synthesized in an excision defective strain of *Escherichia coli* following ultraviolet irradiation. *J. Mol. Biol.* **31**, 291–304.

Rupp, W. D., Wilde, C. E., Reno, D. L., and Howard-Flanders, P. (1971). Exchanges between DNA strands in ultraviolet-irradiated *Escherichia coli*. *J. Mol. Biol.* **61**, 25–44.

Schewe, M. J., Suzuki, D. T., and Erasmus, U. (1971). The genetic effects of mitomycin C in *Drosophila melanogaster* II. Induced meiotic recombination. *Mutat. Res.* **12**, 269–279.

Schultz, R. J. (1971). Special adaptive problems associated with unisexual fishes. *Am. Zool.* **11**, 351–360.

Stubbs, L., and Stern, H. (1986). DNA synthesis at selective sites during pachytene in mouse spermatocytes. *Chromosoma* **93**, 529–536.

Sturtevant, A. H. (1915). No crossing over in the female of the silkworm moth. *Am. Natur.* **49**, 42–44.

Swanson, C. P., Merz, T., and Young, W. J. (1967). "Cytogenetics." Prentice-Hall, Englewood Cliffs, New Jersey.

Tazima, Y. (1964). "The genetics of the Silkworm." Prentice-Hall, Englewood Cliffs, New Jersey.

Thacker, J. (1989). The use of integrating DNA vectors to analyze the molecular defects in ionizing radiation-sensitive mutants of mammalian cells including ataxia telagiectasia. *Mutat. Res.* **220**, 187–204.

Trivers, R. (1988). Sex differences in rates of recombination and sexual selection. *In* "The Evolution of Sex: An Examination of Current Ideas" (B. Levin and R. Michod, eds.), pp. 270–286. Sinauer, New York.

Weismann, A. (1889). "Essays upon Heredity and Kindred Biological Problems," Vol. I. Clarendon Press, Oxford.

West, S. C., Cassuto, E., and Howard-Flanders, P. (1982). Postreplication repair in *E. coli:* Strand exchange reactions of gapped DNA by RecA protein. *Molec. Gen. Genet.* **187**, 209–217.

White, M. J. D. (1973). "Animal Cytology and Evolution," 3rd ed. Cambridge University Press, London.

White, R., and Lalouel, J.-M. (1988). Chromosome mapping with DNA markers. *Sci. Am.* **258 (Feb.),** 40–48.

Whitehouse, H. L. K. (1982). "Genetic Recombination." Wiley, New York.

Wilcox, A. J., Weinberg, C. R., O'Connor, J. F., Baird, D. D., Schlatterer, J. P., Canfield, R. E., Armstrong, E. G., and Nisula, B. C. (1988). Incidence of early loss of pregnancy. *N. Engl. J. Med.* **319**, 189–194.

Chapter Thirteen

The Selective Advantage of Outcrossing

In Chapter 1, Section II, we explained that, according to the DNA repair (and complementation) hypothesis of sex, the selective force maintaining recombination is DNA damage and the selective force maintaining outcrossing in multicellular organisms is deleterious mutation. In Chapters 11 and 12, we argued that meiosis, a central feature of sexual reproduction in eukaryotes, is an adaptation that promotes the recombinational repair of DNA damage. We argue in this chapter that outcrossing in multicellular organisms promotes the masking of mutations through complementation. We introduce the topic of outcrossing by (1) reviewing evidence on the prevalence of sexual reproduction among higher organisms and (2) outlining the sexual cycle of four widely different organisms. This approach illustrates that recombination and outcrossing are very widespread and are the two basic features of sex. We then discuss the costs and benefits of outcrossing, as well as the immediate and long-term costs and benefits of switching to and from various reproduction systems. These reproductive systems include automixis, selfing, outcrossing, endomitosis, apomixis, and vegetative reproduction.

I. *Prevalence of Sexual Reproduction*

Sexual reproduction is very widespread. In Chapter 12, Section IV.B, we noted that, among animals, sexual reproduction is the principal mode of reproduction, while parthenogenesis occurs in less than 0.1% of animal species. As shown in Table XVI, in higher plants

Table XVI
Modes of Reproduction among Higher Plant Species[a]

Reproductive Type[b]	Number Tabulated	Percentage
A	830	55
B	105	7
C	229	15
D	16	1
E	3	0.2
F	2	0.1
G	121	8
H	199	13
Total	1,505	

[a] These data were calculated from Fryxell (1957).

[b] A denotes principally cross-fertilized plants, including self-incompatible, dichogamous, and dioecious species. Dichogamy is the condition in which male and female parts of a flower mature at different times, ensuring that self-pollination does not occur. Dioecious is the condition in flowering plants in which unisexual (male or female) flowers are on separate plants. B denotes partially self-, partially cross-fertilized species. C denotes principally self-fertilized (autogamous) species. D, E, and F denote facultative apomicts, corresponding to the three preceding categories. G denotes obligate apomicts.[c] H denotes those species that have been reported as apomictic, but for which adequate information is not available to enable classification as D, E, F, or G

[c] For the purpose of this table, apomixis includes such vegetative methods of propagation as runners and bulbils, in addition to the more specialized processes of agamospermy. In agamospermy, embryos and seeds are formed by asexual means. Elsewhere in the text, the term apomixis will be used more specifically to refer to the absence of meiosis and fertilization only among organisms that reproduce by eggs.

sexual reproduction is also the most common mode of reproduction. Table XVI shows that about 55% of higher plant species are principally sexual (class A). An additional 7% are partially sexual, i.e., partially self-fertilizing and partially cross-fertilizing (class B). About 15% undergo meiosis but lack significant outcrossing and are principally self-fertilizing (class C). About 1% are principally sexual, but facultatively apomictic (class D). Apomixis means reproduction in the absence of both meiosis and fertilization. Other facultative apomictic classes are E and F, but these are infrequent. Only about

8% of plant species lack both outcrossing and meiosis (obligate apomicts, class G). Meiosis is also observed to occur among numerous species of fungi, algae, and protozoans (Heywood and Magee, 1976) indicating that sex is common among these organisms. Among bacteria and viruses, sexual reproduction is also common (for examples, see this chapter, Section II).

II. *Sex Has Two Basic Features: Recombination and Outcrossing*

Even though variation in the sexual cycle across the broad range of different organisms is considerable, there are certain common features at the level of the genetic material. As previously stated (Chapter 1, Section I), these common features are (1) two haploid genomes, each from a separate parent, come together in a shared cytoplasm (outcrossing); (2) the chromosomes comprising the two genomes pair so that homologous sequences are aligned; (3) exchange of genetic material occurs between the pairs (recombination); and (4) progeny are formed containing the new recombinant chromosomes. To illustrate the commonality of these features, we briefly outline the sexual cycles, as they are presently understood, of four organisms; a mammal, a yeast, a bacterium, and a DNA virus.

In humans, each cell of the body (soma) generally has two sets of chromosomes, one from the father and one from the mother (liver cells are an exception, being polyploid, see Chapter 4, Section IV.D). The two sets are brought together by fertilization. In this process, one set of the father's chromosomes, contained in a haploid sperm cell, enters the mother's haploid egg (outcrossing, step 1, above). The fertilized egg, or zygote, is diploid and gives rise to the approximately 6×10^{12} diploid cells of the soma by successive mitotic cell divisions. In each mitotic division, the chromosomes of a cell first replicate. Then the cell divides, forming two daughter cells, each having the same chromosome constitution as their parent cell (for further description, see Chapter 11, Section I.A). Eggs and sperm, however, are not formed by mitosis, but by meiosis. In meiosis, a diploid cell gives rise to haploid eggs (in females) or haploid sperm (in males). As described in Chapter 11, Section I.A, meiosis is initiated by replication of the chromosomes of the diploid cell to form two copies of each maternal chromosome and two copies of each paternal chromosome. The homologous chromosomes then become aligned (step 2, above). Next, exchange of DNA (recombination) occurs between pairs of homologous chromosomes to form recombinant chromosomes (step 3, above). Two successive cell divisions then occur (with no further replication of the chromosomes), giving rise to haploid eggs or sperm. These can be fertilized by a sperm or egg, respectively, of another individual (outcrossing) to form a diploid zygote (step 4, above). The cycle of fertilization, to form the diploid zygote, and meiosis, to form the haploid egg and sperm, is the sexual cycle of humans.

The fission yeast, *S. pombe,* is single-celled and one of the simplest organisms known to undergo meiosis. Unlike human somatic cells, which are diploid, mitotically active fission yeast cells are haploid. The haploid cell replicates its chromosomes and then divides by mitosis to give rise to two daughter haploid cells. This phase of reproduction is referred to as the vegetative phase to contrast it with the sexual phase. Each vegetative cell has one of two possible mating types. Under certain conditions (e.g., starvation; presence of a DNA-damaging agent [Bernstein and Johns, 1989]), cells of a vegetative culture fuse in pairs (only cells of different mating type fuse) to form diploid zygotes (step 1, outcrossing). The zygote of the fission yeast, in contrast to the zygote in humans, does not undergo further mitotic divisions to form a soma; rather, it undergoes meiosis to form four haploid cells, two of each mating type. Each of these can then enter the vegetative phase. Meiosis in fission yeast is essentially the same as in humans, involving replication of the chromosomes from both parents, an alignment of homologous chromosomes (step 2), recombination between pairs of chromosomes (step 3), two successive divisions without further replication of the chromosomes, and, finally, formation of haploid (i.e., vegetative phase) cells (step 4).

The bacterium *B. subtilis,* like fission yeast, is a single-celled haploid organism. It can reproduce vegetatively by a process in which chromosome replication is followed by cell division to form two haploid cells (binary fission). Although organized much differently than mitosis in eukaryotes, the basic process of partitioning two copies of the replicated chromosome into two daughter *B. subtilis* cells is similar. Sexual reproduction in *B. subtilis,* as in fission yeast, is induced by starvation conditions. The process of mating is referred to as transformation (for further discussion, see Chapter 10, Section II.H). *Bacillus subtilis* cells under starvation conditions enter an altered physiological state called competence. In this state, cells actively take into their cytoplasm chromosome pieces from the surrounding medium that had been released by other *B. subtilis* cells (step 1, outcrossing). Key stages in transformation are the alignment of the chromosome fragment with the resident bacterial chromosome (step 2), and then a physical exchange leading to the formation of a recombinant chromosome (step 3). This recombinant chromosome is then duplicated and passed on to daughter cells by binary fission (step 4). Transformation lacks many of the organized properties of meiosis but shares the feature of chromosomal material from separate individuals coming together in a common cytoplasm (outcrossing) and undergoing physical exchange to form a recombinant chromosome, which is then passed on to progeny.

Phage T4 is a virus that attacks bacterial cells. Its chromosome is composed of DNA, like that of the other organisms considered. Infection of the bacterial host cell is initiated by the virus attaching to the outer surface of the cell and injecting its single chromosome into the cytoplasm through the cell membrane. The vegetative phase of growth consists of multiple rounds of phage DNA replication within the host cell. This is followed by the encapsidation of individual DNA genomes by coat proteins specified by the virus DNA. Mature virus particles are then released by disruption of the cell. Sexual reproduction in phage T4 takes place when two viruses infect the same host cell (step 1, outcrossing). Under these circumstances, the chromosomes contributed by each virus replicate as during vegetative reproduction. However, the chromosomes from the two different viruses may also align with each other (step 2) and exchange DNA (step 3), giving rise to recombinant chromosomes. These can then be encapsidated in progeny virus particles (step 4).

The above four examples are presented to illustrate that, although the sexual cycle varies widely among different organisms, recombination and outcrossing are consistent features. Thus, we consider that sexual reproduction has two fundamental aspects: (1) recombination, in the sense of the breakage and exchange of DNA between two homologous chromosomes, and (2) outcrossing, in the sense that the homologous chromosomes that engage in recombination come from separate parents. This contrasts with asexual reproduction, in which the homologous chromosomes present in any cell are from the same parent.

Although both recombination and outcrossing are ordinarily present together in an organism, it can be argued that recombination is the more basic aspect of sex. This is indicated by the fact that some reproductive systems have abandoned outcrossing but have retained meiotic recombination. For instance, self-fertilization is a common mode of reproduction among higher plant species (Fryxell, 1957). As can be seen in Table XVI, about 22% of plant species are partially or principally self-fertilizing (classes B and C). In contrast, very few reproductive systems exist in which outcrossing is maintained, in the sense that two genomes from different individuals come together in the same nucleus, and yet recombination does not occur. An example of such a system is the diploid hybridogenetic forms of *Poeciliopsis* fishes (Cimino, 1972).

Bernstein *et al.* (1985a; 1987) have proposed that the maintenance of recombination and outcrossing is a consequence of selective forces resulting from the two intrinsic "noise" problems for transmission of genetic information: DNA damage and mutation. As discussed in Chapters 10–12, recombination is probably maintained to repair damage. As discussed below, in multicellular reproductive systems with recombination, outcrossing is maintained because it promotes complementation, or the masking of deleterious mutations. Thus, DNA damage selects for recombination, and mutation in the presence of recombination selects for outcrossing.

III. *Outcrossing*

A basic characteristic of meiosis is the reduction in ploidy that occurs, usually from diploidy to haploidy. This reduction prepares the meiotic products (gametes) for fusion, which restores diploidy. Fusion between two haploid meiotic products can occur by either of two strategies. We will refer to these as open- and closed-system strategies. The open-system strategy is called outcrossing, and it involves fusion between haploid gametes from different individuals. Closed-system strategies involve passage of genetic material from a single parent to progeny. Examples of closed-system strategies are self-fertilization (fusion of gametes from the same individual) and automixis (uniparental production of eggs through meiosis followed by some internal process for restoring diploidy). If recombinational repair of DNA damage were the sole advantage of sexual reproduction, the most cost-effective strategy would be a closed system of asexual reproduction such as self-fertilization or automixis. This would avoid the major short-term costs of sex enumerated in this chapter in Section III.A, below. Thus, we need to explain why the most common strategy in nature, in spite of its costs, is outcrossing, an open-system strategy (see this chapter, Sections III.B, D, and E, below).

A. Outcrossing Has Large Costs, Implying Correspondingly Large Benefits

The large investment of energy, time, and resources devoted to outcrossing by most organisms is impressive. Even without systematic analysis, some of these costs are obvious. For example, in humans, most persons are preoccupied from adolescence with attracting a suitable mate, mating, and keeping a mate. As a second example, many plants make a large investment in elaborate flowers to attract pollinators. Another example is the outlandishly ornamented tail feathers used by peacocks to attract peahens. Systematic analysis of the costs of outcrossing have also been carried out. The major short-term costs of outcrossing have been grouped under the headings of the cost of mating (Bernstein *et al.*, 1985b; Hopf and Hopf, 1985), the cost of males (Maynard Smith, 1978: pp. 2–3), high recombinational load (Shields, 1982), and lower genetic relatedness between parent and offspring (Uyenoyama, 1984, 1985). These costs either stem directly from outcrossing, as in the case of the costs of mating and males, or are enhanced by outcrossing, as in the case of recombinational load and reduced genetic relatedness (Uyenoyama, 1988). A large investment in outcrossing implies that the adaptive benefit must be comparably large. Therefore, any fundamental explanation of why sex is ubiquitous in nature must provide a benefit that is large enough to at least balance the costs. We will now argue that the benefit of masking deleterious mutations accounts for outcrossing (for additional discussion, see Bernstein *et al.*, 1985a, 1987).

B. Complementation and Hybrid Vigor

Both in animals and, particularly, in plants, it is commonly observed that hybrids—i.e., genotypes resulting from the crossing of dissimilar parents (e.g., from two genetically distinct inbred lines)—outperform and are more vigorous than either of their parents. This outcrossing phenomenon, referred to as heterosis, or hybrid vigor, underlies much of the improvement in crop yields achieved in the twentieth century. As noted in Chapter 1, Section III.B, Darwin in 1889 concluded from his studies on the effects of cross- and self-pollination in plants, that the greater vigor of the progeny of out-crosses compared with those of self-fertilization was sufficient to account for sexual reproduction. In this chapter, Sections III.D, and E, below, we will argue that the adaptive advantage of outcrossing is due to complementation, the masking of deleterious recessive alleles. Masking occurs when wild-type genes present in a diploid cell produce normal functional products that compensate for the production of less functional, or nonfunctional, products by homologous mutant genes. We consider that complementation is likely to be the major cause of hybrid vigor and, therefore, regard our explanation for the maintenance of outcrossing as a reformulation and elaboration in modern genetic terms of Darwin's earlier explanation.

C. Why Deleterious Recessive Mutations Occur, How Frequently They Occur, and How They Are Eliminated in *Drosophila* and Humans

In Chapter 2, Section I, we discussed the nature of mutation as distinct from DNA damage. In Chapter 9, Section II.G, we indicated that mutations arise from errors of DNA synthesis associated with either chromosome replication or DNA repair. Improvements in the accuracy of DNA synthesis have costs that include increased energy use (Hopfield, 1974), additional gene products (Alberts *et al.*, 1980), or slower DNA synthesis (Gillen and Nossal, 1976). Thus, there are probably cost-effectiveness limits to indefinite improvement in accuracy, and a finite spontaneous mutation rate is likely intrinsic to DNA replication and repair in nature. Haldane (1937) pointed out that most mutant types are less fit than the normal in the wild state. In the short run, deleterious mutations affect fitness much more than beneficial ones because of their much higher rate of occurrence.

In a population at equilibrium, deleterious mutations are removed by selection at the same rate that they arise by mutation (Haldane, 1937). As shown by Haldane, the average survivorship at equilibrium is solely a function of μ, the rate of deleterious mutation per haploid genome, and is not affected by how individually harmful the mutations are.

The value of μ has been estimated for *Drosophila* by Mukai *et al.* (1972). They estimated that the mutation rate per haploid genome for lethal and nonlethal mutations combined as 0.3. Thus, μ equals 0.3. This gives a survivorship, $e^{-\mu}$, of 0.7. A survivorship of 0.7 implies that 30% of progeny will not survive long enough to reproduce, due to the expression of deleterious mutations. The value of 0.7 is neither so low as to be devastating or so close to 1.0 as to indicate a negligible effect of mutational load. We assume 0.7 to be an approximate general value for survivorship, $e^{-\mu}$, and 0.3 a general value for μ.

Evidence reviewed by Czeizel *et al.* (1988) for humans bears on this approximate value for survivorship. They analyzed epidemiological data on the prevalence of 26 common genetic and partially genetic diseases in Hungary, and from this estimated the detriment to the population in terms of lost and impaired life. Overall, about 16% of all deaths that occur in Hungary every year (in all age groups between 20 and 69 yr) can be attributed to these 29 diseases. Although these age groups are old enough to reproduce, premature death and disability among members of the group can be assumed to reduce significantly reproductive capability. This suggests that the detriment due to the entire load of mutations present in humans, including that in the 0–20-yr age group, is not negligible and is compatible with the above estimate of 70% survivorship for *Drosophila*.

D. Benefit of Changing From a Reproductive System
   That Masks Few Deleterious Mutations to One That
   Masks Many

Table XVII, adapted from Bernstein *et al.* (1989), lists reproductive systems of diploid organisms in the first column. Hopf *et al.* (1988) found, by a generalization of Haldane's argument, that in all of these reproductive systems, survivorship due to expression of deleterious recessive alleles is $e^{-\mu}$ at equilibrium. For haploid organisms (not listed in Table XVII), survivorship at equilibrium is also $e^{-\mu}$, but this case is discussed separately in Section F, below. Because the effect of μ on survivorship at equilibrium is the same for all reproductive systems, μ should not differ among these systems. Thus, we assume μ to be constant and consider below the change in survivorship of switching from one reproductive system to another among those listed in Table XVII.

The second column of Table XVII indicates the number of deleterious recessive mutations accumulated at equilibrium, in terms of μ and of n (the number of functional genes per genome), in the different reproductive systems. The differences among these systems in the number of accumulated mutations are substantial, and vary from μ, which is about 0.3, to n, which is of the order of 100,000. Reproductive systems that are effective at masking deleterious recessive mutations accumulate many; systems ineffective at masking accumulate few.

Because the effect on survivorship of mutational load is the same for all reproductive systems at equilibrium, all such systems should be equally competitive. However, there is a transient benefit to moving downward in Table XVII, i.e., toward greater masking, and a transient disadvantage to moving upward toward diminished

Table XVII
Reproductive Systems of Diploid Organisms[a]

Reproductive System	Masking Ability at Equilibrium	Meiotic Recombinational Repair	Source of Homologous Chromosome
Automixis	Low ($\approx 2\mu$)	Yes	Self
Selfing	Low ($\approx 2\mu$)	Yes	Self
Outcrossing[b]	Intermediate ($\approx \sqrt{n\mu}$)[c]	Yes	Another individual
Endomitosis	High ($\approx n$)	Limited[d]	Self
Apomixis	High ($\approx n$)	No	Not applicable
Vegetative	High ($\approx n$)	No	Not applicable

[a] See Hopf *et al.* (1988) for supporting details.

[b] With some mating between related individuals, as occurs in nature.

[c] n denotes the number of functional genes per genome, which in higher organisms, such as humans, may be about 100,000 (White & Lalouel, 1988); μ is the rate of deleterious mutation per haploid genome which is estimated to be about 0.3.

[d] See Chapter 12, Section IV, B.

masking. To exemplify the transient benefits and disadvantages of shifting between reproductive systems, let us consider a population fixed for selfing. Even though individuals in this population will have accumulated few mutations at equilibrium (Table XVII), new ones occur each generation at a frequency μ. If an outcrossing individual that can mate with the selfers now arises in this population, this individual will be able to mask almost all mutations in its offspring. Complementation in the progeny of this outcrossing individual is nearly complete because deleterious recessive alleles in it and its partner are statistically unrelated to each other. Thus, the fitness of the new outcrosser is not reduced at all by mutational load, and it has a survivorship of almost unity with respect to mutational load.

However, a new outcrosser must pay the costs of outcrossing indicated in Section III.A of this chapter. We can indicate how one might, with enough information, calculate whether or not the expected benefit of the switch to outcrossing is large enough to balance the high costs. The factor by which fitness is reduced, due solely to the costs of outcrossing, can be called C (costs). Here, C is the ratio of the fitness of the selfer (arbitrarily taken as 1.0) to the fitness of the new outcrosser, <1.0 because of the costs of outcrossing. Thus, C is equal to 1.0/<1.0. C is the cost of the shift to outcrossing and can now be compared with the benefit of the shift.

The factor by which fitness is increased due to the masking of deleterious recessive alleles when the shift occurs is given by B (benefits), where B is the ratio of the fitness of the new outcrosser due to masking of mutation (which we said was close to unity, or 1.0) to the fitness of the selfer (which we indicated above would be $e^{-\mu}$) due to the expression of mutations. Thus, $B = 1/e^{-\mu} = e^{\mu}$. If the cost of outcrossing (C) is less than the benefit (B) of masking deleterious recessive alleles (i.e., $C < e^{\mu}$), then a gene for outcrossing will expand in the population. If μ has a value of 0.3, as found for *Drosophila*, above, the benefit $B = e^{\mu}$ has a value of 1.3. This suggests that a shift to outcrossing should occur when the costs (C) are less than 1.3. A benefit of 1.3 is quite large and should cover the costs of outcrossing in many cases.

The value of μ has only been roughly estimated for *Drosophila* and is not known with accuracy for any organism. Thus, a value for μ of 0.7, rather than 0.3, is reasonable. This would lead to a fitness of the selfer ($e^{-\mu}$) of 0.5 and a benefit to shifting to sexual reproduction that would balance a cost as large as twofold ($B = e^{\mu} = 2.0$).

If outcrossing expands in the population at the expense of selfers, then the entire population may become outcrossers (i.e., the outcrossing character becomes fixed). As fixation occurs, deleterious recessive alleles would increase due to the masking effect of outcrossing. As these alleles increase, there is a reduction in fitness due to their expression in the homozygous condition. Eventually, the outcrosser reaches a point where its equilibrium fitness is reduced by both the costs of outcrossing and the costs of mutational load. That is, the long-term effect of the transition from selfing to outcrossing is a net reduction of individual fitness. However, if the outcrosser should shift back to selfing, the many accumulated deleterious alleles would express (i.e., be unmasked). Thus, there is a barrier to a successful shift back to selfing. A shift from outcrossing to selfing may succeed, however, when the costs of outcrossing become very large. In other words, the new selfer, despite its very low fitness, is still fitter than an otherwise very fit outcrosser that can't find a mate. It should be noted that an intermediate level of outcrossing (with some mating between related individuals) is likely preferred over random mating (panmixia) because of the benefit of preserving co-adapted gene complexes (Shields, 1982: p. 2).

E. Switching from Outbreeding to Parthenogenesis or Vegetative Reproduction

When an outcrossing population has reached equilibruim with respect to mutation, a large mutational load has accumulated. If an ameiotic asexual individual arose in this population in which the diploid maternal genome was passed down intact from mother to daughter, then this line would reap the advantage of complementation while avoiding the costs of outcrossing.

However, the absence of meiotic recombinational repair in such asexual systems (see third column of Table XVII) also needs to be considered in evaluating the probability of success of a shift from outcrossing to asexual reproduction. One type of parthenogenetic reproduction—apomixis—is characterized by suppression of meiosis and its replacement by a single mitotic maturation division. In vegetative reproduction, growth is only through mitotic cell divisions and meiosis is absent. As discussed in Chapter 12, Section III.D, recombination between nonsister homologous chromosomes is rare in mitosis although frequent in meiosis. In a shift from outcrossing to apomixis or vegetative reproduction, the advantage of meiotic recombinational repair (Chapters 11 and 12) is largely lost. In particular, double-strand damages present prior to the DNA replication step of cell division cannot be repaired. Thus, apomixis and vegetative reproduction are costly strategies. However, as discussed in Chapter 5, Sections I.C and III.B, the effects of DNA damage may be overcome by cellular selection (i.e., the death of damaged cells and the replication of undamaged ones). Vegetative reproduction and apomixis may be successful strategies when damages are infrequent and/or growth is rapid, but not otherwise.

When an organism shifts from meiotic to mitotic production of eggs, as in the shift from sexual reproduction to ameiotic apomixis, the ability to carry out efficient recombinational repair between nonsister homologous chromosomes might be maintained. This strategy would have the advantage of retaining the ability to efficiently remove double-strand damages in the germ cells. However, the resulting recombination between nonsister homologous chromosomes would rapidly lead, among progeny, to homozygosity and expression of the many deleterious recessive alleles accumulated during the previous sexual phase. Thus, this strategy for switching from sexuality to asexuality is also costly. The above example illustrates a basic problem for nonoutcrossing diploid cells. Physical recombination between homologous chromosomes produces allelic recombination and, hence, creates homozygosity at previously heterozygous loci. Thus, recombinational repair causes expression of deleterious recessive alleles at previously heterozygous loci. Outcrossing avoids this problem and functions to mask the expression of these deleterious recessive alleles.

Another parthenogenetic system involves endomitosis. As described previously in Chapter 12, Section IV.B, this system is characterized by two sequential premeiotic replications followed by an apparently normal meiosis to produce diploid eggs. Double-strand damages that occur before the first premeiotic replication cannot be repaired in this system because all chromatid pairing partners are derived from the same chromosome and there is no intact template corresponding to the damaged site. In conventional meiosis, this problem does not arise because recombination is between nonsister chromatids. Therefore, if double-strand damages are frequent before premeiotic replication, endomitosis is an unsatisfactory alternative.

In conclusion, a shift from outcrossing to other meiotic forms of reproduction should have a large immediate cost due to increased expression of deleterious mutations. This cost is transient in a shift to selfing and automixis, because the few progeny that survive the unmasking of recessive mutations will be those with statistically fewer mutations. The shift from outcrossing to apomixis, endomitosis, or vegetative reproduction, however, should have permanent costs, because these latter reproductive systems would have less capacity for recombinational repair.

Evidence in support of the expected reduced fitness of parthenogens has been summarized by Lynch (1984). Parthenogens most often have lower reproductive rates than their sexual relatives, frequently less than 50%. Lynch noted that these costs of parthenogenesis stem from poor hatching of eggs, due to developmental abnormalities, and in some cases from reduced egg production. Furthermore, newly arisen parthenogens seem to suffer a greater decline in fecundity than established ones. The costs of parthenogenesis predicted by theory and supported by the observed decline in fecundity suggests that parthenogens would only be competitive with outcrossers in circumstances where the costs of outcrossing are larger than the costs of parthenogenesis. One familiar circumstance of this type is in newly created natural habitats such as result from the aftermath of floods and fires, where finding a mate is difficult. Evidence corroborating the generalization that parthenogens are favored where the costs of finding a mate are high has been reviewed by Bernstein *et al.* (1985b).

F. Outcrossing in Multicellular Haploids and Haplodiploids

We have contended that outcrossing is maintained by the transient advantage of masking mutations. However, this explanation cannot apply to mutations in genes that are expressed in the haploid stage of the life cycle, because recessive mutations in such genes are not masked and can be weeded out by selection. Most genes may be of this type in multicellular haploid organisms. If such genes were the only ones of consequence, the advantage of outcrossing would be small and there would be little hindrance to shifting to selfing or automixis. There is a similar difficulty in explaining outcrossing in haplodiploid insects. In the haploid males, recessive mutations are eliminated by selection, and in the diploid females they are largely masked. On first consideration, these cases seem to call into question our hypothesis that the primary function of outcrossing is complementation. Single-celled haploid organisms, in contrast to multicellular haploid organisms, do not present this problem, because they need to outcross simply to bring together homologous chromosomes in the same cell for recombinational repair (for further discussion, see Chapter 14, Section II.B).

To explain the existence of outcrossing in multicellular haploids, Bernstein et al. (1985a) pointed out that in these organisms there are key genes that are only expressed in the diploid stage, such as the genes controlling meiosis. For example, in haploid multicellular fungi, meiosis occurring in the short diploid stage is one of their most complex processes. The multicellular haploid stage, although larger, characteristically has a modular construction with little differentiation. Leslie and Raju (1985) found, in the multicellular haploid fungus Neurospora crassa, that recessive mutations affecting the diploid stage of the life cycle were frequent in natural populations. They estimated that the number of genes affecting the diploid stage was at least 435. These mutations, when homozygous in the diploid stage, formed spores with maturation defects or barren fruiting bodies with few sexual spores. The maintenance of outcrossing can be readily explained by the advantage of masking such mutations.

The conspicuous structures in multicellular mushrooms and bracket fungi are often dikaryons containing nuclei from different parents (Fincham et al., 1979). Because masking of deleterious recessive alleles would be expected to occur in the dikaryons, they are functionally diploid, and, thus, outcrossing is selectively advantageous.

Ferns also have a multicellular haploid stage. However, unlike fungi, they have a large complex diploid sporophyte stage with considerable tissue diversity. Numerous recessive lethal mutations are present in natural populations of at least one genus of fern, Osmunda (Klekowski, 1973). These mutations, when homozygous, prevent normal development of the embryonic diploid sporophyte. Even though these hermaphroditic plants can self-fertilize, outcrossing is thought to be preferred at higher population densities because of the advantage of masking recessive deleterious mutations (Klekowski, 1973).

In haplodiploid insects, some genes are expressed only in the diploid phase (females) and not in the haploid phase (males), and mutations in such genes could select for outcrossing. The estimated percentage of genes whose expression is limited to females was summarized for various haplodiploid species by Kerr (1976). These percentages ranged from 14 to 46% depending on the species and kind of phenotype considered. These data indicate that recessive mutations, limited in their effect to diploid females, are sufficiently common to select for some degree of outcrossing in haplodiploids. The evidence reviewed in this section indicates that in the diploid stages of fungi, ferns, and haplodiploid insects, there is a sufficient advantage to masking deleterious recessive alleles to explain the occurrence of outcrossing in these organisms.

IV. *Conclusions*

Our general hypothesis with regard to the adaptive benefit of sex is that the two intrinsic "noise" problems in replicating genetic information, DNA damage and mutation, are the selective forces that maintain respectively the two main features of sex, recombination and outcrossing. In this chapter, we have presented evidence bearing on the second part of this hypothesis, that outcrossing in multicellular organisms is maintained by the benefit of masking deleterious mutations. We concluded that this second part of the hypothesis explains the predominance of outcrossing sexual reproduction compared with various forms of asexual reproduction. Allelic variation, by this hypothesis, is a by-product of recombination and outcrossing. In focusing on the selective role of DNA damage and mutation in the maintenance of sex, however, we do not wish to imply that the production of allelic variation is an unimportant consequence of sex. Infrequent beneficial allelic variants generated by recombination undoubtedly promote long-term evolutionary success, just as infrequent beneficial mutations do. We consider that the tendency toward randomization of genetic information that occurs with recombination and outcrossing, under general conditions, has a negative effect on fitness in the short run, just as mutations, in general, do. This variation, or recombinational load, probably generally contributes to the immediate cost of sex, and any short-term advantage of creating variation is probably limited to special cases (for further discussion, see Chapter 15, Section II.B). Overall, we consider that DNA damage and deleterious mutation, but not recombinationally produced variation, are the selective forces maintaining sexual reproduction.

References

Alberts, B. M., Barry, J., Bedinger, P., Burke, R. L., Hibner, V., Liu, C.-C., and Sheridan, R. (1980). Studies of replication mechanisms with the T4 bacteriophage *in vitro* systems. *In* "Mechanistic Studies of DNA Replication and Genetic Recombination." ICN-UCLA Symposia on Molecular and Cellular Biology, (B. Alberts, C. F. Fox, and F. J. Stusser, eds.) pp. 449–474. Academic Press, New York.

Bernstein, C., and Johns, V. (1989). Sexual reproduction as a response to H_2O_2 damage in *Schizosaccharomyces pombe*. *J. Bacteriol.* **171**, 1893–1897.

Bernstein, H., Byerly, H. C., Hopf, F. A., and Michod, R. E. (1985a). Genetic damage, mutation and the evolution of sex. *Science* **229**, 1277–1281.

Bernstein, H., Byerly, H. C., Hopf, F. A., and Michod, R. E. (1985b). Sex and the emergence of species. *J. Theor. Biol.* **117**, 665–690.

Bernstein, H., Hopf, F. A., and Michod, R. E. (1987). The molecular basis of the evolution of sex. *Adv. Genet.* **24**, 323–370.

Bernstein, H., Hopf, F. A., and Michod, R. E. (1989). The evolution of sex: DNA repair hypothesis. *In* "The Sociobiology of Sexual and Reproductive Strategies" (A. E. Rasa, C. Vogel, and E. Voland, eds.), pp. 3–18. Chapman and Hall, New York.

Cimino, M. C. (1972). Egg production, polyploidization and evolution in a diploid all-female fish of the genus *Poeciliopsis*. *Evolution* **26**, 294–306.

Czeizel, A., Sankaranarayanan, K., Losonci, A., Rudas, T., and Keresztes, M. (1988). The load of genetic and partially genetic diseases in man. II. Some selected common multifactional diseases: Estimates of population prevalence and of detriment in terms of years of lost and impaired life. *Mutat. Res.* **196**, 259–292.

Fincham, J. R. S., Day, P. R., and Radford, A. (1979). "Fungal Genetics." Blackwell, Oxford.

Fryxell, P. A. (1957). Mode of reproduction in higher plants. *Bot. Rev.* **23**, 135–233.

Gillen, F. D., and Nossal, N. G. (1976). Control of mutation frequency by bacteriophage T4 DNA polymerase I. The tsCB120 antimutator DNA polymerase is defective in strand displacement. *J. Biol. Chem.* **251**, 5219–5224.

Haldane, J. B. S. (1937). The effect of variation on fitness. *Am. Nat.* **71**, 337–349.

Heywood, P., and Magee, P. T. (1976). Meiosis in protists. *Bacteriol. Rev.* **40**, 190–240.

Hopf, F. A., and Hopf, F. W. (1985). The role of the Allee effect on species packing. *Theor. Pop. Biol.* **27**, 27–50.

Hopf, F. A., Michod, R. A., and Sanderson, M. (1988). On the effect of reproductive systems on mutation load and the number of deleterious mutations. *Theor. Pop. Biol.* **33**, 243–265.

Hopfield, J. J. (1974). Kinetic proofreading: A new mechanism for reducing errors in biosynthetic processes requiring high specificity. *Proc. Natl. Acad. Sci. USA* **71**, 4135–4139.

Kerr, W. E. (1976). Population genetic studies in bees. 2. Sex-limited genes. *Evolution* **30**, 94–99.

Klekowski, E. J. (1973). Genetic load in *Osmunda regalis* populations. *Am. J. Bot.* **60**, 146–154.

Leslie, J. F., and Raju, N. B. (1985). Recessive mutations from natural populations of *Neurospora crassa* that are expressed in the sexual diplophase. *Genetics* **111**, 757–777.

Lynch, M. (1984). Destabilizing hybridization, general purpose genotypes and geographic parthenogenesis. *Q. Rev. Biol.* **59**, 257–290.

Maynard Smith, J. (1978). "The Evolution of Sex." Cambridge University Press, London.

Mukai, T., Chigusa, S. I., Mettler, L. E., and Chow, J. F. (1972). Mutation rate and dominance of genes affecting viability in *Drosophila melanogaster*. *Genetics* **72**, 335–355.

Shields, W. M. (1982). "Philopatry, Inbreeding and the Evolution of Sex." State University of New York Press, Albany.

Uyenoyama, M. K. (1984). On the evolution of parthenogenesis: A genetic representation of "the cost of meiosis." *Evolution* **38**, 87–102.

Uyenoyama, M. K. (1985). On the evolution of parthenogenesis. II. Inbreeding and the "cost of meiosis." *Evolution* **39**, 1194–1206.

Uyenoyama, M. K. (1988). On the evolution of genetic incompatibility systems: Incompatibility as a mechanism for the regulation of outcrossing distance. *In* "The Evolution of Sex: An Examination of Current Ideas" (R. E. Michod and B. R. Levin, eds.), pp. 212–232. Sinauer, Sunderland, Massachusetts.

White, R., and Lalouel, J.-M. (1988). Chromosome mapping with DNA markers. *Sci. Am.* **258 (Feb.)**, 40–48.

Chapter Fourteen

Evolutionary Aspects of Sex

In Chapters 10–13, we presented an explanation for the advantages of sex based on the central importance of coping with genome damage and mutation. If, as we assume, genome damage and mutation are fundamental problems in nature that have been present throughout the history of life, sex may have arisen early and been maintained continuously to deal with them. In this chapter, we attempt to show the plausibility of this view by describing evidence that bears on the evolution of sex. We consider, in succeeding sections, (1) the continuity of the evolution of sex, (2) the possible origin of sex in the earliest cells, (3) the selective forces leading to the origin of enzyme-mediated recombinational repair, (4) the advantage of sex in unicellular haploid organisms, and (5) the emergence of organisms with a predominant diploid stage as a modification of the sexual cycle of haploid organisms. In addition to these topics, we consider (6) a major evolutionary consequence of sex—the emergence of species.

I. The Likely Continuous Evolutionary History of Sex: Evidence that Sex Arose prior to the Divergence of Prokaryotes and Eukaryotes

A number of authors have suggested that sex arose very early in the history of life (e.g., Dougherty, 1955; Stebbins, 1960; Baker and Parker, 1973; Maynard Smith, 1978: pp. 6–7; Bell, 1982: p. 84). For instance, Bell commented that it seems certain that mechanisms of genetic exchange and recombination are such ancient features of life that their history traces back to the origin of life itself.

One way of gaining insight into the origin of sex is to consider its presence in the three primary phylogenetic kingdoms: (1) the eukaryotes (which trace their ancestry back to prokaryotes referred to as urkaryotes), (2) the eubacteria, and (3) the archaebacteria (Fox *et al.,* 1980). The common ancestor from which these lineages diverged may have lived over 3 billion years ago. Sexual reproduction is well known among eukaryotes and eubacteria (for examples, see Chapter 13, Section I, and Chapter 10, Section II.H). Mating has also recently been discovered in the archaebacteria (Rosenshine *et al.,* 1989). This suggests that sex could have been present in the early common ancestor of all three major lineages. However, authors have differed on whether or not eukaryotic and prokaryotic sex has a common ancestry. Bell, for instance, found the supposition that prokaryotic sexuality has gradually evolved into eukaryotic sexuality to be improbable. His view was based in large part on a presumed absence of sex in cyanobacteria. Bell (1982: pp. 86–87) considered cyanobacteria to be direct descendants of a species that was the ancestor of green plants. The more common interpretation, however, is that the cyanobacteria are ancestors of only the chloroplasts of higher plants (Margulis, 1981). Even so, evidence for sex among the cyanobacteria is convincing (Singh and Sinha, 1965; Shestakov and Khyen, 1970; Stevens and Porter, 1980), implying that the presumed major gap in the evolutionary sequence of sex does not actually exist.

Dougherty (1955) was the first to propose that sex in all organisms has a common ancestry based on the broad similarity of recombination processes in prokaryotes and eukaryotes. We consider that the large amount of information accumulated on the molecular processes of recombination since Dougherty's proposal reinforces the view that recombination processes are similar in prokaryotes and eukaryotes. Authoritative reviews of such information have been presented by Stahl (1979) and Whitehouse (1982). In a concluding statement at the end of his book, Stahl (1979: p. 245) noted that the similarities in recombination in organisms as diverse as the phage and fungi are impressive. Whitehouse (1982: p. 352) also commented that similarities among recombination mechanisms in diverse organisms are impressive.

Dougherty (1955) proposed that two selective forces were responsible for the origin of sex in primitive cells. First, cells in which an essential DNA species is damaged or destroyed have a potential way of correcting the defect by assimilating a replacement from other cellular compartments. Second, new and desirable combinations of DNA species arising in separate compartments do not have to await independent mutational events in each line of descent. He viewed the first process as making repair and survival possible and the second process as accelerating evolutionary improvement.

Our view is similar to Dougherty's in assuming that sex arose early in evolution as a recombinational repair process and that it has had a continuous evolutionary history (Bernstein *et al.*, 1981, 1984, 1985a, 1985b, 1987, 1989). We differ somewhat with Dougherty in assuming that RNA, rather than DNA, is the primitive genetic material and, more strongly, in considering the acceleration of evolutionary improvement to be a by-product rather than a selective advantage of sex (see Chapter 1, Section III.B). Nevertheless, along with Dougherty, we view genome damage as a primary selective force in the origin of sex.

The questions of whether or not sex arose early and whether or not it has had a continuous evolutionary history also can be approached by examining the molecular machinery of recombination in prokaryotes and eukaryotes for homologies. By examining analogous genes in two different organisms for specific functional and sequence homologies, common ancestry of the two genes can often be inferred. In a review of the literature, Bernstein and Bernstein (1989) found that 13 genes of phage T4, acting in the ubiquitous processes of DNA replication, DNA repair, recombination, and nucleotide synthesis, shared sequence and/or functional homology with genes of bacteria and/or eukaryotes. This suggests that enzymes involved in such fundamental processes probably evolved very early. Adaptive features of early enzymes may have been preserved in the phage T4, bacterial, and eukaryotic lineages because they were already well-tested solutions to basic functional problems by the time these lineages diverged.

Functional homology has been shown between genes specifically required for recombination in widely different organisms. Chen and Bernstein (1988) presented evidence for functional homology, and some sequence homology between genes essential for recombination in a eukaryote and in bacteriophage T4. The *RAD52* gene of the yeast *S. cerevisiae* and the *46/47* gene pair of bacteriophage T4 are essential for most recombinational events in their respective organisms. A functional *RAD52* gene is necessary for meiotic and mitotic recombination in yeast. In addition, it is required for recombinational repair and the production of viable meiotic spores. Genes *46* and *47* are required for recombination, recombinational repair, and production of viable progeny phage. Chen and Bernstein (1988) introduced the *RAD52* gene into plasmid expression vectors that were used to transform *E. coli*. When the expression of *RAD52* was induced, it was observed to complement phage mutants defective in genes *46* and *47* with respect to the three criteria of phage growth, recombination, and recombinational repair. A comparison of the DNA sequence of the *RAD52* gene with each of those of genes *46* and *47* indicated sequence similarity between *RAD52* and gene *46* over a limited region. Thus, gene *RAD52* and genes *46* and *47* seem to have a common ancestry on the basis of both genetic complementation and partial sequence homology. These findings suggest that the basic process of recombination, central to sex, predated the divergence of a bacterial virus and a eukaryote from a common ancestor. Functional similarities have also been found between the *E. coli* RecA protein, a key component of the bacterial recombination machinery, and protein analogues in eukaryotes. In Chapter 12, Section I.A, we described work by Hotta *et al.* (1985) identifying proteins from mouse and lily with enzymatic properties similar to the *E. coli* RecA protein, and in Chapter 10, Section III.A.6, we described work on an *E. coli* RecA analogue in the eukaryotic smut fungus *U. maydis* (Kmiec and Holloman, 1982, 1984). Also, a protein has been isolated from human cells that performs a recA-like function, the invasion of homologous duplex by single-stranded DNA (Hsieh *et al.*, 1986; Cassuto *et al.*, 1987; Fishel *et al.*, 1988). These findings support the suggestion of Dougherty (1955) that sex in all organisms has a common ancestry.

II. *Possible Origin of Sex in the Earliest Cells*

A. Problem of Genome Damage Prior to the Emergence of Sex

The two most basic properties of life may be regarded as the abilities to replicate and to encode information. In accord with the extensive experimental and theoretical work of Eigen and colleagues (Eigen and Schuster, 1979; Eigen *et al.*, 1981), we postulate that life arose as a self-replicating heteropolymer similar to RNA. The first resources for self-replication were probably available in the aqueous environment and were likely similar to ribonucleotides. Natural selection may have arisen among RNA replicators due to competition for ribonucleotides.

Folded configurations of RNA were likely the first functional adaptations (Kuhn, 1972; Bernstein *et al.*, 1983; Darnell and Doolittle, 1986; Orgel, 1987). These configurations were the earliest phenotypes, and they were determined by RNA base sequences, the first genotypes (Michod, 1983). Errors of replication would lead to mutational variants. By the action of natural selection on infrequent beneficial variants, the RNA replicators would then evolve adaptive configurations that promote (1) increased rate and accuracy of replication, (2) protection of replicators from damage both by preventing the occurrence of damage and by coping with damage after it occurs, and (3) improving the ability to incorporate nucleotides from the environment (Bernstein *et al.*, 1983). As the replicators evolved the ability to specify enzymes and other proteins, these three kinds of adaptation became associated with increasingly complex structures. The selective pressures to develop means of protecting against genome damage were probably quite strong. For example, Sagan (1973) determined the flux of solar UV light penetrating the primitive reducing atmosphere of the earth and concluded that this could have posed a major problem for the early evolution of life (see also Chapter 2, Section II.I). Damage to an RNA replicator could block its replication or expression of its encoded information. In a population of independent replicators lacking the ability to repair themselves, badly damaged ones would simply die.

Eigen and colleagues proposed that, although the RNA replicators were at first independent, they evolved mutual dependencies based on joint use of encoded products (e.g., primitive enzymes) (Eigen and Schuster, 1979; Eigen *et al.*, 1981). These authors further suggested an evolutionary progression from free RNA replicators to ''hypercycles'' (sets of mutually dependent replicators) and then to encapsulated hypercycles (for further discussion, see Bernstein *et al.*, 1984). The advantage of encapsulation to form a protocell is that it permits the RNAs to localize their encoded protein products for more efficient use. The problem of damage to the set of RNA replicators comprising a hypercycle becomes more acute when it is encapsulated within a simple membrane (such as a lipid bilayer), because damage in any RNA component of the hypercycle could inactivate the entire protocell. Thus, genome damage, prior to the emergence of sex, may have been an important problem for these early life forms.

B. Most Likely Form of Sex among the Earliest Cells

Protocells and early prokaryotic organisms probably emerged more than 3 billion years ago (Schopf, 1978). Eigen *et al.* (1981) and Woese (1983) have suggested that the genomes of early protocells were composed of single-stranded RNA and that individual genes corresponded to separate RNA segments rather than being linked end-to-end as in extant DNA genomes. For instance, Woese (1983) proposed a universal cellular ancestor that had an RNA genome that was physically disaggregated, comprising a collection of gene size segments. A haploid protocell with such an RNA genome would be vulnerable to damage, because a single damage in any segment is potentially lethal. Although RNA damage has been less well investigated than DNA damage, RNA is susceptible to the same general types of damage as DNA. For instance, UV inactivates both DNA and RNA viruses. The protocells would also be vulnerable to loss of genes because, when the set of segments replicates, one copy of each segment must be passed to each daughter protocell. Failure to segregate properly would lead to inviability of one or both daughters.

Vulnerability to damage or loss could be reduced by maintaining more than one copy of each RNA segment in each protocell, i.e., by maintaining diploidy or polyploidy. Genome redundancy would allow a damaged or lost segment to be replaced by an additional replication of its homologue. For such simple organisms, the resources actually tied up in the genetic material would be a large fraction of the total resource budget. Consequently, under conditions of limiting resources, the protocell reproductive rate would be inversely related to ploidy number. The fitness of the protocell would be reduced by these costs of redundancy. Thus, coping with damaged or lost genetic information while minimizing the costs of redundancy would have been a fundamental problem for early protocells.

A cost–benefit analysis was carried out by Bernstein *et al.* (1984) in which the costs of maintaining redundancy were balanced against the costs of genome damage or loss. It was concluded from this analysis that, under a wide range of circumstances, the selected strategy is for each protocell to be haploid, but to periodically fuse with another haploid protocell to form a transient diploid. The maintenance of the haploid state maximizes the growth rate. The periodic fusions allow mutual reactivation of lethally damaged protocells. If at least one undamaged copy of each gene is present in the transient diploid protocell, viable progeny can be formed. Two viable daughter haploid cells can be produced if there is an extra replication of the intact chromosome homologous to any chromosome that had been damaged or lost prior to the division of the fused protocell. We view the cycle of haploid reproduction, with occasional fusion to a transient diploid state and subsequent splitting to the haploid state, to be the sexual cycle in its most primitive form. The haploid progeny would often have a recombinant genome in the sense that they would have the genes of their two parents in new combinations. Although we consider recovery from genome damage or loss to be the adaptive function of the protocell sexual cycle, allelic recombination would arise as a by-product. In the absence of this sexual cycle, haploid protocells with damaged genomes would simply die.

There may not have been a precise evolutionary moment at which sex evolved. The "ploidies" we have been discussing can be regarded as statistical norms of highly imperfect processes. Despite considerable "sloppiness" in these early systems, the advantages to the protocell of splitting (to decrease the cost of redundancy) and fusion (to reap the benefit of redundancy) would have exerted a continuous pressure on the evolution of the life cycle of the primitive protocell with the sexual cycle as the end product.

C. Segmented Single-Stranded RNA Viruses

The above model for the primitive sexual cycle is hypothetical, but it is very similar to the established sexual behavior of the segmented RNA viruses, which are among the simplest extant organisms known. Influenza virus, whose genome is made up of single-stranded RNA divided into eight physically separate segments (Lamb and Choppin, 1983), is an example of this type of virus. Six of the genome segments of influenza virus code for only one gene apiece, whereas the other two code for two or three genes. When an influenza virus infects a host cell, each of its RNA segments replicates repeatedly, and the viable progeny viruses produced by the infection contain replicas of the original set of eight RNA segments. Should one of the RNA segments of the parental virus be damaged, its replication or expression may be blocked and no viable progeny viruses would be produced.

Influenza viruses can undergo a simple form of sex. If two undamaged viruses infect the same cell, each viable progeny virus will have a complete set of segments, but each set would be likely to contain a mixture of genes from both parents. This exchange of genes is a form of recombination. When influenza virus particles are UV-irradiated, they are able to undergo multiplicity reactivation (Barry, 1961; see also Chapter 10, Section I.A, for a discussion of multiplicity reactivation). In viruses with DNA genomes, multiplicity reactivation is a consequence of enzyme-mediated exchange of DNA between chromosomes leading to progeny that are free of lethal damages. In segmented RNA viruses, however, multiplicity reactivation likely results from replication of undamaged RNA segments of the infecting viruses, and the reassortment of these replicas to form complete, undamaged progeny viruses. We regard this process as an appropriate model for sex in primitive protocells (Bernstein *et al.*, 1984).

III. Selective Forces Leading to the Origin of Enzyme-Mediated Recombinational Repair

We assume that there was an evolutionary progression from segmented, single-stranded RNA genomes to unsegmented, double-stranded DNA genomes. Presumably, as evolution proceeded, more complex genomes arose, each needing mechanisms for overcoming damage. One possible evolutionary sequence of successive stages is (1) segmented single-stranded RNA, in which each segment corresponds to a separate gene; (2) segmented double-stranded RNA, in which each segment corresponds to a separate gene; (3) segmented double-stranded RNA, in which each segment corresponds to several (possibly related) genes; and (4) unsegmented double-stranded DNA. Organisms with each of these types of genome are known to exist in the present day biota, lending plausibility to the assumption that these types existed early in evolution.

Selective pressures that might have led to the progression enumerated above are discussed next. Because an RNA duplex is more stable than an RNA single strand, the primitive single-stranded RNA genomes (stage 1) may have evolved into double-stranded RNA genomes (stage 2). These remained in a segmented form with each gene on a separate segment. As with the single-stranded RNA genomes, these double-stranded RNA genomes presumably engaged in sexual union and recombination to form damage-free genomes. Reovirus, a present-day organism, is an example of a segmented double-stranded RNA virus that undergoes multiplicity reactivation of damaged genomes (McClain and Spendlove, 1966). Thus, our proposed model for stage 2 genomes has an analogue in the current biota.

A difficulty with having a segmented genome, however, is ensuring that, upon replication, a complete set of genes is passed to each progeny. In fact, influenza viruses frequently form defective progeny viruses with incomplete sets of genes because of unreliable segregation. At stage (3), the segregation problem could be partially solved by clustering genes with related functions on individual duplex RNA segments. In the current biota, in fact, bacteriophage phi 6 has three duplex RNA segments, which contain at least four, three, and three genes, respectively (Cuppels *et al.,* 1980). At this stage, presumably, the problem of genome damage could still be solved by sexual union and recombination to form undamaged genomes.

As protocells with segmented genomes evolved increasing numbers of genes, however, the segregation problem must have become increasingly acute. A simple way of solving this problem is to link all genes end-to-end. Thus, we think that the segmented genome of protocells evolved into the most common present day type of genome, one where genes are linked end-to-end, in order to solve the segregation problem. At this stage (4), the duplex RNA genome may have evolved into duplex DNA genomes because of the greater stability of DNA than RNA. This stage presumably involved two steps: the change of the segmented genome into an unsegmented genome, and the change of RNA, as the genetic material, into DNA. When genes became linked end-to-end, the problem of creating undamaged genomes from damaged ones needed a new solution; i.e., recombination to form undamaged genomes can no longer occur merely by assortment of undamaged replicas. Special enzymes had to evolve to permit the replacement of damaged segments of DNA by the splicing-in of undamaged segments. Thus, enzyme-mediated breakage and exchange between DNA molecules, as presently occurs in DNA viruses, bacteria, and eukaryotes (Chapter 10), must have evolved.

Alternative plausible evolutionary sequences that go from segmented single-stranded RNA to unsegmented double-stranded DNA could also be proposed. For instance, an alternative intermediate might be an unsegmented single-stranded RNA genome. As reviewed by King (1987), several unsegmented RNA viruses have been shown to undergo recombination (e.g., foot-and-mouth disease virus, poliovirus, mouse hepatitis virus). Whatever the exact pathway, the joining of genes end-to-end, we think, was an adaptive response to solve the segregation problem. We consider that this joining gave rise to the need to evolve enzymes to carry out recombinational repair.

IV. Advantage of Sex in Unicellular Haploid Organisms

In Chapter 13, Section III, we argued that outcrossing in multicellular organisms is maintained by the benefit of complementation. However, we also argued in this chapter, Section II.B, that outcrossing arose in primitive cells as a means of bringing two haploid genomes together to allow recombinational repair. Because the diploid stage of such primitive unicellular organisms was probably transient, complementation was less likely to be an advantage than repair in maintaining outcrossing.

This reasoning suggests that in extant haploid unicellular organisms, outcrossing still may be primarily an adaptation for DNA repair. This idea can be experimentally tested in the simple unicellular haploid fission yeast *S. pombe*. In this organism, introduction of DNA damage during the predominant haploid phase might stimulate outcrossing, leading to meiosis and recombinational repair. *Schizosaccharomyces pombe* is facultatively sexual–asexual. Any wildtype population is a mixture of two mating types, because a cell of one mating type can give rise to a cell of the opposite mating type every two generations (Klar, 1987). A vegetative cell can enter the mitotic cycle and continue to grow vegetatively, or it may associate with a cell of the opposite mating type, fuse, undergo meiosis, and form sexual spores. If vegetative cells are grown in medium rich in nitrogen, mating is infrequent and sexual spores constitute $<10^{-3}$ of the exponential phase cell population. When the cells are transferred to a medium without nitrogen and incubation is continued for 24 hr, about 20% of the cells form zygotic structures or sporulate (Nurse, 1985). If, as we propose, meiosis is a adaptation for DNA repair, vegetative cells growing in a medium with adequate nitrogen may enter the sexual cycle more frequently when treated with a DNA-damaging agent than when untreated. As discussed in Chapter 6, Section I, endogenous hydrogen peroxide (H_2O_2) is a likely important natural DNA-damaging agent. As reviewed by Bernstein and Johns (1989), the major lethal effect of H_2O_2 appears to be on DNA, rather than on other molecular or cellular constituents. Therefore, Bernstein and Johns (1989) tested whether or not treatment of vegetative cells with H_2O_2 causes an increase in sexual reproduction. They compared populations of vegetative cells that had been treated with H_20_2 in late exponential phase and then grown to stationary phase in adequate nitrogen medium with similar populations that had not been treated with H_2O_2. Among untreated S. pombe populations that reached stationary phase, the sexual spores produced by meiosis represented about 1% of the total cells. However, treatment of late exponential-phase vegetative cells with H_2O_2 increased the percentage of meiotic spores in the stationary phase by an average of about 8.4-fold. These observations indicate that oxidative damage induces sexual reproduction in S. pombe. These results can be explained by the hypothesis that outcrossing, in this organism, promotes DNA repair. Furthermore, the data obtained are consistent with the idea that if an S. pombe cell, after treatment with H_2O_2, is able to grow vegetatively, it does so; however, if it has DNA damage that prevents vegetative growth, it may mate, undergo recombinational repair, and generate viable spore progeny.

V. Emergence of Diploid Organisms

In Sections II and IV of this chapter, we indicated that the sexual cycle was selected for because it promotes repair of DNA damage. In this section, we consider the evolutionary emergence of predominantly diploid organisms as a modification of the sexual cycle. We include in this explanation a brief summary of our previous argument (Chapter 13, Section III) that in multicellular organisms it is mutations, rather than damages, that play the dominant selective role in maintaining outcrossing sexual reproduction.

A. Improving Replication Accuracy as a Way of Avoiding Mutation

As haploid organisms evolved more complex phenotypes, their genomes expanded to encode the additional information needed. This increase in information made the organisms intrinsically more vulnerable to deleterious mutations. In other words, a large genome with many essential gene functions can go wrong in more possible ways than a genome with few essential gene functions. This problem was probably dealt with by improving the accuracy of the enzymatic machinery that replicates DNA. This would allow the mutation rate per genome per replication to remain low. Evidence from present-day haploid organisms indicates that accuracy of replication increases in direct proportion to the size of the genome. Drake (1974) compared five haploid organisms (phages T4 and lambda, the bacteria *E. coli* and *S. typhimurium,* and the fungus *N. crassa*) and showed that with increasing genome size over a 1,000-fold range, mutation rate per base pair per replication declines in such a way that the mutation rate per genome per replication remains roughly constant at about 0.001–0.003 for all five organisms. This supports the inference that an increase in genetic information selects for increase in replication accuracy.

B. Masking Mutations by Genetic Complementation and Diploidy

Bernstein *et al.* (1981) have suggested that as the haploid genome expanded, a point of diminishing returns was reached at which the cost of improvements in the replicative machinery (loss of speed in replication, etc.) no longer was balanced by the benefit of increased accuracy. A new strategy could be used at this stage to allow for further expansion of the genome. This is to permit diploidy, previously a transient stage of the sexual cycle, to become the dominant stage. Thus, a stage of the sexual cycle, previously used to promote genome repair, would be much prolonged to allow masking of mutational errors. Diploidy allows the expression of deleterious mutations to be masked through complementation (see also Chapter 13, Section III.B). When diploidy first became predominant, genome size and, hence, information content could expand without the constraint of improving accuracy of replication. The opportunity to increase information content at low cost would be advantageous by allowing new adaptations to be encoded.

The shift to diploidy, however, entails costs associated with readjusting the life cycle and also with the synthesis of an extra genome per cell. These costs presumably delayed the transition to diploidy. But the transition to diploidy may have happened when the costs of maintaining an extra genome within the cell became small relative to the costs of maintaining the total cellular metabolism. We argued in Section II.B of this chapter that outcrossing arose in primitive protocells because of the cost of maintaining more than one set of chromosomes within a protocell. These costs stem from the simple fact that for a cell to duplicate, all its chromosomes must replicate. For primitive protocells, the costs of replicating the genome takes a large part of its available resources. As a result, for protocells, the costs of duplicating the genome would be prohibitive, and for the diploid cell would approach twice that of duplicating the haploid cell. Later, with the evolution of complex intracellular structures and functions, the costs of replicating the genome would become a smaller portion of the total resource budget. As the relative costs of diploidy decreased, diploidy would have emerged as the predominant stage of the life cycle because of the advantages of complementation. As discussed in Chapter 13, Section III, at this point the advantage of complementation (in the presence of recombination) would select for the maintenance of outcrossing.

The benefit of diploidy cannot be exploited indefinitely, however, because deleterious mutations are protected from being weeded out by natural selection and will consequently accumulate (Muller, 1932). This accumulation will continue until a new balance is reached between the occurrence of new mutations and the loss of deleterious homozygous recessive mutations through natural selection. When this equilibrium point is attained, many more deleterious recessive mutations are present in each individual than were previously present in the haploid. Furthermore, at this point, the expression of deleterious mutations in the predominant diploid is as frequent as it was in the original predominant haploid. Because of the accumulation of deleterious recessive mutations, the advantage of shifting from predominant haploidy to predominant diploidy is only temporary (except for the presumed increase in genome size and the new beneficial adaptations this allows, as discussed above). It is clear, however, that in the short run diploidy, through complementation, provides advantages over haploidy. Of course, the long-range consequences of evolutionary events are irrelevant to whether or not the events will occur and those events are selected that are immediately beneficial. The switch to predominant diploidy was probably irreversible, because, once equilibrium was reached, a shift back to haploidy would express all the deleterious alleles, which are now at a higher frequency. However, if the switch to predominant diploidy was exploited to expand the genome (before homozygous recessive deleterious mutations became a problem), then among extant organisms one might expect little or no overlap in genome size from haploid prokaryotes and haploid eukaryotes as compared with diploid eukaryotes. According to the data compiled by Sparrow and Nauman (1976), this seems to be the case, even though genome size for all organisms varies over eight orders of magnitude.

After organisms became diploid, the balance of selective forces maintaining sex probably changed. While recombinational repair remained necessary for removing germ line DNA damage, outcrossing, which was initially required just for bringing haploid genomes together for recombinational repair, acquired an additional role. This new role was to maintain the masking of mutations.

VI. *Sex and the Emergence of Species*

A. Darwin's Dilemma

In "The Origin of Species," Darwin (1859) explained evolution in terms of natural selection but was puzzled by the clustering of life into species. At the beginning of his chapter VI, entitled "Difficulties of the Theory," Darwin noted that long before the reader has arrived at this part of his work, "a crowd of difficulties" will have occurred to him. Darwin commented that some of them were so serious that he could hardly reflect on them without to some degree being staggered, but to the best of his judgment, most were only apparent, and those that were real were not fatal to the theory. Darwin then went on to describe several difficulties, but only the first concerns us here. Darwin asked why, if species have descended from other species by fine gradations, do we not see innumerable transitional forms everywhere? Why is not all life in confusion, instead of the species being as we see them, well defined? We can refer to this dilemma as the absence or rarity of transitional varieties in space. Related to this dilemma is the the absence or rarity of transitional varieties in time. Darwin noted that innumerable transitional forms must have existed, and asked why do we not find them embedded in countless numbers in the crust of the earth? (Darwin, 1859: p. 125). That clearly defined species actually do exist in nature in both space and time suggests that some fundamental characteristic of natural selection operates to generate and maintain them.

B. Species Emerge as a Result of the Dynamics of Sexual Reproduction

Division of the biota into species on the basis of morphological distinctions is noted by humans in all natural languages, and this perception preceded any notion of evolution. In this century, the concept of a species has come to be based on the reproductive isolation of a collection of interbreeding organisms as the key idea, in contrast to "essential" morphological similarities. An example of a contemporary definition is that of Mayr (1982: p. 273), who commented that a species is a reproductive community of populations (reproductively isolated from others) that occupies a specific niche.

Keeping the modern definition of species in mind, we can now address Darwin's dilemmas. Bernstein et al. (1985c) argued that the resolution of Darwin's dilemmas lies in the fact that outcrossing sexual reproduction has an intrinsic cost of rarity. The cost of rarity stems from the simple fact that it takes two individuals to mate. At low population density, the fitness of outcrossing individuals declines because of increased costs of finding a mate. A cost–benefit analysis of sexual species splitting was carried out by Bernstein et al. (1985c) and Hopf and Hopf (1985). The benefit of splitting was the finer adaptation of the smaller species resulting from the split. Using this approach, it was shown that when a cost of rarity is introduced, a point of diminishing returns is reached with increasing splitting. At this point, sexual species are stable and will not proliferate further into smaller species. For asexual organisms, there is no cost of rarity; as a result, there are only benefits to fine-scale adaptation. This analysis suggested that transitional sexual species with few members cannot compete with larger established species, even if moderately better adapted, because of the cost of rarity. Thus, the dilemma raised by Darwin concerning the absence or rarity of transitional forms in space can be explained by the dynamic of sexual reproduction.

C. Evidence in Nature for a Cost of Rarity

A number of lines of evidence support a cost of rarity in sexual species. For many populations, mating success is density-dependent over wide ranges of density. Examples include many wind-pollinated plants, aquatic organisms such as sexual zooplankton (Gerritsen, 1980), other invertebrates that depend on random collision for mating, and insects (Milne, 1950; Park, 1933; Kort, 1962). Andrewartha and Birch (1954: pp 334–343) discuss further examples. With respect to the difficulties of mating in sparse populations, Andrewartha and Birch (1954: p. 342) note that it probably occurs frequently in natural populations without having been noticed very often.

The disadvantageous effects on mating success of low population density can be overcome to some extent by communication and directed movement. Thus, many species in the current biota occur at low absolute density and, yet, individuals in these species manage to find mates. However, communication and directed movement introduce indirect costs of rarity. Direct costs are measured in terms of decreases in mating success. Indirect costs are measured roughly by the amount of time and energy devoted to attracting or finding a mate. Such investments will detract from other components of fitness such as ability to obtain food or to protect against predators. Therefore, the cost of rarity need not be manifested solely as a decrease in ability to find a mate.

Other lines of evidence suggesting that sexual reproduction is costly under conditions of low population density come from studies of facultatively sexual organisms and from studies of the distribution of parthenogens in nature. Facultatively sexual organisms can alternate between parthenogenetic reproduction and sexual reproduction. If there is a cost of rarity, then at high population densities sexual reproduction should tend to occur and at low densities asexual reproduction should be favored. Such a pattern among facultative sexual organisms is, in fact, frequently found in nature. Bernstein *et al.* (1985c) have reviewed examples of this pattern among rotifers, cladocerans, fresh water coelenterates, and ciliate protozoans. In addition, studies of the distribution of animal parthenogenesis suggest that it predominates in novel or disturbed habitats such as temporary bodies of water. Parthenogenetic species predominate in small temporary ponds, streams, and rock pools, while sexual species predominate in larger, more permanent bodies of water (Bell, 1982: pp. 359–361). In temporary bodies of water, population densities should be low initially, implying a high cost of rarity. Other examples of empty, newly colonized habitats in which parthenogenesis is favored are the habitats created by floods, fires, or receding glaciers. Thus, numerous observations suggest that sexual reproduction has an associated cost of rarity.

Additional support for the thesis that the cost of rarity imposed by sexual reproduction accounts for the emergence of species can be found in studies of asexual populations. Because asexual reproduction does not incur a cost of rarity, we would expect more fine-grained adaptation among asexuals. Thus, a contrast should exist between organisms reproducing sexually with those reproducing asexually. Maynard Smith (1983) argued that discontinuities only exist between species if there is sexual reproduction, although some workers (e.g., Holman, 1987; Mishler, 1990) contend that asexual species can be as distinct as sexual species. Reviews of apomictic clonal populations by Bell (1982: pp. 48–55) and Parker (1979) concluded that sympatric polytypic clones are the norm among obligate parthenogenic animals. A sympatric distribution is one in which the different types coincide or overlap. A proliferation of overlapping types seems to be typical of clonal apomictic populations. Thus, the segregation of sexual organisms into coarse-grained groups (species) contrasts with the fine-grained profusion of types characteristic of asexual organisms. The presence of a cost of rarity among sexuals and the absence of such a cost among asexuals, we think, is sufficient to explain the contrasting distributions (see also Bernstein *et al.,* 1985a, 1985c; Hopf and Hopf, 1985).

D. The Species Concept

Our explanation of why separate species exist supports Mayr's view that species classification properly reflects not a common essence as in the traditional notion, but the fact of interbreeding. Thus, the underlying evolutionary cause of morphological similarities is the sexual dynamic of natural selection, rather than similarity of individuals being the fundamental cause of interbreeding. The concept of a niche, which Mayr included in his definition of species (see above), we think actually depends conceptually on prior species differentiation. We think it is species differentiation (resulting from the sexual dynamic) that defines niches rather than vice versa.

There are two aspects to the question "what is the origin of species?" These are (1) what are the evolutionary mechanisms of speciation, and (2) what accounts for the individuality and separateness of species in the biota. So far we have focused on the second aspect. However, since Darwin's time, efforts to understand the nature of species have primarily focused on the first aspect, namely mechanisms by which established species give rise to new species. It is now widely accepted that the underlying concept needed to understand the origin of new species from old ones is reproductive isolation (Mayr, 1982: p. 274).

E. Punctuated Evolution

We began this section with the question of why organisms are distributed into species rather than into a continuum of finely adapted types. As we have argued, the dynamic of sexual reproduction with its associated cost of rarity generates species. Just as this dynamic gives rise to a distribution of distinct types in space, i.e., species, it also should contribute to distinctness in time (Bernstein *et al.,* 1985a, 1985c). The cost of rarity leads to an intrinsic stability of moderately adapted species that are already established. Such species would only be replaced by new, *substantially* better adapted species. Potential new species, that are only marginally better adapted, would have a high cost of rarity when they arise. Therefore, the few individuals in the better adapted new species could not compete unless their new adaptive features could compensate for the high cost of rarity in the beginning. Consequently, displacement of one species by another would have a quantum character. In general, this view of species evolution is consistent with that of Eldridge and Gould (1972), who regard evolution as occurring by long periods of stasis punctuated by relatively rapid changes in species and community compositions.

VII. *Summary and Conclusions*

In this chapter, we have argued that sexual reproduction arose very early in the evolution of life as a way of overcoming genome damage. The earliest organisms, primitive protocells, had genomes comprised of a small number of unconnected genes. The initial sexual process brought together protocells with damaged genomes, allowing reassortment of an undamaged set of genes. This process was a primitive form of genome repair. As genetic information increased, genes became connected, end-to-end, as an adaptation for reliably assorting larger numbers of genes during protocell division. This, we think, led to selection for enzyme-mediated recombinational repair as a way of continuing to repair genome damage. As organisms acquired further adaptations and genome information content continued to expand, diploidy, at first transient, became the predominant way of coping with the increased vulnerability to mutation. This change allowed further genome expansion. The switch from predominant haploidy to predominant diploidy is essentially irreversible, because reversion to predominant haploidy would lead to expression of accumulated deleterious recessive mutations. Outcrossing, at this point, becomes necessary to accomplish continued masking of deleterious recessive mutations.

A consequence of sex is an intrinsic cost of rarity. At low population density, fitness declines due to increased costs of finding a mate. This fundamental constraint can inhibit evolutionary success of the best-adapted species if it is small in numbers. This can account for the existence of discrete, long-lasting species. In general, we consider that the existence of discrete species, and their characteristic cohesion and stability over time, are direct consequences of sex, and sex in turn is a consequence of the need to cope with genome damage through recombinational repair, while at the same time masking deleterious mutations through complementation.

References

Andrewartha, H. G., and Birch, L. C. (1954). "The Distribution and Abundance of Animals." University of Chicago Press, Chicago.

Baker, R. R., and Parker, G. A. (1973). The origin and evolution of sexual reproduction up to the evolution of male-female phenomenon. *Acta Biotheoretica* **22(2)**, 1–77.

Barry, R. D. (1961). The multiplication of influenza virus II. Multiplicity reactivation of ultraviolet irradiated virus. *Virology* **14**, 398–405.

Bell, G. (1982). "The Masterpiece of Nature: The Evolution and Genetics of Sexuality." University of California Press, Berkeley.

Bernstein, C., and Johns, V. (1989). Sexual reproduction as a response to H_2O_2 damage in *Schizosaccharomyces pombe*. *J. Bacteriol.* **171**, 1893–1897.

Bernstein, H., and Bernstein, C. (1989). Bacteriophage T4 genetic homologies with bacteria and eucaryotes. *J. Bacteriol.* **171**, 2265–2270.

Bernstein, H., Byers, G. S., and Michod, R. E. (1981). Evolution of sexual reproduction: Importance of DNA repair, complementation and variation. *Am. Nat.* **117**, 537–549.

Bernstein, H., Byerly, H. C., Hopf, F. A., Michod, R. E., and Vemulapalli, G. K. (1983). The Darwinian dynamic. *Q. Rev. Biol.* **58**, 185–207.

Bernstein, H., Byerly, H. C., Hopf, F. A., and Michod, R. E. (1984). Origin of sex. *J. Theor. Biol.* **110**, 323–351.

Bernstein, H., Byerly, H. C., Hopf, F. A., and Michod, R. E. (1985a). The evolutionary role of recombinational repair and sex. *Int. Rev. Cytol.* **96**, 1–28.

Bernstein, H., Byerly, H. C., Hopf, F. A., and Michod, R. E. (1985b). DNA repair and complementation: The major factors in the origin and maintenance of sex. *In* "Origin and Evolution of Sex" (H. O. Halvorsen, ed.), pp. 29–45. Alan R. Liss, New York.

Bernstein, H., Byerly, H. C., Hopf, F. A., and Michod, R. E. (1985c). Sex and the emergence of species. *J. Theor. Biol.* **117**, 665–690.

Bernstein, H., Hopf, F. A., and Michod, R. E. (1987). The molecular basis of the evolution of sex. *Adv. Genet.* **24**, 323–370.

Bernstein, H., Hopf, F. A., and Michod, R. E. (1989). The role of DNA repair in sexual reproduction. *In* "Sociobiology of Reproduction, Strategies in Animal and Man" (E. Voland and C. Vogel, eds.), pp. 3–18. Chapman and Hall, New York.

Cassuto, E., Lightfoot, L. A., and Howard-Flanders, P. (1987). Partial purification of an activity from human cells that promotes homologous pairing and the formation of heteroduplex DNA in the presence of ATP. *Mol. Gen. Genet.* **208**, 10–14.

Chen, D. S., and Bernstein, H. (1988). Yeast gene *RAD52* can substitute for phage T4 gene *46* or *47* in carrying out recombination and DNA repair. *Proc. Natl. Acad. Sci. USA* **85,** 6821–6825.

Cuppels, D. A., van Etten, J. L., Burbank, E. E., Lane, L. C., and Vidaver, A. K. (1980). In vitro translation of the three bacteriophage phi 6 RNAs. *J. Virol.* **35,** 249–251.

Darnell, J. E., and Doolittle, W. F. (1986). Speculation on the early course of evolution. *Proc. Natl. Acad. Sci. USA* **83,** 1270–1275.

Darwin, C. (1859). "The Origin of Species by Means of Natural Selection," pp. 1–386. The Modern Library, New York. [1936 Facsimile of the first edition, 1859.]

Dougherty, E. C. (1955). Comparative evolution and the origin of sexuality. *Syst. Zool.* **4,** 145–190.

Drake, J. W. (1974). The role of mutation in bacterial evolution. *Symp. Soc. Microbiol.* **24,** 41–58.

Eigen, M., and Schuster, P. (1979). "The Hypercycle, a Principle of Natural Self-Organization." Springer-Verlag, Berlin.

Eigen, M., Gardiner, W., Schuster, P., and Oswatitsch, P. (1981). The origin of genetic information. *Sci. Am.* **244 (April),** 88–118.

Eldridge, N., and Gould, S. J. (1972). Punctuated equilibria: An alternative to phyletic gradualism. *In* "Models in Paleobiology" (T. J. M. Schopf, ed.), pp. 82–115. Freeman, Cooper and Company, San Francisco.

Fishel, P. A., Detmer, K., and Rich, A. (1988). Identification of homologous pairing and strand exchange activity from a human tumor cell line based upon DNA affinity chromatography. *Proc. Natl. Acad. Sci. USA* **85,** 36–40.

Fox, G. E., Stackebrandt, E., Hespell, R. B., Gibson, J., Maniloff, J., Dyer, T. A., Wolf, R. S., Balch, W. E., Tanner, R. S., Magrum, L. J., Zablen, L. B., Blakemore, R., Gupta, R., Bonen, L., Lewis, B. J., Stahl, D. A., Luehrsen, K. R., Chen, K. N., and Woese, C. R. (1980). The phylogeny of procaryotes. *Science* **209,** 457–463.

Gerritsen, J. (1980). Sex and parthenogenesis in sparse populations. *Am. Nat.* **115,** 718–742.

Holman, E. W. (1987). Recognizability of sexual and asexual species of rotifers. *Syst. Zool.* **36,** 381–386.

Hopf, F. A., and Hopf, F. W. (1985). The role of the Allee effect on species packing. *Theor. Pop. Biol.* **27,** 27–50.

Hotta, Y., Tabata, S., Bouchard, R. A., Pinon, R., and Stern, H. (1985). General recombination mechanisms in extracts of meiotic cells. *Chromosoma* **93,** 140–151.

Hsieh, P., Meyn, M. S., and Camerini-Otero, R. D. (1986). Partial purification and characterization of a recombinase from human cells. *Cell* **44**, 885–894.

King, A. M. Q. (1987). RNA viruses do it. *Trends Genet. Anal.* **3**, 60–61.

Klar, A. J. S. (1987). Differentiated parental DNA strands confer developmental asymmetry on daughter cells in fission yeast. *Nature (London)* **326**, 466–470.

Kmiec, E. B., and Holloman, H. K. (1982). Homologous pairing of DNA molecules promoted by a protein from *Ustilago. Cell* **29**, 367–374.

Kmiec, E. B., and Holloman, W. K. (1984). Synapsis promoted by *Ustilago rec-1* protein. *Cell* **36**, 593–598.

Kort, J. (1962). Effect of population density on cyst production in *Heterodera rostochiensis woll. Nematologica* **7**, 305–308.

Kuhn, H. (1972). Self-organization of molecular systems and evolution of the genetic apparatus. *Angew. Chem. Int. Ed.* **11**, 798–820.

Lamb, R. A., and Choppin, P. W. (1983). The gene structure and replication of influenza virus. *Annu. Rev. Biochem.* **52**, 467–506.

Margulis, L. (1981). "Symbiosis in Cell Evolution." W. H. Freeman, San Francisco.

Maynard Smith, J. (1978). "The Evolution of Sex." Cambridge University Press, London.

Maynard Smith, J. (1983). The genetics of stasis and punctuation. *Annu. Rev. Genet.* **17**, 11–25.

Mayr, E. (1982). "The Growth of Biological Thought: Diversity, Evolution, and Inheritance." Harvard University Press, Cambridge.

McClain, M. E., and Spendlove, R. S. (1966). Multiplicity reactivation of reovirus particles after exposure to ultraviolet light. *J. Bacteriol.* **92**, 1422–1429.

Michod, R. E. (1983). Population biology of the first replicators. *Am Zool.* **23**, 5–14.

Milne, A. (1950). The ecology of the sheep tick, *Ixodes ricinus* L., spatial distribution. *Parasitology* **40**, 35–45.

Mishler, B. D. (1990). Reproductive biology and species distinctions in the moss genus *Tortula*, as represented in Mexico. *Syst. Bot.* **15**, 86–97.

Muller, H. J. (1932). Some genetic aspects of sex. *Am. Nat.* **66**, 118–138.

Nurse, P. (1985). Mutants of the fission yeast *Schizosaccharomyces pombe* which alter the shift between cell proliferation and sporulation. *Mol. Gen. Genet.* **198**, 497–502.

Orgel, L. E. (1987). Evolution of the genetic apparatus: A review. *Cold Spring Harbor Symp. Quant. Biol.* **52**, 9–16.

Park, T. (1933). Studies in population physiology. II. Factors regulating initial growth of *Tribolium confusum* populations. *J. Exp. Zool.* **65**, 17–42.

Parker, E. D., Jr. (1979). Ecological implications of clonal diversity in parthenogenetic morphospecies. *Am. Zool.* **19,** 753–762.

Rosenshine, I., Tchelet, R., and Mevarech, M. (1989). The mechanisms of DNA transfer in the mating system of an archaebacterium. *Science* **245,** 1387–1389.

Sagan, C. (1973). Ultraviolet selection pressure on the earliest organisms. *J. Theor. Biol.* **39,** 195–200.

Schopf, J. W. (1978). The evolution of the earliest cells. *Sci. Am.* **239 (Sept),** 110–139.

Shestakov, S. V., and Khyen, N. T. (1970). Evidence for genetic transformation in blue-green alga *Anacystis nidulans. Mol. Gen. Genet.* **107,** 372–375.

Singh, R. N., and Sinha, R. (1965). Genetic recombination in a blue-green alga, *Cylindrospermum majus* Kuetz. *Nature (London)* **207,** 782–783.

Sparrow, A. H., and Nauman, A. F. (1976). Evolution of genome size by DNA doublings. *Science* **192,** 524–529.

Stahl, F. W. (1979). "Genetic Recombination: Thinking about It in Phage and Fungi." W. H. Freeman, San Francisco.

Stebbins, G. L. (1960). The comparative evolution of genetic systems. *In* "The Evolution of Life" (S. P. Tax, ed.), pp. 197–226. University of Chicago Press, Chicago.

Stevens, S. E., and Porter, R. D. (1980). Transformation in *Agmenellum quadruplicatum. Proc. Nat. Acad. Sci. USA* **77,** 6052–6056.

Whitehouse, H. L. K. (1982). "Genetic Recombination." Wiley, New York.

Woese, C. R. (1983). The primary lines of descent and the universal ancestor. *In* "Evolution from Molecules to Men" (D. S. Bendall, ed.), pp. 209–233. Cambridge University Press, London.

Chapter Fifteen

Other Theories of Aging and Sex

In the previous chapters, we have presented the evidence bearing on the DNA damage hypothesis of aging and the DNA repair (and complementation) hypothesis of sex. At least 10 theories of aging and 3 of sex have been described in the scientific literature based on other hypotheses. Eight of the theories of aging, with some changes of emphasis or small alterations of content, can be shown to be compatible with, or complementary to, the DNA damage hypothesis of aging. We discuss these eight theories of aging first. The ninth theory of aging, the error catastrophe theory, has been extensively tested experimentally, but most of the results obtained have tended to refute, rather than support, the theory. We describe some of these experiments, because the error catastrophe theory has had a substantial influence in recent years. The tenth theory of aging is that of programmed aging. Although, a number of workers have presented arguments in the literature to refute it, this theory still retains some influence. We present this theory and review the arguments bearing on it.

Of the three theories of sex in the literature based on other hypotheses, one has similarities to the DNA repair hypothesis of sex, which we present in this book. We first review this similar theory of sex. A second type of theory is based on the hypothesis, originally proposed by Weisman in 1892, that the benefit of sex is the production of genetic variation. A number of difficulties with this idea have been pointed out, especially in the last 15 years, and we discuss some of these. The third, more recent theory of sex, based on the selfish-DNA hypothesis, has been developed since 1982. It, too, has serious difficulties, which we discuss.

I. Aging Theories

A. The Free Radical Theory of Aging

In Chapter 6, we reviewed evidence indicating that oxidative free radical-induced DNA damage is probably a major cause of aging in mammals. DNA, in contrast to other macromolecules, is particularly vulnerable because it is present in only two copies per diploid cell, instead of the many copies typical of mRNA and protein species. We have argued (Chapter 5, Section I.A) that because DNA contains the master code, loss of its ability to transcribe vital genes is lethal to the cell. Other workers have taken the broader view that free radical damage to cellular constituents in general, not just to DNA, is the primary factor in mammalian aging.

The free radical theory of aging as usually expressed does not emphasize DNA damage. This theory was originally formulated by Harman (1956, 1957) and has been reviewed recently by Pryor (1987) and Sohal (1987). According to Harman (1981), accumulating evidence indicates that the sum of deleterious free radical reactions occurring continuously throughout the cells and tissues constitutes the aging process or is a major contributor to it. In mammalian systems, the free radical reactions are mainly those involving oxygen.

An often-cited example of a direct effect of free radical damage on a biomolecule is the accumulation of age pigment (lipofuscin). Age pigment is formed by oxidative polymerization of lipids (probably largely mitochondrial) and of proteins (Miquel et al., 1977). However, age pigments do not appear to be life-shortening or pathological (Pryor, 1987). Direct oxidative damage to proteins may also occur, but this is unlikely to be critical as suggested by the following observation. Treatment with hydrogen peroxide (which should accelerate natural oxidative free radical effects) can damage proteins as well as DNA. However, the damaged proteins, in contrast to the DNA, rapidly turn over (Davies and Goldberg, 1987).

In Chapter 3, Section I, we presented evidence that DNA damage interferes with RNA transcription, and in Chapter 5, Section II.A, we reviewed evidence that in several mammalian tissues mRNA transcription and protein synthesis decline with age. We argued that this decline is likely due to the accumulation of damage in DNA. In general, the DNA damage hypothesis and the free radical theory are compatible if one assumes that the most important target of free radical damage is DNA, and that other oxidatively damaged molecules are efficiently replaced as long as the DNA is intact.

B. Neuroendocrine Theory of Aging

Finch (1972) hypothesized that certain brain cells (present in the hypothalamus and limbic system, which control endocrine function via the pituitary) may act as pacemakers of aging in mammals. This hypothesis assumes that deficiencies in such cells may be the mechanism underlying the age-associated decline in homeostasis, leading to eventual death. Neurotransmitter systems are considerably modified with age as evidenced by the decrease of brain catecholamines (Finch, 1978; McGeer and McGeer, 1975) and the reduction in catecholamine receptors in the corpus striatum and cerebellum of old rats (Greenberg and Weiss, 1978). Age-related reductions in receptor concentrations have been implicated in many of the postmaturational reductions of hormone response (reviewed by Roth, 1979). The mechanisms responsible for hormone receptor alterations during aging are unknown. As discussed in Chapter 5, however, the accumulation of DNA damage in nondividing cells likely leads to reduced gene expression. This reduced expression might explain both the decreased production of hormone receptors and the decline in general neuroendocrine function. Thus, the neuroendocrine theory of aging is compatible with the DNA damage hypothesis of aging.

C. Evolutionary Theory of Aging

Medawar (1952) and Williams (1957) proposed an evolutionary theory of aging. They pointed out that in a hypothetical population of organisms that did not undergo senescence, death due to such factors as accident, predation, disease, and starvation would still occur. In such a population, chronologically older individuals would necessarily be less well represented than chronologically younger individuals. As a consequence, the force of natural selection acting on chronologically older populations would progressively decline. Because of this decline in selective force, genes that are beneficial in youth would tend to be selected for even if they were deleterious later in life.

Genes with multiple effects, such as these, are referred to as pleiotropic. The expression of such pleiotropic genes would then result in senescence. Sacher (1968), on the other hand, considered that such pleiotropic theories have serious flaws, the main one being that such age-dependent genes have not been shown to exist.

In answer to this criticism, we think that an example of the age-dependent pleiotropic "genes" postulated by the Medawar–Williams theory might be the genetic elements that regulate DNA repair in somatic cells. The studies of Gensler (1981) and Karran *et al.* (1977) on the brain and of Lampidis and Schaiberger (1975) and of Stockdale (1971) on muscle suggest that the transition from the mitotic to the postmitotic condition is accompanied by a reduction in DNA repair. This turnoff of repair is presumably selected for because, as discussed in Chapter 4, Section II.D, the resources of the cell that were previously devoted to repair, as well as to replication and cell division, could now be used more effectively for critical neuronal or muscular functions. The harmful effect of this genetically controlled turnoff is to allow enhanced accumulation of DNA damages. This, in turn, causes more frequent transcriptional impairment and progressive loss of cell and tissue function. However, because the harmful effects are cumulative, and most severe in chronologically older individuals (whose numbers are depleted by causes of death unrelated to senescence), the beneficial effects of the genetic elements that control turnoff of repair during youth would predominate. Thus, regulatory genetic elements that turn off DNA repair in postmitotic cells may be important examples of the postulated pleiotropic "genes," which are beneficial in youth but deleterious at a later age.

Edney and Gill (1968) noted that the theory of Medawar and Williams predicts that with a decline in the rate of mortality due to reductions in such factors as accidents, predation, disease, or starvation, the rate of senescence would also tend to diminish. A decline in the rate of aging could evolve by at least two mechanisms. Somatic DNA repair processes might become more efficient or vulnerable cells might be produced in greater numbers, with greater redundancy, to enable functional compensation for losses due to damage. We have already reviewed evidence that shows a correlation between life span and DNA repair capacity among various mammalian species (Chapter 7, Section I), a correlation that suggests that improvements in the efficiency of DNA repair increase longevity. We have also described the correlation between relative brain weight (an approximate measure of the number of neurons present relative to body weight) and longevity (Chapter 5, Section III.A), a correlation that suggests the evolution of neuronal redundancy.

The increase in DNA repair and neuronal redundancy would be costly during youth in terms of energy and resources diverted from other activities but would be beneficial in slowing the rate of senescence. The genetic elements controlling these changes would also be examples of the pleiotropic "genes" postulated by Medawar and Williams. In general, natural selection should be able to act on such genes either to increase or decrease the rate of aging.

Recently, Rose and Hutchinson (1987) reviewed the Medawar–Williams theory and suggested that it is corroborated by available evidence. In general, the Medawar–Williams theory is compatible with the DNA damage hypothesis of aging. The Medawar–Williams theory deals with the general evolutionary process that determines the rate of aging, whereas the DNA damage hypothesis deals with the underlying molecular mechanism of aging.

D. Theory That Aging Results from Cell Differentiation

Miquel *et al.* (1979) reviewed evidence bearing on the hypothesis, originally proposed by Minot (1907), that cellular aging and death of metazoan animals are the result of cell differentiation. This idea has also been discussed by Strehler (1977: p. 247).

Miquel *et al.* (1979) reviewed data on the fruit fly *Drosophila* indicating that aging is accompanied by a gradual disorganization of its postmitotic cells, probably resulting from deleterious reactions associated with oxidative metabolism. At the physiological level, this disorganization is manifested as a decline in such abilities as negative geotaxis and mating ability. At the structural level, it is manifested by such features as loss of ribosomes and mitochondria and by age pigment accumulation. They also theorized that the primary aging changes in mice occur in the fixed postmitotic cells and extracellular components. This hypothesis is compatible with the DNA damage hypothesis of aging because, by the DNA damage hypothesis, it is in postmitotic cells specifically that DNA damage should be found to accumulate. In Chapter 5, Section II, we presented evidence that cellular aging was most prominent in postmitotic differentiated cells of the brain and muscle.

E. Disposable Soma Theory

The disposable soma theory of aging was proposed by Kirkwood (1977, 1981) and Kirkwood and Holliday (1979). According to this theory, there is a trade-off between maintenance of somatic parts to secure the potential for longer life and the immediate investment by the individual in reproduction (Kirkwood, 1988). Kirkwood (1985) modeled the effects of this trade-off and showed that the strategy of maximum fitness balances somatic maintenance against reproductive effort in such a way that the investment of resources and energy in maintenance and repair of the soma is always less than what is required for indefinite somatic longevity (see also Chapter 5, Section I.B). This leads to two predictions: (1) aging is due to an accumulation of unrepaired somatic defects, and (2) the rate of aging can be modulated by altering the level of key maintenance and repair processes. This theory is clearly compatible with the DNA damage hypothesis of aging if the "unrepaired somatic defects" are taken to be primarily DNA damages, and "maintenance and repair processes" are taken to be primarily DNA repair processes.

F. Immunological Theory of Aging

Some gerontologists, notably Walford (1969), have maintained that the whole pattern of aging is a reflection of the immune system. Burnet (1978: pp. 84–87) thought that progressive accumulations of genetic errors in the immune system are responsible for many of the manifestations of old age, including the characteristic vulnerability of the old to disease. Old people die more frequently from infection because their immune systems have lost their former capacity to respond briskly and to generate a long-lasting immunity against most of the infections they encounter. Aging in mice and humans is accompanied by atrophy of the thymus, diminution of the circulating levels of lymphocytes, and diminution of the peripheral lymphoid tissues (Burnet, 1970: p. 94). In Chapter 4, Section V.B, we reviewed evidence for accumulation of DNA damage in lymphocytes with age, and in Chapter 5, Sections II.A.4 and II.B.4, we reviewed evidence for decline in lymphocyte function with age. We think it unlikely that the decline in the immune system is responsible for the whole pattern of aging because in most humans immunity remains effective at least until the fifties. According to Burnet (1978: p. 85), serious deficiency in tests of immune responses is rarely seen in healthy old people until after 75 yr, although the tests that are used almost certainly will miss minor degrees of immune deficiency. We do not dispute Burnet's view that progressive accumulation of "genetic errors" in the immune system are responsible for many features of old age. However, on the basis of arguments presented in Section H, below, we think the type of genetic errors that are most relevant are DNA damages rather than the somatic mutations emphasized by Burnet.

G. Wear-and-Tear Theory of Aging

The wear-and-tear theory of aging was discussed previously in Chapter 8, Section III.D. Accumulation of damage in any biological structure that might contribute to aging, except for damage to DNA, can be regarded as a process of wear and tear. We consider damage to DNA to be a special case because DNA encodes the genetic information of the cell. We previously cited bacterial decay of teeth, and the pitting and eroding of the articular surfaces where bones contact each other, as examples of wear and tear in humans. We will now describe two further cases of wear and tear in vertebrates: the deterioration of red blood cells and the cross-linking of macromolecules. We will then speculate on the relationship between wear and tear and DNA damage in the overall aging process.

Kay *et al.* (1990) have proposed a specific mechanism for the aging of vertebrate red blood cells. This mechanism involves the generation of a specific red blood cell membrane antigen, referred to as senescent cell antigen. Senescent cell antigen is generated by the modification, probably by oxidation, of a major membrane protein. Exposure of senescent cell antigen on a cell's surface marks the cell for death by initiating the binding of a specific autoantibody and subsequent removal by phagocytes. Because red blood cells lack DNA, they should be incapable of carrying out structural repairs that depend on encoded genetic information. Therefore, these cells should be uniquely vulnerable to wear and tear. It seems reasonable that a mechanism for removing worn out red blood cells should have evolved.

The specific membrane protein, which, upon modification, gives rise to the senescent cell antigen is ubiquitous, having been identified on a variety of types of cells. This protein is a major transmembrane structural protein and is also important in anion transport. Kay *et al.* (1990) have suggested that exposure of senescent cell antigen may occur on old cells generally and, therefore, that this exposure may provide a universal mechanism for removal of senescent cells in vertebrates.

We now suggest a biological rationale for such a general mechanism of targeting senscent cells for destruction. Inactive forms of numerous specific enzymes have been shown to accumulate with age in animals, and this has been interpreted as being due, in part, to inefficient disposal by the cellular protein degradation system (Resnick *et al.*, 1981). We think that this loss of ability to degrade inactive proteins may be due to accumulation of DNA damage with age. DNA damage may not only reduce a cell's capacity to degrade defective proteins but may also reduce a cell's capacity for self-renewal generally. The oxidized form of the transmembrane protein, the likely precursor to senescent cell antigen, may provide a signal that a cell's general capacity for self-renewal has been critically reduced. Immunological mechanisms may have evolved to recognize the antigens exposed in these oxidized proteins and to mark the cell for destruction. This interpretation couples the signal for cell destruction with loss of capacity for self-renewal due to DNA damage. Therefore, it provides a biological rationale for the targeting of senescent cells for destruction.

Another wear-and-tear process is the progressive linking together of macromolecules such as proteins. The idea that this process is the underlying cause of aging is referred to as the cross-linking theory of aging. This theory has recently been reviewed by Bjorksten and Tenhu (1990). We have previously discussed cross-linking of the two strands of DNA as a significant kind of DNA damage (see Figure 5, Chapter 2, Section I, and Chapter 2, Section II.H) and have also noted the occurrence of DNA–protein cross-links (Chapter 2, Section II.F). In the sense that macromolecular cross-linking includes specific types of DNA damages, the cross-linking theory and the DNA damage hypothesis of aging overlap. However, we think that other types of DNA damages, in addition to those involving cross-links, are important in aging (see Chapter 2, Section II). Cross-linked macromolecules, other than DNA, can be potentially replaced or repaired if the genetic information in DNA remains intact. This suggests that cross-linking by itself is insufficient to explain aging.

Human brain proteins from older individuals appear to be more highly cross-linked than proteins from young individuals (Bjorksten and Tenhu, 1990). Many changes occur in brain cells with age in addition to the cross-linking of proteins (see Chapter 5, Sections II.A.1 and II.B.1). On the basis of the evidence that DNA damage increases in brain cell DNA (Chapter 4, Section III.B) and gene expression declines (Chapter 5, Section II.A.1), we think that the primary defect leading to aging in the brain is DNA damage. Thus, we interpret the increased cross-linking of brain proteins to mean that gene functions necessary for replacing or repairing cross-linked proteins decline with age due to accumulating DNA damage.

In general, we do not dispute the idea that wear and tear of biological structures is pervasive. On the other hand, it is clear that cells have a considerable capacity for self-renewal involving replacement and repair of worn out components. This capacity appears to be encoded in the cell's DNA, as indicated by the fact that cells without DNA, such as red blood cells, deteriorate rapidly. Thus, we think that the wear-and-tear theory, by itself, does not explain aging in most organisms. One exception may be trees, which are composed largely of nonliving tissue. Here, wear and tear seems to be the major factor in determining longevity (for further discussion, see Chapter 8, Section III.D). We consider that aging can be accounted for by the progressive loss, due to DNA damage, of the program for renewal and repair, in concert with the constant toll of wear and tear on biological structures. Thus, we think that the DNA damage hypothesis of aging and the wear and tear theory of aging are complementary rather than competing ideas.

H. Somatic Mutation Theory of Aging

Szilard's (1959) theory of aging, a theory based on somatic mutation, has had an important influence on later thinking. Szilard hypothesized that the elementary step in the process of aging is an "aging hit," which "destroys" a chromosome of the somatic cell in that it renders inactive all genes carried on the chromosome. A cell becomes nonfunctional when two homologous chromosomes have each suffered a hit (somatic mutation) or when one of a pair of homologues has suffered a hit and the other carries a "fault" (deleterious recessive allele). Maynard Smith (1959) criticized this theory and maintained that Szilard's work has made a theory of aging by somatic mutation less, rather than more, promising. He noted that the implausible assumption that a hit renders ineffective all the genes carried by the chromosome was not gratuitous. He suggested that the assumption was made because, as Szilard shows, if it were assumed that the target of a hit were an individual gene, it would be necessary also to assume that each individual carried a load of faults so high as to be inconsistent with the actual fertility of consanguineous marriages. During the past three decades, there has been a lack of evidence for the prevalence of mutations that inactivate whole chromosomes, and that lack reinforces Maynard Smith's criticism. Basic difficulties with Szilard's theory were also discussed by Sacher (1968) and more recently by Kirkwood (1988). At present, the theory appears unsupportable. Weaknesses of mutation theories of aging, in general, were analyzed by Maynard Smith (1962). One specific weakness he stressed was that if rates of gene mutation in somatic cells are similar to those in germ line cells, then the rate at which somatic cells would express deleterious recessive mutations in a homologous pair of genes would be exceedingly low and not sufficient to account reasonably for aging.

Alexander (1967) also presented experimental evidence tending to refute a major role of somatic mutation in aging. The monofunctional alkylating agent, EMS, is highly mutagenic at dose levels that show no toxicity. EMS has been shown to produce mutations in *Drosophila, Neurospora,* barley, bacteria, and bacteriophage. Yet, when administered to mice, even at very highly mutagenic doses, it had no effect on longevity. Despite difficulties with the early somatic mutation theories of aging, the basic concept remained attractive to some workers, and Burnet (1974) has presented a comprehensive discussion of aging from this perspective.

Clones of cells that arise by somatic mutation may accumulate with age in rapidly dividing tissue, such as mouse colon intestinal epithelium (Winton *et al.,* 1988). Somatic mutations may have a variety of effects including allowing the reactivation of genes that have been silenced by a normal regulatory process (Wareham *et al.,* 1987). In Chapter 8, Section III.C, we considered the role in aging of focal lesions caused by somatic mutation. We also noted that somatic mutations are of central importance in aging of filamentous fungi. The incidence of focal lesions increases with age in mammals. Although some types of focal lesions such as ulcers, cancer, and atherosclerotic plaque have major health consequences, these lesions do not appear to be the primary cause of aging in mammals. They may, however, have a contributory role. We consider the DNA damage hypothesis of aging to be compatible with the somatic mutation theory of aging, but with the proviso that somatic mutation is a substantially less significant factor than somatic cell DNA damage in mammals.

I. Error Catastrophe Theory of Aging

The somatic mutation theory retained its appeal in part because of a hypothesis presented by Orgel (1963) based on the idea that inaccuracy of the protein-synthetic apparatus leads to an "error catastrophe." Errors in the enzymes employed in transcription and translation could accelerate synthesis of further defective proteins. This amplification of mistakes could snowball until it reached catastrophic proportions. One of the consequences of inaccurate protein synthesis would be development of an inaccurate DNA synthesis system (Orgel, 1963), which in turn would cause somatic mutations. The somatic mutations would, thus, be part of the sequence of events that constitute the cellular aging process.

In support of the error catastrophe hypothesis, various DNA polymerases isolated from fibroblasts aged in culture are reported to be less faithful in copying synthetic polynucleotides than the same DNA polymerases isolated from early passage cells (Linn et al., 1976; Murray and Holliday, 1981; Krauss and Linn, 1982). On the other hand, the fidelity of DNA synthesis was not different when comparing DNA polymerase-alpha purified from peripheral lymphocytes from young versus old (60–77 yr) humans (Agarwal et al., 1978). It has also been shown that DNA polymerases-alpha and -beta from regenerating mouse liver exhibited no alterations in fidelity with respect to age of the mouse (Silber et al., 1985). DNA polymerase-beta, the mammalian DNA polymerase used in repair, is the major DNA polymerase present in the neuronal nuclei of adult rodents (see

Chapter 4, Section II.C). Rao *et al.* (1985) found that the fidelity of this enzyme was not significantly different in the neurons of young and old (40 mo) mice. These results are in accord with previous studies using livers of mice of different ages; chromatin-associated DNA polymerase-beta did not show decreased fidelity with age (Fry *et al.*, 1981).

Rao *et al.* (1985) offered an explanation for the contrasting decreased fidelity of DNA polymerases in aged cultured cells and the unchanged fidelity of DNA polymerases taken from old animals. They commented that the observation of Krauss and Linn (1982), that the fidelity of DNA polymerases is markedly reduced as early-passage cells become confluent, suggests that the fidelity of DNA polymerases observed in late-passage cells may be more related to confluence in culture than to aging. Rao *et al.* (1985) considered that their own finding of unaltered fidelity of DNA polymerase-beta isolated from neurons of very aged mice is strong evidence against the error catastrophe theory.

Implicit in the error catastrophe theory is the idea that miscoding leading to random amino acid substitutions is the main mechanism by which dysfunctional protein molecules are formed. An accumulation of altered enzymes with age has been shown in a number of different animal species (see Gershon, 1979, for review). Gershon (1979) and Rothstein (1979), however, both concluded that the alteration of enzymes is not based on amino acid substitutions but, rather, on modifications that occur subsequent to protein synthesis.

Harley *et al.* (1980) showed that errors in protein synthesis do not increase during the aging of cultured human fibroblasts. Furthermore, they observed no correlation between error frequency, donor age, and maximal cellular life span *in vitro*. In other experiments, Edelman and Gallant (1977) grew the bacterium *E. coli* in the presence of streptomycin, an antibiotic that causes an increase in translational errors during protein synthesis. The resulting modification of proteins proved not to be a self-perpetuating process involving accelerated decrease in fidelity of protein synthesis. That the fidelity of translation did not deteriorate progressively argues against the error catastrophe theory. Bozcuk (1976) reached a similar conclusion on the basis of experiments in which *Drosophila* were fed amino acid analogues. These analogues are incorporated into proteins and would be expected to increase errors; however, the analogues did not cause accelerated aging.

In a more recent study, Johnson and McCaffrey (1985), using two-dimensional gel electrophoresis, found no evidence that the fidelity of protein synthesis decreases with increasing age. The accuracy of poly(U) translation by ribosomes isolated from the liver and brain of 7- and 33-mo-old Fischer F344 rats was measured by Filion and Langhera (1985). They found that the translational fidelity of the ribosomes did not change with increasing age. This agrees with previous studies on the translational fidelity of ribosomes as reviewed by Richardson et al. (1985). Numerous further experiments have been carried out to test the error catastrophe theory. These have been reviewed by Harley et al. (1980), Davies (1985), Rothstein (1987), Gershon (1987), and Richardson and Semsei (1987). The majority of this evidence appears to be inconsistent with the error catastrophe theory. For instance, Rothstein (1987) commented that almost all of the evidence bearing on the error catastrophe hypothesis, both direct and indirect, is strongly negative, and some of it is unequivocally so. Richardson and Semsei (1987) pointed out that in the decade between 1977 and 1987 a large number of studies measured the fidelity of protein synthesis in young and old organisms. These studies were carried out to determine if the error catastrophe hypothesis of aging is a viable hypothesis. They concluded that the general consensus was that age-related change in the fidelity of protein synthesis does not occur. As noted by Rothstein (1987), taken together, the data make it impractical to accept the idea of an error catastrophe, or even the idea of a lesser degree of errors occurring in proteins as a result of aging.

DNA damages block general mRNA transcription (Chapter 3, Section I), which in turn blocks general protein synthesis. Thus, the DNA damage hypothesis of aging predicts an age-related reduction in protein synthesis. However, this hypothesis does not predict that the fidelity of protein synthesis should be reduced. The considerable evidence for a reduction in mammalian gene expression at the level of mRNA synthesis and protein synthesis was reviewed in Chapter 5, Sections II.A.1–5). Thus, the available evidence on gene expression with age indicates that the quantity, but not the accuracy, of protein synthesis declines. This supports the DNA damage hypothesis of aging and tends to refute the error catastrophe theory.

J. Programmed Aging

Comfort (1968) has pointed out that if molecular information loss is the timekeeper of aging, two hypotheses can explain its occurrence: (1) unprogrammed random accumulation of chemical noise in the cellular information systems, and (2) programmed differentiation, caused by the switching-off of synthetic processes, which cannot be switched on again without the loss of differentiation. Hayflick (1987) characterized the theory of developmentally programmed aging as age changes resulting from a continuation of the same kind of genetic program of sequential, orderly gene expression that presumably instructs the fertilized egg to develop into a mature organism.

Because the DNA damage hypothesis of aging assumes that aging is a nonadaptive consequence of DNA damage, it is incompatible with any theory that assumes aging to be a programmed adaptation (see also Chapter 1, Section III.A). In previous decades, the idea that aging is a programmed adaptation was supported explicitly in the aging literature. For instance, Finch (1972) concluded that the general physiological sequence of aging in mammals is carefully regulated genetically, and that a basic program of aging has persisted in placental mammals since their descent from a common ancestor at least 70 million years ago. In recent years, however, there has been little explicit support for the idea that aging is a genetically programmed adaptation, although as noted by Kirkwood (1984) this idea is still implicit in some of the current discussions of aging. In Chapter 1, Section III.A, we referred to the arguments by which Kirkwood (1984) refutes this idea. His main argument is that the idea of programmed aging depends on the implausible assumption that the benefit of aging to the species is a stronger selective force than the benefit to individual organisms derived from the reproductive advantages of a longer life.

Hayflick (1987) reviewed the arguments bearing on the idea that there is a purposeful sequence of events written into the genome that constitutes developmentally programmed aging. He considered that the conceptual simplicity of this idea is part of its allure, but that it cannot be defended in light of the driving force of natural selection and evolution. Hayflick's argument, in brief, was that the prevalence of aged organisms is an aberration of civilization. Aged organisms, during most of evolutionary history, were rapidly culled by predation or accident before humans protected their own and a few other species. He stated that if aging occurs at all in wild animals, its expression is quantitatively small and temporally brief. He reasoned that because few feral animals ever become old, natural selection could not have favored the development of a genetically programmed aging process.

The idea that aging is developmentally programmed as an adaptation is often confused with the idea that aging is under genetic control. There is broad agreement that aging is under genetic control. On the DNA damage hypothesis of aging, genes that specify enzymes that inactivate DNA-damaging agents (Figure 9, Chapter 6, Section I) or repair DNA damages (Chapter 9) should affect the rate of aging. However, this does not imply that the progressive impairments of function that define aging are developmentally programmed as an adaptive benefit to the organism or the group (species).

II. *Theories of Sex*

A. Reprogramming the Developmental Clock

Holliday (1988) proposed that an important function of meiosis is to reprogram the DNA prior to the formation of germ cells and the fertilized egg. By this proposal the control of gene activity during development in higher organisms depends partly on the presence or absence of 5-methylcytosine in specific controlling DNA sequences. The methylation of cytosine is heritable from one cell generation to the next through the activity of a maintenance DNA transmethylase. Ordered changes occur in the pattern of methylation during the unfolding of the genetic program for development. However, a constant pattern must be transmitted through the germ line, and maintenance of this pattern is essential for normal development from the fertilized egg. Defects in the DNA methylation patterns can arise in both somatic and germ cells. The former might contribute to the aging of the soma; the latter can be transmitted to the next generation and this would be deleterious to proper development. Abnormal development would be avoided if the loss of methylation in important controlling sequences produced a signal for recombination. This would result in formation of heteroduplex DNA as an intermediate step in recombination (see Figure 13, Chapter 11, Section I.B).

Within the heteroduplex region, the activity of the maintenance methylase would allow repair of the demethylated site, i.e., restoration of methylation. Loss of a cystosine methylation is neither a mutation nor a damage but, rather, is referred to as an epigenetic defect. Thus, sexual reproduction and outbreeding maximize the opportunities for the correction of heterozygous epigenetic defects. Asexual reproduction and inbreeding might often lead to accumulation of homozygous epigenetic defects that could not then be removed at meiosis. Holliday suggests that his hypothesis provides a possible explanation for the importance of sexual reproduction and outbreeding and the short-term harmful effects of asexual reproduction and inbreeding. However, he also noted that the general argument is in many ways similar to that presented by Bernstein *et al.* (1988) but the basic difference is in the type of DNA abnormalities recognized at meiosis. We think that resetting the developmental clock may be an additional function of meiosis, but it is unlikely to be the general selective force maintaining recombination and outcrossing. This is because the vast majority of recombination events in some organisms occur in somatic cells (Chapter 12, Section III.B) and a great deal of the benefit of outcrossing can be attributed to complementation (hybrid vigor) (Chapter 13, Section III.B).

B. Hypothesis That the Primary Advantage of Sex is the Production of Heritable Variation

The ideas presented in this section are mainly from Bernstein *et al.* (1987, 1988). In general, progeny arising from sexual reproduction differ genetically from their parents, whereas progeny arising by asexual reproduction are genetically the same as their sole parent. The progeny of sexual reproduction not only differ from their parents but also from each other (except for identical twins). Consequently, sex creates new genetic variants at each generation. Since first proposed by Weisman (1892) (see Chapter 1, Section III.B), it has been widely thought that the benefit of sex is the genetic variation (new combination of alleles) it produces. For convenience, we refer to this classical idea as the "variation hypothesis." For most of the period since it was first proposed, the hypothesized advantages of variation were ones that acted at the level of the group or species and not at the level of the individual organism. It was only in the early 1960s that group selection started to be criticized (see below), and today its use in evolutionary explanation is limited and requires careful analysis of factors relating to population structure. Since the early 1960s, however, the variation hypothesis has been the basis for numerous population genetic models, which attempt to explain the benefit of sex by selection at the level of the individual organism (for reviews, see Williams, 1975; Maynard Smith, 1978; Bell, 1982).

1. *Group Selection for Sex*   Fisher (1958: p. 50) held the general view that characters evolve as a result of their benefit (or lack thereof) to individuals. However, in sharp contrast to his general view, he singled out sexual reproduction as the possible exception that could be interpreted as having evolved for the species rather than the individual advantage. Weisman (1892: pp. 199), Fisher (1958: p. 160), and Muller (1932) each proposed ways in which the genetic variation generated by sex might benefit species in the long-term by increasing their rate of evolution. Group selection explanations for the adaptive benefit of sex were largely based on competition between populations differing in extent of recombinationally generated variation. This type of group selection explanation has, even recently, been supported by Nunney (1989).

Dawkins (1982: pp. 100–109), however, presented an argument relevant to whether or not such a complex process as sexual reproduction could have arisen by group selection. Dawkins considered that group selection is only meaningful at the level of competition between species. He contended that a lower level of population grouping than species is not stable and unitary enough to be selected as a unit of natural selection in preference to another population. The "generation time" of higher animal and plant species (i.e., the time from one speciation event to the next) ranges from thousands to hundreds of thousands of years compared with the generation time of individual organisms, which typically ranges from months to decades. Therefore, natural selection processes based on species competition are several orders of magnitude slower than those based on individual competition. It seemed implausible to Dawkins that the multiple parallel trends that make up any complex adaptive process (such as sexual reproduction) could happen within a reasonable evolutionary time span by species selection, given the "generation time" of species.

The idea that sex is beneficial at the level of the species, as distinct from the level of the individual organism, has also been discussed and criticized by Bell (1982: pp. 46–47, 91–103). The major difficulty for the maintenance of sex by group selection is the large short-term costs imposed by sex at each generation on individual organisms. These short-term costs of sex were largely ignored by early writers but have been emphasized in recent studies (for enumeration of these costs, see Chapter 13, Section III.A). On theoretical grounds, it is difficult for selection, operating at the species level, to overcome large short-term costs to individual organisms. The issue of whether or not sex provides any long-term benefits through group selection has still not been resolved. Rather, it has been superceded, for most investigators, by the more pressing issue of short-term benefits.

2. *Short-Term Benefit of Sex* Numerous theoretical studies have approached the problem of the short-term benefit of sex by assuming that selection for sex is synonymous with selection for higher rates of exchange of alleles at linked loci. This idea is modeled by assuming the existence of a gene that promotes recombination but has no direct effect itself on fitness. This neutral modifier gene causes exchange of alleles at other loci, which are themselves assumed to have direct effects on fitness (see, for example, Kimura, 1956; Nei, 1967, 1969; Turner, 1967; Eshel and Feldman, 1970; Feldman and Balkau, 1972, 1973; Teague, 1976; Strobeck *et al.*, 1976; Charlesworth *et al.*, 1979; Feldman *et al.*, 1980; Holsinger and Feldman, 1983; Kondrashov, 1984). A neutral modifier gene presumably affects the induction of chromosome breaks (i.e., damages), leading to the repair of these breaks by rejoining the broken chromosome with its homologue. There are two specific problems with modeling the evolution of recombination using neutral modifier genes.

First, it is well established that gene products that catalyze recombination are not neutral (as is assumed by the neutral modifier theory) but have direct effects on fitness because they are employed in a major form of repair (for review, see Chapter 10). Second, neutral modifier models depend on hypothetical conditions, which are not broadly applicable in nature, to select increased rates of allelic recombination. The ubiquity of sex in the biota implies that the advantages of sex are general, yet the neutral modifier models depend on restrictive conditions to work. It is beyond the scope of this book to review the particular conditions used in each of the models (see Maynard Smith [1978] and Bell [1982] for comprehensive discussions of these models). However, it is evident from the discussions of Maynard Smith and Bell that despite considerable theoretical work, no model based on the neutral modifier hypothesis is both general in scope and consistent with available evidence. There is now widespread skepticism that any particular variation argument can provide a general explanation for the evolution of sex. Despite this skepticism, variation theories remain predominant in the literature as an explanation of sex.

Recently, Crow (1988) summarized the three major advantages of sex according to variation theories: (1) adjusting to a changing environment, (2) incorporating beneficial mutations, and (3) getting rid of deleterious mutations. As an example of the controversy concerning these ideas we note recent comments by Williams (1988) from his retrospective article in "The Evolution of Sex: An Examination of Current Ideas" in which he discusses the views of the other contributors to the same volume. First, he commented on the *disadvantages* (or costs) of recombination and sexual reproduction. According to Williams, the contributor Ghiseln described, and the contributor Crow provided, a formal list of major and widely applicable disadvantages to recombination, and these authors supported them with convincing and straightforward arguments. Williams commented that these disadvantages were mainly factors understandable as immediately important to natural selection at the individual level. Williams then contrasted these to the *advantages,* which have been proposed for the genetic variation produced by sex. According to Williams, by contrast to the disadvantages of sex, the advantages in Crow's list seemed weak and restricted in scope, the arguments in their favor rather devious, and their applications most readily understood as long-term effects at the group level. A similar conclusion was reached by Shields (1982: p. 105). He commented that all explanations so far advanced for the evolution of sex based on the advantage of variation to the individual are restricted by a lack of generality, a reliance on admittedly unrealistic assumptions, or on infrequent conditions as primary selection forces.

3. *Other Approaches to Choosing Between the Variation Hypothesis and the Repair (and Complementation) Hypothesis* A choice between the variation hypothesis and the repair (and comple-

mentation) hypothesis of sex might be made by examining the mechanism of meiotic recombination. We take the general view that by analyzing a structure something can often be inferred about its function. While we do not think that a biological structure need be optimally adapted for its function, we consider it appropriate, given alternative functional hypotheses, to ask whether or not the structure in question has evolved such that it is better adapted to carry out one function rather than another. Therefore, we focus on meiotic recombination as a central feature of the sexual cycle and we ask whether, on the basis of its mechanical design, it is an adaptation for DNA repair or allelic recombination or both. We next consider two fundamental features of meiosis, which, we believe, allow a choice to be made. These are premeiotic replication and the frequent occurrence of physical recombination without allelic recombination of flanking alleles.

The first feature, premeiotic replication, has been discussed in some detail in Chapter 12, Section II, and we will only briefly review this subject here. Premeiotic replication is a general characteristic of meiosis giving rise to four homologous chromosomes, which are then able to recombine in pairs. On the repair hypothesis, premeiotic replication might have the following advantage. It may lead to the formation of gaps in the newly replicated complementary strand opposite single-strand damages, many of which may be difficult for excision repair enzymes to recognize. These gaps may then act as general signals for recombinational repair of the damages. On the hypothesis that meiosis is an adaptation for promoting variation, however, premeiotic replication is difficult to explain because it allows the occurrence of sister-chromatid exchanges, a type of recombination that generates no genetic variation. Thus, a very general aspect of meiosis, premeiotic replication, is unlikely to be an adaptation for promoting recombinational variation but can be explained by the hypothesis that meiotic recombination is an adaptation for DNA repair. We will refer to physical recombination events that do not generate genetic recombination, such as sister-chromatid exchanges, as cryptic recombination.

We next consider the observation that, during meiosis, physical recombination occurs frequently between nonsister homologous chromatids without exchange of regions flanking the recombination event. This observation tends to refute the hypothesis that sex is selected to promote recombinational variation. Numerous studies have been performed in fungi where meiotic recombination between nonsister homologous chromatids was detected by gene conversion (aberrant segregation at a genetically marked site), and genetic markers at flanking loci were scored as to whether or not they were exchanged (for further discussion, see Chapter 11, Section II.B). In the majority of these studies, the flanking markers were exchanged in <50% of the recombination events detected by gene conversion (Whitehouse, 1982: p. 321). In eight individual studies involving four different species, the ratio of recombination events with nonexchanged flanking markers to those with exchanged flanking markers was on average 64:36 (for summary of these data, see Whitehouse, 1982: p. 320). The average frequency of exchange of flanking markers varied from species to species, from about 26% in *Sordaria brevicollis* to about 50% in *S. cerevisiae*. In addition to the data obtained from fungal species, data from half-tetrad analyses using compound autosomes in *D. melanogaster* showed that only one-third of recombination events at the rosy cistron resulted in exchange of flanking markers (Chovnick *et al.*, 1970).

Those physical recombination events occurring between nonsister homologous chromatids, which do not generate recombination of flanking alleles, can also be referred to as cryptic recombination. We can ask: Why are most recombination events that are detected by gene conversion cryptic? The preponderance of cryptic recombination can be explained either as an adaptation or as an unavoidable consequence of the basic mechanism of meiotic recombination. On the hypothesis that sex is selected to promote genetic variation, cryptic recombination events are nonadaptive and wasteful, because they use up resources but generate no benefit. Thus, they should be eliminated by natural selection unless they are unavoidable. If it can be shown that cryptic recombination is avoidable, the high level of cryptic recombination observed would be evidence against the variation hypothesis.

Bernstein *et al.* (1988) have shown that in the double-strand break repair model, the currently favored model for meiotic recombination, the cryptic recombination generated could readily be avoided by eliminating a complex and costly step. This step is the formation of the second Holliday junction (see Figure 13, Step F, Chapter 11, Section I.B). Without this step, physical recombination would produce outside marker exchange all the time (see fig. 2 in Bernstein *et al.*, 1988). Consequently, the function of this step appears to be to reduce exchange of alleles outside the small region of physical recombination. Thus, we conclude that cryptic recombination is not an unavoidable consequence of physical recombination. The fact that recombination is usually cryptic suggests that variation is not the selected-for advantage of recombination. The extensive occurrence of cryptic recombination also suggests that the common approach, in evolutionary biology, of modeling recombination in terms of the modification of the linkage relationship of genes (for review, see Brooks, 1988) is, at best, an incomplete description of the phenomenon to be explained. At worst, it is misleading with respect to the evolutionary forces operating on recombination.

Another approach to choosing between the variation hypothesis and the repair (and complementation) hypothesis of sex was discussed previously in Chapter 12, Section III.B. This approach was based on the fraction of physical recombination events that result in allelic recombination. We calculated that for the human female this fraction is infinitesimally small, less than 2.6×10^{-6}. This small value results from the fact that sister-chromatid exchanges in somatic cells, in aggregate, are far more numerous than meiotic recombination events. That such a small fraction of recombination results in allelic exchange is consistent with our argument that the fundamental function of all recombination events is DNA repair, whether these events occur in the germ line or in the somatic line, whether they are between sister or nonsister homologues, and whether, in the case of nonsister exchange, they lead to allelic recombination or no allelic recombination. Because the fraction of total recombination events that give rise to allelic recombination is so tiny, at least in this one case where it can be calculated, allelic recombination cannot plausibly be regarded as the general adaptive function of recombination.

The variation hypothesis can be evaluated from still another approach based on the independent assortment of chromosomes at meiosis. The physical linking of genes to form chromosomes is an almost universal feature of life. Most eukaryotes have multiple chromosomes and during meiosis, genes on separate chromosomes assort independently of each other. Crow (1988) has argued that independent assortment and the genetic variation it produces is an adaptive feature of meiosis; however, we do not share this view.

Linkage probably evolved initially to facilitate the assortment of complete genomes into daughter cells (Bernstein *et al.*, 1984; see also Chapter 14, Section III). Before genes became linked in chromosomes, the problem of damage to a gene could be dealt with by replication of an undamaged homologous gene. The homologous gene was available because the primitive sexual cycle presumably involved haploid cell fusion to form a transient diploid. The replacement of a damaged gene by replication of an undamaged copy was the earliest form of "repair." Once linkage evolved, this strategy could no longer work because the whole chromosome would have to be discarded if it contained even a single damaged gene. The only option for repairing damage at this point is to remove the damaged information and replace it using the homologous information from a chromosome containing an undamaged copy of the gene. We believe the mechanism that evolved to carry this out is physical recombination. Depending on the frequency of recombinational repair, there is a chromosome length beyond which increased linkage would have no effect on variation because the genes at the ends of the chromosomes already behave as if they are physically unlinked. In other words, the organization of the genome into multiple chromosomes, rather than a single long chromosome, does not necessarily generate more variation than already exists because of the effect of recombinational repair on linkage.

4. Natural Selection Reduces Recombinational and Mutational Variation   The occurrence of genetic variation is an essential condition for evolution. Mutations and genetic recombination are the sources of this variation. Mutations arise largely from replication errors during either chromosome replication or repair synthesis. Despite the importance of mutational variation for evolution, the replicative machinery has clearly been selected to be extraordinarily accurate, so that the error rates per replicated nucleotide are of the order of 10^{-8} to 10^{-11}, depending on the organism (Drake, 1974). In the absence of selectivity by the replicative machinery, however, complementary base pairing is, by itself, relatively inaccurate, having an error rate of approximately 10^{-2} (Loeb *et al.*, 1978). The enzymatic machinery that catalyzes replication is complex and this complexity is considered by enzymologists studying these structures to result from adaptations for promoting accuracy (Alberts *et al.*, 1980).

Just as mutational variation can be viewed as a by-product of DNA replication, recombinational variation can be considered to be a by-product of recombinational repair of DNA. Furthermore, just as the replicative machinery has been selected to reduce mutational variation, we think that the recombinational repair machinery of meiosis has been selected to reduce recombinational variation (see preceding section). We consider that the main selective pressure operating on recombinational variation, like that operating on mutation, is the immediate reduction in fitness caused by random changes in genetic information (for further discussion, see Shields, 1982: p. 67). We think that genetic variation is, fundamentally, a kind of informational noise produced as an incidental by-product of genome replication and repair, processes essential for life. The view that genetic variation is informational noise does not conflict with the concept that infrequent beneficial variants, either mutational or recombinational, are essential for adaptive evolution. That genetic variation is necessary for evolution does not mean it is selected as an adaptation (for further discussion, see Michod, 1986). Indeed, selection apparently has reduced the genetic variation that would otherwise be spontaneously produced by both DNA replication and repair.

One can ask, if allelic recombination is a form of genetic noise, why hasn't natural selection reduced its frequency to a very low level or to zero? This question has been considered in detail by Bernstein et al. (1988). They concluded that the observed level of recombination of outside markers associated with gene conversion (about 36% on average) probably reflects a balance between the costs of reducing it further (in terms of modifying the mechanism of recombination) and the benefits of such a reduction. One benefit of reducing allelic recombination arises from avoiding the breakup of coadapted gene complexes (for further discussion, see Shields, 1982: pp. 66–102). However, it is difficult to judge how strongly such a benefit would counterbalance the costs of modifying the recombination mechanism.

5. *Overview of the Variation Hypothesis of Sex*   Historically, explanations for the advantage of sex have been based almost exclusively on the advantage of genetic variation, at first at the level of the group or species and more recently at the level of the individual

organism. The variation hypothesis currently has the status of dogma in evolutionary biology. For instance, in a review of the field, Bell (1982: p. 336) observed that the consequence of increased genetic diversity of progeny is a category that includes all hypotheses of interest. We noted difficulties with this dogma. First, maintenance of sex by group selection faces the problem of large short-term costs. Efforts to find a short-term benefit focus on the evolution of a gene that promotes allelic recombination, but is otherwise neutral. However, we have pointed out that the gene products that are known to catalyze recombination are not neutral and the specific models of the evolution of sex based on the neutral modifier approach are not broadly applicable to nature. Next, we showed that two general features of meiosis, premeiotic replication and the cryptic recombination associated with gene conversion, are best explained by the hypothesis that meiosis is an adaptation for recombinational repair of DNA, not genetic variation. In addition, only a tiny fraction of recombination events in the human female are associated with allelic recombination. Thus, allelic recombination cannot be the general adaptive function of recombination in humans. We argued that recombinational variation and mutation are two forms of genetic noise that are selected against, although we noted that infrequent beneficial variants, either recombinational or mutational, are essential for adaptive evolution.

C. Selfish-DNA Hypothesis for the Evolution of Sex

Hickey (1982) developed a model for the evolution of transposable genetic elements. Such elements have been referred to as examples of "selfish DNA" or "parasitic genes" because their spread throughout the genome of a higher organism is driven by the intrinsic properties of the element even though they might provide no advantage, or even a disadvantage, to the "host" organism. This type of process has been termed molecular drive (Dover, 1982). In developing the consequences of his model, Hickey suggested that sex itself, and especially outbreeding, is a consequence of parasitic genes. Dawkins (1982: p. 160) also raised this possibility. He referred to replicating "engineers" of meiosis, which achieve their own replicative success as a consequence of forcing meiosis upon the organism.

Could sex have arisen and been maintained, not as an adaptation, but rather as an unavoidable parasitic disease? We consider this idea to be implausible, primarily because it attributes no advantage to sex to balance the costs. For most species, the costs of sexual reproduction are considerable. In many organisms, there is the twofold cost of producing males (Maynard Smith, 1978: pp. 2–3). In addition, there are the energy and material costs of forming sexual structures, maintaining their function, and all of the activities that comprise mating behavior (for other costs, see Chapter 13, Section III.A). If these costs were imposed at each generation without benefit to the organism by parasitic genes, some effective method would surely have evolved to eliminate them. Bacteria can deal with foreign invading DNA through the production of restriction enzymes that recognize and degrade this DNA (Kornberg, 1980: pp. 333–340). Additional mechanisms exist in microorganisms for inhibiting competing DNA at the level of gene expression. We think it likely that analogous devices for avoiding the huge costs of sex would have evolved in higher organisms if sex was maintained merely for the benefit of parasitic genes. We are not arguing here that processes should evolve to eliminate all parasitic genes. Parasitic genes that are relatively innocuous to their host might be able to persist by molecular drive; however, genes governing sexual reproduction clearly are not innocuous.

In the previous section, where we compared the variation hypothesis with the repair (and complementation) hypothesis, our approach was to consider whether certain general features of meiosis were adaptations for promoting variation or repair, these being presumed benefits to the organism. We cannot use this approach to compare the parasitic DNA hypothesis with the repair (and complementation) hypothesis, because the former assumes that sex is not adaptive to the organism. This inherent aspect of the parasitic DNA hypothesis makes it difficult to test experimentally. However, as discussed above, the large costs, without compensating benefit to the organism, make the molecule drive model implausible as the major force explaining the origin and maintenance of sex.

[A summary of the evidence presented in this chapter on other theories of aging and sex is given in the next chapter (Chapter 16, Section I.C) as part of our general overview of the material covered in this book.]

References

Agarwal, S. S., Tuffner, M., and Loeb, L. A. (1978). DNA replication in human lymphocytes during aging. *J. Cell. Physiol.* **96,** 235–243.

Alberts, B. M., Barry, J., Bedinger, P., Burke, R. L., Hibner, U., Liu, C.-C., and Sheridan, R. (1980). Studies of replication mechanisms with the T4 bacteriophage *in vitro* system. *In* "Mechanistic Studies of DNA Replication and Genetic Recombination." ICN-UCLA Symposia on Molecular and Cellular Biology (B. Alberts, C. F. Fox, and F. J. Stusser, eds.), pp. 449–474. Academic Press, New York.

Alexander, P. (1967). The role of DNA lesions in processes leading to aging in mice. *Symp. Soc. Exp. Biol.* **21,** 29–50.

Bell, G. (1982). "The Masterpiece of Nature: The Evolution and Genetics of Sexuality." University of California Press, Berkeley.

Bernstein, H., Byerly, H. C., Hopf, F. A., and Michod, R. E. (1984). Origin of sex. *J. Theor. Biol.* **110,** 323–351.

Bernstein, H., Hopf, F. A., and Michod, R. E. (1987). The molecular basis of the evolution of sex. *Adv. Genet.* **24,** 323–370.

Bernstein, H., Hopf, F. A., and Michod, R. E. (1988). Is meiotic recombination an adaptation for repairing DNA, producing genetic variation, or both? *In* "The Evolution of Sex: An Examination of Current Ideas" (R. E. Michod and B. R. Levin, eds.), pp. 139–160. Sinauer, New York.

Bjorksten, J., and Tenhu, H. (1990). The cross-linking theory of aging— Added evidence. *Exp. Gerontol.* **25,** 91–95.

Bozcuk, A. N. (1976). Testing the protein error hypothesis of aging in *Drosophila. Exp. Gerontol.* **11,** 103–112.

Brooks, L. D. (1988). The evolution of recombination rates. *In* "The Evolution of Sex: An Examination of Current Ideas" (R. E. Michod and B. R. Levin, eds.), pp. 87–105. Sinauer, New York.

Burnet, F. M. (1970). "Immunology, Aging and Cancer." W. H. Freeman, San Francisco.

Burnet, F. M. (1978). "Endurance of Life." Cambridge University Press, Cambridge.

Burnet, M. (1974). "Intrinsic Mutagenesis: A Genetic Approach to Aging." Wiley and Sons, New York.

Charlesworth, D., Charlesworth, B., and Strobeck, C. (1979). Selection for recombination in partially self-fertilizing populations. *Genetics* **93,** 237–244.

Chovnick, A., Ballantyne, G. H., Baille, D. L., and Holm, D. G. (1970). Gene conversion in higher organisms: Half-tetrad analysis of recombination within the rosy cistron of *Drosophila melanogaster. Genetics* **66,** 315–329.

Comfort, A. (1968). Feasibility in age research. *Nature (London)* **217,** 320–322.

Crow, J. F. (1988). The importance of recombination. *In* "The Evolution of Sex: An Examination of Current Ideas" (R. E. Michod and B. R. Levin, eds.), pp. 56–73. Sinauer, New York.

Davies, I. (1985). Theories of aging. *In* "Textbook of Geriatric Medicine and Gerontology," Chapter 4 (J. C. Brocklehurst, ed.), pp. 62–104. Churchill Livingstone, New York.

Davies, K. J. A., and Goldberg, A. L. (1987). Proteins damaged by oxygen radicals are rapidly degraded in extracts of red blood cells. *J. Biol. Chem.* **262**, 8227–8234.

Dawkins, R. (1982). "The Extended Phenotype." W. H. Freeman, San Francisco.

Dover, G. (1982). Molecular drive: A cohesive mode of species evolution. *Nature (London)* **299**, 111–117.

Drake, J. W. (1974). The role of mutation in bacterial evolution. *Symp. Soc. Gen. Microbiol.* **24**, 41–58.

Edelman, P., and Gallant, J. (1977). On the translation error theory of aging. *Proc. Natl. Acad. Sci. USA* **74**, 3396–3398.

Edney, E. B., and Gill, R. W. (1968). Evolution of senescence and specific longevity. *Nature (London)* **220**, 281–282.

Eshel, I., and Feldman, M. W. (1970). On the evolutionary effect of recombination. *Theor. Pop. Biol.* **1**, 88–100.

Feldman, M. W., and Balkau, B. (1972). Some results on the theory of three gene loci. *In* "Population Dynamics" (T. N. E. Greville, ed.), pp. 357–383. Academic Press, New York.

Feldman, M. W., and Balkau, B. (1973). Selection for linkage modification. II. A recombination balance for neutral modifiers. *Genetics* **74**, 713–726.

Feldman, M. W., Christiansen, F. B., and Brooks, L. D. (1980). Evolution of recombination in a constant environment. *Proc. Natl. Acad. Sci. USA* **77**, 4838–4841.

Filion, A. M., and Laughrea, M. (1985). Translation fidelity in the brain, liver, and hippocampus of the aging Fischer 344 rat. *In* "Molecular Biology of Aging: Gene Stability and Gene Expression" (R. S. Sohal, L. S. Birnbaum, and R. G. Cutler, eds.), pp. 257–261. Raven Press, New York.

Finch, C. E. (1972). Cellular pacemakers of aging in mammals. *In* "Cell Differentiation" (R. Harris, P. Allin, and C. Viza, eds.), pp. 259–262. Munksgaard, Copenhagen.

Finch, C. E. (1978). The brain and aging. *In* "The Biology of Aging" (J. A. Behnke, C. E. Finch, and G. B. Moment, eds.), pp. 301–309. Plenum Press, New York.

Fisher, R. A. (1958). "The Genetical Theory of Natural Selection." Dover, New York.

Fry, M., Loeb, L. A., and Martin, G. M. (1981). On the activity and fidelity of chromatin-associated hepatic DNA polymerase-beta in aging murine species of different lifespans. *J. Cell Physiol.* **106**, 435–444.

Gensler, H. L. (1981). Low level of UV-induced unscheduled DNA synthesis in postmitotic brain cells of hamsters: Possible relevance to aging. *Exp. Gerontol.* **16**, 199–207.

Gershon, D. (1979). Current status of age altered enzymes: Alternative mechanisms. *Mech. Ageing Dev.* **9**, 189–196.

Gershon, D. (1987). The "error catastrophe" theory of aging and its implications. *In* "Modern Biological Theories of Aging" (H. R. Warner, R. N. Butler, R. L. Sprott, and E. L. Schneider, eds.), pp. 135–137. Raven Press, New York.

Greenberg, L. H., and Weiss, B. (1978). Beta-adrenergic receptors in aged rat brain: Reduced number and capacity of pineal gland to develop supersensitivity. *Science* **201**, 61–63.

Harley, C. B., Pollard, J. W., Chamberlain, J. W., Stanners, C. P., and Goldstein, S. (1980). Protein synthetic errors do not increase during aging of cultured human fibroblasts. *Proc. Natl. Acad. Sci. USA* **77**, 1885–1889.

Harman, D. (1956). A theory based on free radical and radiation chemistry. *J. Gerontol.* **11**, 198–300.

Harman, D. (1957). Prolongation of the normal life span by radiation protection chemicals. *J. Gerontol.* **12**, 257–263.

Harman, D. (1981). The aging process. *Proc. Natl. Acad. Sci. USA* **78**, 7124–7128.

Hayflick, L. (1987). Origins of longevity. *In* "Modern Biological Theories of Aging" (H. R. Warner, R. N. Butler, R. L. Butler, R. L. Sprott, and E. L. Schneider, eds.), pp. 21–34. Raven Press, New York.

Hickey, D. A. (1982). Selfish DNA: A sexually transmitted nuclear parasite. *Genetics* **101**, 519–531.

Holliday, R. (1988). A possible role for meiotic recombination in germ line reprogramming and maintenance. *In* "The Evolution of Sex: An Examination of Current Ideas" (R. E. Michod and B. R. Levin, eds.), pp. 45–55. Sinauer, New York.

Holsinger, K. E., and Feldman, M. W. (1983). Linkage modification with mixed random mating and selfing: A numerical study. *Genetics* **103**, 323–333.

Johnson, T. E., and McCaffrey, G. (1985). Programmed aging or error catastrophe? An examination by two dimensional polyacrylamide gel electrophoresis. *Mech. Ageing Dev.* **30**, 285–297.

Karran, P., Moscona, A., and Strauss, B. (1977). Developmental decline in DNA repair in neural retinal cells of chick embryos. *J. Cell Biol.* **74**, 274–286.

Kay, M. M. B., Lin, F., Marchalonis, J. J., Schluter, S. F., and Bosman, G. (1990). Human erythrocyte aging: Molecular and cell biology. *Transfus. Med. Rev.,* in press.

Kimura, M. (1956). A model of a genetic system which leads to closer linkage by natural selection. *Evolution* 10, 278–287.

Kirkwood, T. B. L. (1977). Evolution of aging. *Nature* 270, 301–304.

Kirkwood, T. B. L. (1981). Repair and its evolution: Survival versus reproduction. *In* "Physiological Ecology: An Evolutionary Approach to Resource Use" (C. R. Townsend and P. Calow, eds.), pp. 165–189. Blackwell Scientific, Oxford.

Kirkwood, T. B. L. (1984). Towards a unified theory of cellular aging. *Monogr. Devel. Biol.* 17, 9–20.

Kirkwood, T. B. L. (1985). Comparative and evolutionary aspects of longevity. *In* "Handbook of the Biology of Aging," 2nd ed. (C. E. Finch and E. L. Schneider, eds.), pp. 27–44. Van Nostrand Reinhold, New York.

Kirkwood, T. B. L. (1988). DNA, mutations and aging. *Mutat. Res.* **pilot issue,** 7–13.

Kirkwood, T. B. L., and Holliday, R. (1979). The evolution of aging and longevity. *Proc. R. Soc. Lond. B* 205, 531–546.

Kondrashov, A. S. (1984). Deleterious mutations as an evolutionary factor. I. The advantage of recombination. *Genet. Res.* 44, 199–217.

Kornberg, A. (1980). "DNA Replication." W. H. Freeman, San Francisco.

Krauss, S. W., and Linn, S. (1982). Changes in DNA polymerases-alpha, -beta, and -delta during the replicative lifespan of cultured human fibroblasts. *Biochemistry* 21, 1002–1009.

Lampidis, T. J., and Schaiberger, G. E. (1975). Age-related loss of DNA repair synthesis in isolated rat myocardial cells. *Exp. Cell Res.* 96, 412–416.

Linn, S., Kairis, M., and Holliday, R. (1976). Decreased fidelity of DNA polymerase activity isolated from aging human fibroblasts. *Proc. Natl. Acad. Sci. USA* 73, 2818–2822.

Lange, V. E. (1988). Aus zellulären vorgängen abgeleitete hinweise zum sexual täts problem. *Biol. Zent. bl.* 107, 489–503.

Loeb, L. A., Weymouth, L. A., Kunkel, T. A., Gopinathan, K. P., Beckman, R. A., and Dube, D. K. (1978). On the fidelity of DNA replication. *Cold Spring Harbor Symp. Quant. Biol.* 43, 921–927.

Maynard Smith, J. (1959). A theory of aging. *Nature (London)* 184, 956–958.

Maynard Smith, J. (1962). Review lectures on senescence. I. The causes of aging. *Proc. R. Soc. Lond. B* 157, 115–127.

Maynard Smith, J. (1978). "The Evolution of Sex." Cambridge University Press, London.

McGeer, E. G., and McGeer, P. L. (1975). Age changes in the human for some enzymes associated with metabolism of catecholamines, GABA and acetylcholine. *In* "Neurobiology of Aging" (J. M. Ordy and K. R. Brizzee, eds.), pp. 287–305. Plenum Press, New York.

Medawar, P. B. (1952). "An Unsolved Problem in Biology." H. K. Lewis, London.

Michod, R. E. (1986). On fitness and adaptedness and their role in evolutionary explanation. *J. Hist. Biol.* **19**, 289–302.

Minot, C. S. (1907). The problem of age, growth and death. *Pop. Sci. Mon.* **71**, 509–523.

Miquel, J., Oro, J., Bensch, K. G., and Johnson, J. E., Jr. (1977). Lipofuscin: Fine structural and biochemical studies. *In* "Free Radicals in Biology," Vol. 3 (W. A. Pryor, ed.), pp. 133–182. Academic Press, New York.

Miquel, J., Economos, A. C., Bensch, K. G., Atlan, H., and Johnson, J. E., Jr. (1979). Review of cell aging in *Drosophila* and mouse. *Age* **2**, 78–88.

Muller, H. J. (1932). Some genetic aspect of sex. *Am. Nat.* **66**, 118–138.

Murray, V., and Holliday, R. (1981). Increased error frequency of DNA polymerases from senescent human fibroblasts. *J. Mol. Biol.* **146**, 55–76.

Nei, M. (1967). Modification of linkage intensity by natural selection. *Genetics* **57**, 625–641.

Nei, M. (1969). Linkage modification and sex difference in recombination. *Genetics* **63**, 681–699.

Nunney, L. (1989). The maintenance of sex by group selection. *Evolution* **43**, 245–257.

Orgel, L. E. (1963). The maintenance of the accuracy of protein synthesis and its relevance to aging. *Proc. Natl. Acad. Sci. USA* **49**, 517–521.

Pryor, W. A. (1987). The free-radical theory of aging revisited: A critique and a suggested disease-specific theory. *In* "Modern Biological Theories of Aging" (H. R. Warner, R. N. Butler, R. L. Sprott, and E. L. Schneider, eds.), pp. 89–112. Raven Press, New York.

Rao, K. S., Martin, G. M., and Loeb, L. A. (1985). Fidelity of DNA polymerase-beta in neurons from young and very aged mice. *J. Neurochem.* **45**, 1273–1278.

Resnick, A. Z., Lavie, L., Gershon, H. E., and Gershon, D. (1981). Age associated accumulation of altered FDP aldolase B in mice. *FEBS Lett.* **128**, 221–224.

Richardson, A., and Semsei, I. (1987). Effect of aging on translation and transcription. *In* "Review of Biological Research in Aging," Vol. 3 (M. Rothstein, ed.), pp. 467–483. Alan R. Liss, New York.

Richardson, A., Roberts, M. S., and Rutherford, M. S. (1985). Aging and gene expression. *In* "Review of Biological Research in Aging," Vol. 2 (M. Rothstein, ed.), pp. 395–419. Alan R. Liss, New York.

Rose, M. R., and Hutchinson, E. W. (1987). Evolution of aging. *In* "Review of Biological Research in Aging," Vol. 3 (M. Rothstein, ed.), pp. 23–32. Alan R. Liss, New York.

Roth, G. S. (1979). Hormone action during aging: Alterations and mechanisms. *Mech. Ageing Dev.* **9,** 497–514.

Rothstein, M. (1979). The formation of altered enzymes in aging animals. *Mech. Ageing Dev.* **9,** 197–202.

Rothstein, M. (1987). Evidence for and against the error catastrophe hypothesis. *In* "Modern Biological Theories of Aging" (H. R. Warner, R. N. Butler, R. Sprott, and E. L. Schneider, eds.), pp. 139–154. Raven Press, New York.

Sacher, G. A. (1968). Molecular versus systemic theories on the genesis of aging. *Exp. Gerontol.* **3,** 265–271.

Shields, W. M. (1982). "Philopatry, Inbreeding and the Evolution of Sex." State University of New York Press, Albany.

Silber, J. S., Fry, M., Martin, G. M., and Loeb, L. A. (1985). Fidelity of DNA polymerases isolated from regenerating liver chromatin of aging *Mus musculus*. *J. Biol. Chem.* **260,** 1304–1310.

Sohal, R. S. (1987). The free radical theory of aging: A critique. *Rev. Biol. Res. Aging* **3,** 431–449.

Stockdale, F. E. (1971). DNA synthesis in differentiating skeletal muscle cells: Initiation by ultraviolet light. *Science* **171,** 1145–1147.

Strehler, B. L. (1977). "Time, Cells and Aging." Academic Press, New York.

Strobeck, C., Maynard Smith, J., and Charlesworth, B. (1976). The effects of hitchhiking on a gene for recombination. *Genetics* **82,** 547–558.

Szilard, L. (1959). On the nature of the aging process. *Proc. Natl. Acad. Sci. USA* **45,** 30–45.

Teague, R. (1976). A result on the selection of recombination altering mechanism. *J. Theor. Biol.* **59,** 25–32.

Turner, J. R. G. (1967). On supergenes. I. The evolution of supergenes. *Am. Nat.* **101,** 195–221.

Walford, R. L. (1969). "The Immunologic Theory of Aging." Williams and Wilkins, Baltimore.

Wareham, K. A., Lyon, M. F., Glenister, P. H., and Williams, E. D. (1987). Age related reactivation of an X-linked gene. *Nature* **327,** 725–727.

Weisman, A. (1892). "Essays upon Heredity and Kindred Biological Problems." Clarendon, Oxford.

Whitehouse, H. L. K. (1982). "Genetic Recombination." Wiley, New York.

Williams, G. C. (1957). Pleiotropy, natural selection, and the evolution of senescence. *Evolution* **11,** 398–411.

Williams, G. C. (1975). "Sex and Evolution." Princeton University Press, Princeton, New Jersey.

Williams, G. C. (1988). Retrospect on sex and kindred topics. *In* "The Evolution of Sex: An Examination of Current Ideas" (R. E. Michod and B. R. Levin, eds.), pp. 287–298. Sinauer, New York.

Winton, D. J., Blount, M. A., and Pounder, B. A. J. (1988). A clonal marker induced by mutation in mouse intestinal epithelium. *Nature* **333,** 463–466.

Chapter Sixteen

Overview, Future Directions, and Implications

In this chapter, we first give an overview of the ideas on aging and sex that we presented in the previous 15 chapters. Then we discuss the implications of the DNA damage hypothesis of aging for slowing the rate of aging in humans, and what consequences this might have. Next, we discuss the further work needed, on the basis of the ideas presented here, to fill in the gaps in our knowledge and to test the validity of these ideas. Lastly, we summarize our basic view—that DNA damage and mutation are forms of informational noise arising from the intrinsic chemical nature of DNA. We argue that aging and sex reflect the compromises made through the course of natural selection to cope with the intrinsic instability of DNA.

I. *General Overview*

Until this point in the book, we have referred to our two key ideas as hypotheses, i.e., the DNA damage hypothesis of aging and the DNA repair (and complementation) hypothesis of sex. The word *hypothesis* implies an inadequacy of evidence in support of an explanation that is tentatively inferred, often as a basis for further experimentation. However, as we have shown and will review in this section, considerable evidence already exists in support of these ideas. Therefore, we consider it appropriate to elevate the two hypotheses to the status of *theories*. Consequently, in the remaining text we refer to the DNA damage *theory* of aging and the DNA repair (and complementation) *theory* of sex.

The underlying assumption in this book has been that damage to the genetic material is a universal problem for life. The first half of this book deals mainly with the theory that DNA damage causes aging. The second half deals mainly with the theory that the adaptive advantages of sex are DNA repair and (in multicellular organisms) the masking of mutations. Aging and sex are related in that aging reflects the accumulation of DNA damage and sex reflects, in part, the removal of DNA damage. In Section A, below, we review the major facts and ideas used to support the DNA damage theory of aging and in Section B we review those that support the DNA repair (and complementation) theory of sex. We avoid breaking up the flow of the ideas presented with references, but the arguments are structured in approximately the order they were given, in detail, in the first 14 chapters of the book; references are accessible there. In Section C, below, we review the other 10 theories of aging and 3 theories of sex, which we discussed in detail in Chapter 15. Again, we avoid breaking the flow of the ideas with references, which can be found in Chapter 15.

A. Aging

Aging can be defined as inherent, progressive impairments of function. Although it has been most well studied in mammals, aging occurs in other species as well, including some unicellular organisms. The DNA damage theory assumes that aging in mammals is due to the accumulation of DNA damages in somatic cells. This idea was preceded by the somatic mutation theory of aging. The two ideas are related in that both attribute aging to "errors" in the genetic material. However, DNA damages are very different from mutations. DNA damages, unlike mutations, cannot be replicated and therefore cannot be inherited. Also, DNA damages, unlike mutations, can be repaired. Thus, the DNA damage theory of aging leads to different predictions than the somatic mutation theory.

DNA damages appear to occur frequently in somatic cells. Apparently tens of thousands of DNA damages occur per average mammalian cell per day. DNA damages, if unrepaired, interfere with gene transcription and DNA replication. This interference can cause progressive impairment of cell function and/or cell death. In cells where the rate of occurrence of DNA damages exceeds the rate of repair, DNA damages accumulate.

In mammals, long-lived neurons, differentiated muscle cells, and other differentiated cell types do not divide or divide only slowly, and they accumulate DNA damage with age. These cells appear to govern the rate of mammalian aging. The brain is composed in large part of nondividing neurons. In the brain, the level of DNA repair is low, endogenous DNA damages accumulate with age, mRNA synthesis declines, and protein synthesis is reduced. Furthermore, cell loss occurs, tissue function declines, and functional impairments directly related to the central processes of aging occur. Thus, for the brain, there appears to be a direct relationship between the accumulation of DNA damage and important features of mammalian aging. Similarly, in muscle, the level of DNA repair is low, DNA damage accumulates with age, transcription decreases, protein synthesis declines, cellular structure deteriorates, and cells die. These declines parallel a reduction in muscle strength and speed of contraction. Again, a direct cause-and-effect relationship between DNA damage and decline in muscle performance seems a reasonable explanation of the data. Also, some evidence relates DNA damage to functional decline in liver and lymphocytes. In contrast to nondividing or slowly dividing cell populations, at least some types of rapidly dividing cell populations appear to cope with DNA damage by replacing lethally damaged cells through replication of undamaged ones. Examples include duodenum and colon epithelial cells and hemopoietic cells of the bone marrow.

Oxidative DNA damage is a likely major cause of aging in mammals. Normal respiratory metabolism produces free radicals from molecular oxygen and these reactive molecules damage DNA. Oxidative DNA damages occur at a high rate in mammals, at least a thousand per average cell per day. In comparisons of human, monkey, rat, and mouse, a higher incidence of oxidatively altered DNA bases per cell correlates with shorter life span. Oxidative DNA-damaging agents are likely to be a problem in the brain because of the high rate of oxygen utilization in this organ. The brain of individuals with Down's syndrome may be especially vulnerable to oxidative DNA damage because of an extra copy of the CuZn superoxide dismutase gene. This leads to elevated intracellular CuZu superoxide dismutase resulting in excess production of hydrogen peroxide and the potent hydroxyl radical. This excess of reactive oxygen species could account, in part, for the accelerated aging phenotype of Down's syndrome.

In numerous studies comparing different mammalian species, correlations were found between life span and the capacity to perform DNA repair. These observations fulfill an expectation of the DNA damage theory of aging. Another expectation of this theory is that exposure of an animal to sublethal doses of a DNA-damaging agent might shorten life span and accelerate at least some aspects of normal aging. Treatment with either ionizing radiation or chemical DNA-damaging agents has been found to induce life-shortening with some similarity to normal aging.

According to the DNA damage theory of aging, any human genetic condition that increases the incidence of unrepaired DNA damage might have features of premature aging. In humans, 11 genetic syndromes have both features of premature aging and a possibly increased incidence of DNA damage. In nine of these syndromes, evidence also indicates neuropathology as would be expected if neurons are particularly susceptible to DNA damage.

Among different animal and plant species, several distinct strategies appear to have evolved for responding to DNA damage. Even within a multicellular organism, different tissues may use different strategies. A strategy of continuous replication accompanied by replacement of lethally damaged cells may allow a cell population to cope with DNA damage and avoid aging. This strategy seems to be used by bacteria and by plants under some circumstances. It also appears to be used by some mammalian cell populations such as hemopoietic cells. A difficulty with this strategy in multicellular organisms is that mutational variants that proliferate abnormally may arise. Examples of these are senescence plasmids in filamentous fungi and focal lesions such as cancer in humans.

The main strategy that has evolved in mammals and insects for coping with DNA damage apparently involves maintaining major classes of cells of the adult organism in a nonreplicating state. This circumvents the problem of DNA damage blocking chromosomal replication but not the problem of blockages in the transcription of essential genes. Cellular redundancy partially compensates for the cell attrition due to the lethal effects of DNA damage in nondividing populations. This strategy avoids some of the costs of the strategy of rapid proliferation, for instance the cost of cell turnover and the cost of mutations that cause abnormal proliferation. However, a price is paid in terms of accumulation of DNA damage and the general aging of the nondividing cell population.

The level of DNA damage in a cell is determined by the balance between production of damage and its repair. Damages are of two basic types: single-strand damages and double-strand damages. Single-strand damages are those that affect only one of the two strands of DNA. These can be repaired by a number of different repair processes. The type of repair known as excision repair appears to be the predominant type in mammalian somatic cells for removing single-strand damages. Consequently, this type of repair may be most important in resisting aging. All forms of excision repair involve the removal of the damaged region from one strand of the DNA and accurate replacement of the lost information by copying from the undamaged complementary strand. Currently, seven different specific types of DNA damage in mammals are known to be removed by excision repair enzymes and this number is likely to increase as more such enzymes are discovered. Some of these excision repair enzymes remove specific types of altered bases caused by oxidative damage to DNA.

Double-strand damages are ones in which the two strands of the DNA duplex are altered at, or near, the same position. Excision repair is probably ineffective at removing this kind of damage because it depends on the ability of the DNA strand opposite the damaged one to act as an intact template for repair synthesis. Double-strand damages include double-strand breaks and interstrand cross-links. Such damages are formed by oxidative reactions and heat, although at a lower frequency than single-strand damages. Because double-strand damages are less easily repaired than single-strand damages, they may contribute substantially to the DNA impairments leading to aging despite their lower incidence. Informational redundancy is necessary for repair of damaged information. The source of such redundancy in repairing a double-strand damage must be a second DNA molecule with intact information homologous to that lost in the first molecule. Repair can occur if the intact DNA molecule donates a single-stranded section to the damaged molecule to replace the lost information. Such a process of exchange between two DNA molecules is referred to as recombination and, in this case, recombinational repair.

B. Sex

Sexual reproduction (sex) occurs in many species. It is the most frequent form of reproduction in higher animals and plants. It also occurs among simple organisms such as protozoans, bacteria, and viruses. Sexual reproduction has two key features, which appear to be universal and therefore can be regarded as defining characteristics. The first is recombination, which is the exchange of genetic material between pairs of chromosomes. The second is outcrossing, which means that the two chromosomes that exchange information come from different parents. According to the repair (and complementation) theory of sex, the function of recombination is the repair of DNA.

Recombinational repair has been shown to be efficient at overcoming DNA damages, especially double-strand damages. It is versatile, being able to handle a wide variety of different types of damage, and it is prevalent in nature, occurring in a broad range of organisms. Recombinational repair has been intensively studied in both bacterial viruses and bacteria. In the bacterium *E. coli,* there is efficient recombinational repair of three kinds of double-strand damage: cross-links, double-strand breaks, and gaps opposite pyrimidine dimers. Other types of damage are also repaired by this process in both bacteria and bacterial viruses. The bacterium *B. subtilis* carries out a sexual process, called transformation, which is also common to other bacterial species. Transformation involves the uptake of free DNA released by other members of a bacterial population. This DNA is then incorporated into the chromosome of the recipient bacterium by recombination. The function of transformation appears to be to allow the recipient bacterium to use the donor DNA as a template in the recombinational repair of its own damaged chromosome.

The fungi, including yeast, are often used as simple model systems for studying molecular processes in eukaryotes. Recombinational repair appears to be common in the fungi, as it is in viruses and bacteria. Recombinational repair of double-strand breaks and cross-links is highly efficient in yeast. The fruit fly *D. melanogaster* is used as a model for studies of the genetics of multicellular eukaryotes. Genes in *Drosophila* have been identified to have a likely role in recombinational repair, suggesting that this is a significant process in multicellular organisms. In mammalian cells, evidence suggests that double-strand breaks and cross-links, as well as other types of damage, are overcome by recombinational repair. Also, some evidence indicates recombinational repair in plants. In general, these findings provide support for the first part of the repair (and complementation) theory of sex—that recombination is an adaptation for dealing with DNA damage.

Meiosis is a key stage of the sexual cycle in eukaryotes. Recombination is a conspicuous feature of meiosis. Meiosis is initiated in a diploid cell, i.e., a cell that has two sets of chromosomes. One set of chromosomes traces its information back to one parent (e.g., for humans, the mother or the father) and the other set traces its information back to the other parent. In the diploid cell, for each chromosome from one parent there is a second homologous chromosome from the other parent. Meiosis occurs in four steps: (1) replication of the chromosomes, (2) lining up and intimate pairing of homologous chromosomes, leading to recombination, (3) cell division, and (4) another cell division. Overall, meiosis is a process by which a diploid cell produces four haploid cells. Recombination involves exchange of DNA segments between two of the four homologous chromosomes (chromatids). Any two of the four chromatids may be involved in the exchange.

The problem of how recombination occurs is one of the classical problems of genetic research. Much work has been performed over the last 40 years to achieve a molecular model of recombination that satisfactorily explains the large amount of experimental data available. The molecular model that dominates current thinking is referred to as the double-strand break repair model. This model is solidly based on current data from fine-structure genetic analyses as well as knowledge of the physical characteristics of DNA and the enzymatic reactions involved in its processing. The most significant features of the model from the perspective of the repair (and complementation) theory of sex are the first steps. These are the formation of a double-strand break that is extended to a double-strand gap. The chromatid containing the double-strand gap then receives information to fill the gap from the homologous chromatid with which it is paired. The initial formation of a double-strand gap is consistent with the idea that meiotic recombination is designed to remove any damages, particularly double-strand damages, in a simple, direct fashion. The double-strand break repair model thus provides a general mechanism for repairing all types of DNA damage at meiosis.

Much evidence indicates that recombination during meiosis functions to repair DNA. In both plants and animals, DNA repair synthesis during meiosis is maximal at the time when recombination occurs. This repair synthesis increases after treatment with DNA-damaging agents. Enzymes necessary for DNA repair increase specifically during the period of recombination. In the fruitfly *Drosophila* and in yeast, genes that are employed jointly in meiotic recombination and DNA repair have been identified.

Meiosis is limited to the cell divisions preceding gamete formation. We think this placement of meiosis reflects the very high cost of producing gametes, which are nonfunctional because they contain unrepaired DNA damages. A parent contributes only a single gamete to each progeny and this gamete must be free of damages. No other cell in a multicellular organism has such a direct effect on reproductive success. Even with meiotic recombinational repair, attrition of fertilized eggs in mammals is still rather high, with miscarriages ranging from 31 to 78%. This suggests that repair processes may be just managing to cope with damage.

Organisms that undergo vegetative reproduction or some forms of parthenogenesis can apparently deal with DNA damage without the necessity of meiosis. These organisms may maintain themselves through cellular selection, in which lethally damaged cells are replaced by replication of undamaged ones. Less than 0.1% of animal species reproduce parthenogenetically. The majority of plants reproduce sexually, undergoing both meiosis and outcrossing. Another substantial fraction undergo meiosis but are partially or principally self-fertilizing. Less than 10% appear to be limited strictly to vegetative reproduction or parthenogenesis. Thus, asexual reproduction is not a generally successful strategy among animals, and it is of limited success as a sole strategy among plants.

We noted above that sex has two fundamental aspects: recombination and outcrossing. Recombination appears to be the more fundamental aspect of sex, because reproductive systems that have abandoned outcrossing but have retained meiotic recombination are a substantial minority. One example is self-fertilization in higher plants. On the other hand, reproductive systems in which outcrossing is retained and meiosis is abandoned are rare.

According to the DNA repair (and complementation) theory of sex, the selective force maintaining recombination is DNA damage, and the selective force maintaining outcrossing in multicellular organisms is mutation. Above, we reviewed the evidence bearing on the first part of this theory. We now review the argument that the selective force maintaining outcrossing is mutation.

In addition to recombination, a second fundamental feature of meiosis is the reduction in chromosome sets from two per cell (diploidy) to one per cell (haploidy). This prepares the gametes for fertilization, which restores diploidy. In general, there are two different ways of restoring diploidy. The first involves fusion of haploid cells from the same individual, as in self-fertilization or automixis. The second involves fusion of gametes from separate individuals, as in outcrossing. If recombinational repair were the sole advantage of sexual reproduction, the processes of self-fertilization or automixis would be more cost-effective than the process of outcrossing. The former strategies avoid most of the major costs of sex. These costs include the obvious costs of structures and functions involved in the mating process as well as more subtle genetic costs. Thus, we need to explain why the most common strategy in nature, in spite of its costs, is outcrossing sex.

In 1889, Darwin argued that the greater vigor of the progeny of outcrosses (hybrid vigor), compared with the progeny of self-fertilization, was sufficient to account for outcrossing. Although this explanation was long ignored, it has been recently reformulated in modern genetic terms. In a diploid cell, the expression of a mutant gene with defective function can be masked by the expression of the corresponding wild-type gene in the second homologous chromosome. When there are mutations in different genes in each of the homologous chromosomes within a diploid cell, each mutation can be masked by the homologous wild-type gene in the partner chromosome. This mutual masking of recessive deleterious mutations, when two homologous chromosomes are present in a common cytoplasm, is referred to as complementation.

We consider complementation to be the main explanation for hybrid vigor. Hybrid vigor is a very well-known phenomenon, underlying much of the improvement in crop yield in the twentieth century. In a population of multicellular diploid organisms, unrelated individuals are less likely to have mutations in the same gene than related individuals. Therefore, outcrossing or mating between unrelated individuals will be favored over mating with close relatives because of the advantage of complementation. Self-fertilization, which involves union of gametes from the same individual, is the most extreme form of inbreeding and consequently does not benefit significantly from complementation.

Why has hybrid vigor been largely ignored as an explanation for the maintenance of outcrossing? The probable reason is that the evolutionary advantage of outcrossing is transient in a long-term evolutionary sense. The masking of recessive mutations in the progeny of outcrossers prevents these mutations from being weeded out efficiently by natural selection. If a self-fertilizing organism were to switch to outcrossing, recessive mutations would accumulate over successive generations in the descendants of the outcrossers. Eventually, an equilibrium would be reached between the occurrence of new mutations and the loss of mutations by natural selection. At this point, the outcrossers have no advantage through complementation over their original self-fertilizing ancestor. In fact, the outcrossers bear the costs of mating without compensating benefit. However, long-term consequences are irrelevant to the operation of natural selection, which is influenced only by the selective forces present at each generation. So long as the benefits of complementation outweigh the costs of outcrossing, outcrossing will be selected.

When equilibrium is reached there are many more recessive mutations in the genome of the outcrosser than in the genome of the original self-fertilizing ancestor. Thus, if an outcrosser, at this stage, should switch back to being a self-fertilizer, the many accumulated recessive mutations would be immediately expressed in the progeny. These progeny, then, would be inviable or have reduced fitness, so that there is a severe penalty for an outcrosser at equilibrium that switches back to selfing. In summary, there is a substantial advantage for a self-fertilizing organism to switch to outcrossing, due to the benefit of complementation, whereas there is a substantial penalty when an outcrosser at equilibrium switches to self-fertilization, due to the unmasking of mutations.

An alternative strategy for avoiding the costs of sex as well as the costs of unmasking mutations would be to produce diploid eggs by mitosis rather than haploid eggs by meiosis. This parthenogenetic process is referred to as ameiotic apomixis. This strategy, however, permanently foregoes the benefit of recombinational repair between non-sister homologous chromosomes. If this type of recombinational repair is unavailable, double-strand damages present before pre-meiotic replication cannot be removed. Thus, it seems that abandoning outcrossing sex for either self-fertilization or ameiotic apomixis ordinarily has severe short-term costs, whereas a switch in the opposite direction has short-term benefits because of enhanced repair and/or complementation. The selective forces sketched here, we think, are sufficient to explain the maintenance of outcrossing in multicellular diploid organisms.

The evolution of sex can be viewed as a continuum, with the main selective forces driving it being DNA damage and mutation. Sex is presumed to have arisen in primitive RNA-containing protocells. We speculate that the earliest form of sex was similar to that occurring in extant segmented single-stranded RNA viruses, which are among the simplest known organisms. We think that this early form of sex, involving fusion of haploid protocells, arose as an adaptation for coping with genome damage. A haploid protocell with lethal damage to one of its RNA genome segments might produce viable progeny by fusing with another damaged haploid protocell. In the diploid fusion cell, nonenzymatic reassortment of replicas of undamaged RNA segments would give rise, upon division, to viable haploid progeny. Such a primitive form of recombinational repair is thought to occur in influenza virus-infected cells. We believe this primitive process of repair evolved into the currently common form of recombinational repair, which involves enzyme-mediated breakage and exchange between multigenic DNA molecules.

As some lines of descent became more complex, their genome information increased. This led to greater vulnerability to mutation and, therefore, to increased selection pressure to cope with mutations. This selection pressure led to the diploid stage of the sexual cycle, which was initially transient, becoming the predominant stage in some lines of descent because it allows complementation. As discussed above, the advantage of complementation probably also maintains outcrossing. However, outcrossing can be abandoned in favor of parthenogenesis or self-fertilization when the costs of mating are very high. Thus, parthenogenesis might be favored in recently disturbed habitats in which population densities are low and finding a mate is difficult.

Darwin, viewing evolution as a gradual process, wondered why we find the biota segregated into distinct species rather than being "blended together in inextricable chaos." This problem was approached by a cost–benefit analysis, using a computer simulation. A sexual population was allowed to split successively into smaller and smaller species, each more finely adapted to a continuous resource gradient. After each split, the new species formed had fewer individual organisms. Finding a mate is more costly when the number of individuals in a species declines. Thus, sex was assumed to have an intrinsic cost of rarity. The cost–benefit analysis showed that, with a cost of rarity, splitting does not proceed indefinitely but leads, rather, to segregation into distinct species. The current concept of species is tied intimately to sexual reproduction because, by convention, individual organisms are allocated to species on the basis of sexual compatibility. The cost–benefit analysis described above clarifies why this convention is natural. We consider that the answer to Darwin's dilemma as to why the biota is not in "an inextricable chaos" lies in the fact that most organisms are sexual and that sex results in an intrinsic cost of rarity, which creates distinct species.

C. Other Theories of Aging and Sex

The following theories of aging are partial explanations and can be viewed as consistent with the DNA damage theory of aging: (1) the free radical theory, (2) the neuroendocrine theory, (3) the Medawar–Williams evolutionary theory, (4) the theory that aging results from cell differentiation, (5) the disposable soma theory, and (6) the immunological theory. All of these theories are incomplete in that they do not offer a specific explanation for the molecular basis of aging. The DNA damage theory of aging provides such an explanation and the above theories of aging can be shown, with some modification, to be consistent with this theory.

In addition, there is the wear-and-tear theory of aging. This assumes that general deterioration of biological structures accounts for aging. However, cells have a considerable capacity for self-renewal, which is encoded in the genome. This capacity allows cells to replace or repair non-DNA structures. We argue that only when the capacity for self-renewal declines due to DNA damage do cells experience irreversible wear and tear. Thus, the wear-and-tear theory and the DNA damage theory are basically complementary rather than competing theories. The somatic mutation theory of aging also offers an explanation for the molecular basis of aging. It attributes aging to errors in the genetic material, as is also true of the DNA damage theory of aging. This similarity, in fact, has led to confusion between the two ideas. However, on the basis of available evidence, and also on logical grounds, the somatic mutation theory does not explain the central phenomena of aging in mammals. On the other hand, it does explain the increased incidence of focal lesions with age, a significant problem of human aging. Thus, this theory can be considered to supplement the DNA damage theory of aging.

The error catastrophe theory of aging also offers an explanation for the molecular basis of aging. However, its key prediction, an accelerating decline in the accuracy of protein synthesis with age, is in conflict with most experimental evidence. Another theory is that aging is programmed as an adaptation and is selected for its benefit to the species. One of the major arguments against this is the implausibility of the benefit of aging to the species being a stronger selective force than the benefit to an individual from a longer reproductive life.

Three explanations of sex, not based on the DNA repair (and complementation) theory of sex, have been proposed. The first alternative explanation for sex postulates that an important function of meiosis is to reprogram the developmental clock prior to the formation of germ cells. Outcrossing, on this explanation, permits correction during meiosis of epigenetic errors that have accumulated in the germ line DNA. Although this type of error correction may occur in meiosis, we think it unlikely to be the major selective force maintaining recombination and outcrossing, because it does not deal with the major problems of DNA damage and mutation.

The second and most frequently discussed alternative explanation for sex is the variation hypothesis. During most of the past century, it was thought that sex evolved because it produces genetic variation. Recombinational variation has been regarded as beneficial for (1) adjusting to a changing environment, (2) incorporating beneficial mutations, and/or (3) getting rid of deleterious mutations. Until recently, the presumed advantages of recombinational variation were considered to act at the level of the group or species, rather than at the level of the individual organism. Explanations for the evolutionary advantage of sex based on group selection depend on competition between populations differing in extent of recombinationally generated variation. The major difficulty in explaining the advantage of sex by group selection is the large short-term costs imposed by sex at each generation on individual organisms. For selection operating among groups or species to overcome large short-term costs to individual organisms would be very difficult. For this reason, advocates of the variation hypothesis have, in recent years, sought to find short-term advantages of sex that explain its selective maintenance.

Numerous theoretical studies have approached the problem of the short-term advantages by assuming that selection for sex is synonymous with selection for higher rates of allelic recombination. In general, it has not been possible to devise a theoretical model that selects for increased recombinational variation under conditions that are broadly applicable to nature. It now seems doubtful that any particular model based on the variation hypothesis can provide a general explanation for the evolution of sex.

The following argument also suggests that genetic variation is not the main function of recombination. In humans, the average number of recombination events per cell during meiosis and mitosis, respectively, are about 43 and 5.5. If the function of recombination is repair, these values imply that recombinational repair is significant for cells generally, but that it is especially important in the production of germ cells. In mitotically dividing cells of humans the most common kind of recombination is that between the two sister chromosomes formed upon chromosome replication. This type of recombination is conventionally referred to as sister-chromatid exchange. In women, for example, the total number of mitotically dividing somatic cells making up the body is at least 1 million times greater than the number of egg cells produced. As a consequence, the total number of recombination events occurring by sister-chromatid exchange in all of the mitotically dividing cells of the female body is vastly greater than the number of meiotic recombination events occurring in just the egg cells. Sister chromatids are copies of each other; therefore, recombination between them generates no genetic variation. Consequently, the vast majority of recombination events in females are of a type that produces no genetic variation. Production of genetic variation is thus unlikely to be the general adaptive function of recombination in at least this one case where a rough estimate of total meiotic and mitotic recombination events can be made.

Another way of choosing between the variation hypothesis and the repair (and complementation) theory of sex is to examine meiotic recombination and ask: for which function is it adapted? The majority of physical recombination events during meiosis (i.e., events where DNA is exchanged between homologous chromosomes) occur in a way that either yields no allelic recombination or minimizes it. This evidence suggests that meiotic recombination is an adaptation for repair rather than for producing genetic variation.

The main challenge to the DNA repair (and complementation)

theory of sex comes from the proponents of the variation hypothesis. These proponents are not convinced that DNA damage is a sufficient problem for life to explain the maintenance of meiotic recombination, nor that the masking of mutations is a sufficient benefit to explain outcrossing in multicellular diploid organisms. On the other hand, we and other proponents of the repair (and complementation) theory point to the lack of convincing theoretical and experimental support for the variation hypothesis. The main attraction of the variation hypothesis appears to be the idea that sex should be selected because genetic variation is necessary for evolution. Although we agree, of course, that genetic variation is necessary for evolution, we do not think that this means that natural selection operates to create variation. Our view is that the two main kinds of genetic variation, mutation and allelic recombination, are forms of genetic noise produced as unavoidable by-products of processes essential for life. Mutation is principally a by-product of DNA replication and DNA repair synthesis. Allelic recombination is produced as a by-product of recombinational repair of DNA damage.

The third alternative explanation for sex is the selfish-DNA hypothesis. This supposes that sexual processes are specified by selfish-DNA. Such DNA, also termed "parasitic genes," can spread throughout the genome of higher organisms. This spread is driven by the intrinsic properties of the element even though the element might provide no advantage, or even a disadvantage, to the "host" organism. This idea seems to be implausible because it attributes no advantage to sex to balance the considerable costs.

II. Medical Implications

Ludwig (1980) made the following general observations on the future of research on aging. Contemporary medicine has proven successful at curing or preventing diseases caused by the environment. Medical practice has achieved this by neutralizing pathogens or by compensating for the lack of something the environment normally supplies. The thrust of contemporary medicine is ecological. Altering the intrinsic makeup of the human being has been beyond medicine's reach. However, with increasing age, the cause of disease shifts away from the environment to originate more and more in the human makeup itself. A reasonable goal for medical care is to forestall intrinsic pathogenesis as effectively as that originating in the environment. The technical means of forestalling aging, we think, are at hand or soon will be. We outline these means in the next section. Then, we discuss some of the ethical issues involved in proceeding in this direction.

A. Approaches to Forestalling Aging

According to the DNA damage theory of aging, there are two general approaches to forestalling aging: (1) decrease the incidence of DNA damage and (2) increase the level of DNA repair.

1. *Reducing the Incidence of DNA Damage* We have assumed that the DNA damages leading to aging arise largely from intrinsic sources (Chapter 2, Section II). One intrinsic source that appears to be important is oxidative free radicals produced as by-products of metabolism (Chapter 6). The rate of metabolism may be slowed by dietary restriction, and this may lead to a slower rate of production of oxidative-free radicals. Weindruch and Walford (1988) reviewed evidence that long-term dietary restriction increases life span in mammals and other species. These authors concluded (p. 306) from the numerous almost uniformly positive animal studies, coupled with the fact that diverse (albeit limited) human studies point in the same direction, that dietary restriction will likely retard the rate of aging and extend maximum life span in humans. They recommended (p. 336) dietary restriction as an ethical option for human use under appropriate conditions.

Another dietary approach to forestalling aging could be to increase the intake of free radical scavaging (antioxidant) compounds. Experimental evidence from animal studies bearing on this idea was discussed in Chapter 6, Section III.A. It is still unclear in the animal studies whether or not increases in longevity in animals fed antioxidants is due to a slowing of the intrinsic aging process. Therefore, it is also uncertain whether or not adding antioxidants to the human diet would be beneficial in slowing aging.

A further hypothetical approach for reducing the incidence of DNA damage would be to increase the level or efficiency of enzymes that inactivate DNA-damaging free radicals. Examples of such enzymes, and the reactions they carry out, are shown in Figure 9,

(Chapter 6, Section I). These enzymes include superoxide dismutase, catalase, glutathione peroxidase, and glutathione reductase. In principle, these enzymes might be increased by techniques of genetic engineering (to be discussed in the next section). Other methods, not invasive to the human genome, might be found to increase the activity of these enzymes on a long-term basis, although none is known at present.

Even if a prolonged increase in an enzyme that defends against active oxygen species or other DNA-damaging agents could be brought about, there might be unwanted side effects. As discussed in Chapter 6, Section IV, individuals with an extra copy of a small region of chromosome 21, which encodes the gene for superoxide dismutase, suffer from Down's syndrome. At least some aspects of Down's syndrome may be due to the excess generation of hydrogen peroxide from the superoxide radical, the reaction catalyzed by superoxide dismutase. This example may reflect a general problem for any effort to decrease the incidence of DNA damage by increasing the level of an enzyme. When only one component of a battery of defensive enzymes is increased, imbalances that have unwanted side effects may be created.

2. *Increasing the Level of DNA Repair* There are two possible approaches to increasing the level of DNA repair in humans: (1) dietary and (2) genome manipulation. We will discuss these in turn.

Naturally occurring flavorings such as anisaldehyde, cinnamaldehyde, coumarin, and vanillin are widely used in our daily food. Vanillin reduces the frequency of chromosomal aberrations induced by the DNA-damaging agent mitomycin C in mammalian cells both *in vitro* (Sasaki *et al.*, 1987) and *in vivo* (Inouye *et al.*, 1988). Sasaki *et al.* (1990a) showed that induction of chromosome aberrations by UV light or X-rays in cultured mammalian cells was suppressed by anisaldehyde, cinnamaldehyde, coumarin, and vanillin. After UV- or X-irradiation, cell survival was higher when these flavorings were present than when they were not. Further examination of the effects of vanillin suggested that it could be promoting repair of DNA strand breaks (Sasaki *et al.*, 1990a). When vanillin, cinnamaldehyde, or p-anisaldehyde were given orally to mice after X-ray treatment, the number of chromosome aberrations induced by X-rays was suppressed compared with controls in which mice were not fed these supplements (Sasaki *et al.*, 1990b). The suppressing effect of these supplements was demonstrated using the mouse micronucleus test, which is a well-established and sensitive method to detect chromosome aberrations in bone marrow cells. The authors suggested that these flavorings may promote repair of DNA strand breaks, which then reduces the formation of chromosome aberrations.

We can speculate that, if specific flavorings promote DNA repair under the experimental conditions tested, they might also promote DNA repair in the cells of the body generally. If so, they should also promote longevity. The very fact that certain molecules have an attractive taste suggests that natural selection has acted to promote their ingestion on the basis of some fitness benefit. A slower rate of aging could contribute to fitness. If these speculations are valid, then a dietary approach to increasing DNA repair and longevity should be feasible.

The second approach to increasing DNA repair involves genome manipulation. In recent years, impressive advances have been made in (1) cloning mammalian genes, (2) inserting them into the chromosomes of mammalian cells, and (3) demonstrating expression of the cloned gene within the whole animal (for a nontechnical review, see Montgomery, 1990). Although therapeutic use of such genetic engineering techniques has not yet been achieved, successful application is likely in the near future (Culliton, 1990). Perhaps the first genetic disease that will yield to gene therapy will be adenosine deaminase deficiency (ADA), a rare but severe immune disorder that often leads to death in children who are born without a fully functioning gene for ADA. An approach to treating this condition currently being pursued (Culliton, 1990) is to insert the normal form of the ADA gene into the bone marrow stem cells, which give rise to all blood cells, including the white cells of the immune system. Armed with the normal ADA gene, which they otherwise lack, the stem cells would go on to produce immune cells with plenty of ADA. This gene therapy approach, if successful, will likely be applied to other inherited conditions. As currently conceived, gene therapy would involve altering the genetic makeup of groups of cells of the body (somatic cells) to achieve some beneficial result for a particular individual.

A more ambitious level of genetic engineering would be to manipulate the gametes (egg or sperm) or the fertilized egg to achieve a desired genetic result for the offspring arising from the fertilized egg, as well as for the descendants of this individual. Although this type of gene therapy is not yet a practical possibility, it may become feasible within the next decade or so.

In the preceding section, we mentioned a number of enzymes that metabolize reactive oxygen species and thus render them harmless. In addition to the genes that specify these enzymes, human genes also specify DNA repair enzymes. Examples of such genes are the ones that encode DNA glycosylases and apyrimidimic endonucleases (Chapter 9, Section II.B), DNA polymerase-beta (Chapter 4, Section II.C), and the recA-like proteins that are probably involved in recombinational repair (Chapter 12, Section I.A). Identification of the genes that encode repair enzymes in mammals and humans is an active field of research (e.g., see book edited by Friedberg and Hanawalt, 1988: pp. 279–324). Therefore, such genes are likely to be available for gene therapy in the coming decades.

Cutler (1975) studied the evolution of longevity in the human ancestral line and estimated the genetic complexity governing aging rate. He concluded that only a few genetic changes are necessary to decrease uniformly the aging rate of many different physiological functions. He also suggested that aging, indeed, might be controlled at the gene regulatory level. This suggests that changes in the regulatory elements of a relatively few genes might be responsible for the increase in longevity over recent primate evolution leading to humans. Among these longevity-determining genetic elements may be ones involved in DNA protection and repair. The possible existence of a small number of genetic elements controlling the rate of human aging makes it feasible to contemplate a gene therapy approach. The aim of such an approach would be to insert genes involved in DNA protection and repair into the fertilized egg. In principle, the genetic regulatory elements controlling the gene's activity could be adjusted to ensure a high level of production of the encoded enzyme. Such an approach could be used to decrease the incidence of DNA damage or increase the level of DNA repair and thus forestall aging. The special danger in such an approach is that any harmful side effects would not only affect the individual that developed from the original manipulated egg cell, but also the descendants of that individual. On the other hand, the inherited alterations introduced by genetic engineering are not fundamentally different than those that arise by chance mutation. Thus, it could be argued that the genetic engineer would be merely accelerating a process that might also occur by evolution.

So far, in discussing the genetic engineering approach to fore-stalling aging, we have focused on manipulation at the level of the fertilized egg. This approach, however, could only be used for future generations. Is there any reasonable expectation of using genetic engineering to forestall aging among people already living? We think not. The basic problem, we think, is the impracticality of genetically altering large numbers of existing postmitotic cells. For instance, if genetic engineers set as their goal to increase the low DNA repair capability of brain neurons (Chapter 4, Section II.B), a substantial fraction of the trillion or so neurons would have to be genetically modified. The technology for doing this is not on the horizon. In addition, the benefit of slowing aging in one organ is doubtful if other tissues age at the usual rate. Unless most postmitotic and slowly dividing cells could be genetically altered, the overall benefits would be marginal. Thus, for the foreseeable future, the genetic engineering approach offers no significant promise for slowing the aging process in living people.

B. Ethical Issues in Slowing the Rate of Aging by Gene Therapy

Let us assume that it becomes a practical possibility to use genetic engineering on human germ cells to reduce DNA damage and thus slow the rate of aging. Then the question arises as to whether or not it is ethical to forge ahead with such gene therapy. To answer this question it is important, first, to anticipate the quality of the extended period of life. Slowing the rate of aging implies slowing the general loss of mental and physical vigor. Major causes of debility and misery among the elderly in developed countries include chronic diseases such as cardiovascular disease, cancer, arthritis, and Alzheimer's disease. Each of these diseases occur in only a portion of the elderly population, and the incidence of some depend on life-style. Thus, they may have separate causes apart from that of the intrinsic aging process. It is not obvious, on the DNA damage theory of aging, whether or not slowing the rate of aging would affect the incidence of these specific diseases. A reduction would be expected if the under-lying cause of a particular disease is DNA damage. We have pre-viously discussed the role of DNA damage in cancer and in athero-sclerosis leading to cardiovascular disease (Chapter 8, Section III.C) and in Alzheimer's disease (Chapter 7, Section III.H). If reduction in such diseases could reasonably be expected, it would strengthen the argument for gene therapy. However, if little or no promise of a reduction in such diseases exists, even though intrinsic aging had been slowed, gene therapy would be less worthwhile.

In addition to the immediate effects on individuals of slowing the rate of aging, long-term consequences can also be expected. A longer, healthy average life span would undoubtedly have social, economic, and political impacts. The potential use of genetic engineering for slowing aging also raises the general issue of whether or not humans should tinker with the genetic information in their germ cells. Is it wise to create alterations that may change the nature of our species? Side effects of a seemingly beneficial genetic alteration may be difficult to anticipate and yet have profound implications. For instance, increasing the portion of cellular resources devoted to DNA repair may detract from other cellular functions resulting in subtle losses of human capabilities. As a conceivable example, there might be a trade-off between longevity and mental capacity.

III. *Future Research Directions*

Progress in science depends on the development of a theoretical framework within which experimental and observational evidence can be understood. As Darwin noted, all observation must be for or against some view if it is to be of any service (see Barlow, 1958: p. 161). In the preceding chapters, we presented a theoretical framework for understanding aging and sex based on the DNA damage theory of aging and the DNA repair (and complementation) theory of sex. This theoretical framework should also make clear what further work needs to be done. For convenience, we have listed, as a series of problems, the main issues we think need to be resolved.

A. Gaps in Our Knowledge Concerning the DNA Damage Theory of Aging

Resolution of the following problems would fill in major gaps in our understanding of aging from the perspective of the DNA damage theory of aging.

Problem 1. The distribution of types of DNA damage that occur naturally in mammalian cells is not known. We have reviewed evidence that damages caused by heat (Chapter 2, Sections II.A–C) and oxidation (Chapter 2, Sections II.F and H, and Chapter 6) are major types, but we do not know if other types of damages (e.g., alkylation damages) are important. Furthermore, we have no knowledge of relative rate of occurrence of the different kinds of damages.

Problem 2. The DNA damages likely to be most important in aging are the ones that accumulate over the life span of the organism. The rate of accumulation of a particular type of damage depends on the balance between its rate of occurrence and its rate of repair. In Chapter 4, Sections III.A and IV.B, we reviewed evidence that DNA single-strand breaks accumulate in muscle, brain, and liver. Also, evidence indicates that other types of DNA damage accumulate (e.g., methylated guanine in muscle, and oxidized bases, and methyl and other adducts in liver). However, we have not identified the full spectrum of damages that accumulate, nor do we know which types accumulate most rapidly.

Problem 3. In Chapter 3, we reviewed evidence that when mammalian cell DNA is damaged by an exogenous source, this causes blockage of transcription and cell death. We have also reviewed evidence that as mammals age, their brain, muscle, and liver accumulate DNA damage, mRNA synthesis declines, synthesis of specific proteins is reduced, cell loss occurs, and tissue and organ function declines (Chapters 4 and 5). However, it has not been shown in the aging mammal that the spontaneous DNA damages that accumulate are the specific cause of the loss in gene function and higher-order function. For the DNA damage theory of aging to be valid such a cause-and-effect relationship would have to exist.

Problem 4. In Chapter 7, Section III, we reviewed evidence indicating that 11 human genetic syndromes with features of premature aging also may have elevated DNA damage (see Table XII). However, in some cases the evidence for elevated DNA damage was weak, and in no specific case was there compelling evidence that elevated DNA damage caused the features of accelerated aging. In most of these genetic syndromes, our understanding would be substantially advanced if the nature of the gene function that is altered were established. Recently, major advances have been made in the technology for locating on the human genome map those genes causing specific genetic diseases. Once located, it may be feasible to determine the gene's function. Progress in identifying the genetic functions that are defective in accelerated aging syndromes could provide key information linking DNA damage to aging.

Problem 5. In Chapter 9, we reviewed current knowledge of mammalian DNA repair processes with possible relevance to aging. Study of DNA repair in mammalian cells is a new field and, as yet, our understanding is very incomplete. These pathways of repair need to be defined in humans and their relative importance in promoting longevity worked out.

B. Gaps in Our Knowledge Concerning the DNA Repair (and Complementation) Theory of Sex

Problem 1. In Chapter 10, we reviewed substantial evidence showing that recombinational repair processes are prevalent in nature, that they are effective against a range of DNA damage types, and that they are particularly efficient against double-strand damages. However, most of this evidence came from studies of viruses, bacteria, and fungi. The evidence for the existence of recombinational repair in higher plants and animals was mainly indirect. Because the DNA repair (and complementation) theory of sex predicts the existence of recombinational repair in all sexual organisms, it is important to seek more direct evidence for this process in higher animals and plants.

Problem 2. This problem is related to Problem 1 above. The DNA repair (and complementation) theory of sex assumes that meiotic recombination is a DNA repair process. Substantial evidence from work on microorganisms indicates that recombination is involved in repair (Chapter 10); there is also good evidence in higher plants and animals that during the period of meiosis when recombination happens, DNA repair also occurs (Chapter 12, Sections I.A and B). Furthermore, as reviewed in Chapter 12, Section III.C, repair and protective processes may be just barely able to cope with DNA damage, as judged by the low survival of fertilized eggs and the high frequency of chromosomal abnormalities present in spontaneously aborted fetuses. Nevertheless, it has not been shown that meiotic recombination functions to repair spontaneous DNA damages that would otherwise be deleterious to the fertilized egg. Thus, more work is needed to determine if meiotic recombination is specifically an adaptation for DNA repair.

Problem 3. The most serious challenge to the DNA repair (and complementation) theory of sex comes from the proponents of the variation hypothesis of sex (Chapter 15, Section II.B). It might be possible to distinguish between these two ideas by elucidating the detailed molecular mechanism of meiotic recombination. An understanding of this mechanism may reveal whether meiotic recombination is an adaptation for repair or to promote variation. We have already argued that our present understanding suggests that meiotic recombination is an adaptation to reduce recombinational variation rather than increase it (Chapter 15, Section II.B.4). Despite half a century of work on the mechanism of meiotic recombination and the steady progress that has been made, we are still far from fully understanding the process.

Problem 4. In Chapter 12, Section IV.C, we discussed the finding that frequencies of meiotic recombination are often lower in the male than in the female of the species. We attributed this to the higher level of metabolism during oogenesis than during spermatogenesis. We reasoned that the greater metabolism could produce a higher level of oxidative free radicals and thus a higher level of DNA damage leading to recombination. However, we also noted that in some species, such as silkworm moths, it is the female that has reduced recombination rather than the male. These exceptions represent a challenge to our theory, and, thus, we consider it important to better understand this phenomenon.

IV. *General Implications*

We have assumed throughout that genome damage is an intrinsic problem for life. We believe that genome damage both causes aging and generates the selective pressure (along with mutation) for sex. These ideas are in fundamental conflict with any explanation of aging or sex that assumes that loss of genetic information is selected as a beneficial adaptation. Thus, we disagree with the idea that aging is a programmed adaptation (Chapter 1, Section III.A, Chapter 15, Section I.J) and the idea that sex is an adaptation for generating recombinational variation, a form of informational noise (Chapter 1, Section III.B, Chapter 15, Section II.B).

We wish to point out now that the two ideas with which we disagree have a fundamental similarity. The first assumes that life tends to be too well designed, and the second assumes that this design tends to be too reliably transmitted to descendants; as a consequence, natural selection operates to reduce these tendencies. Aging, by the first view, evolved as an adaptation to overcome the tendency for individuals to last longer than is good for the group (species) (see argument by Weismann, Chapter 1, Section III.A). Sex, by the second view, evolved as an adaptation because present genetic design ordinarily does not cope with problems of the future as well as with those of the present. Therefore, a dose of recombinational variation is good for one's descendants. However, because we view DNA damage and mutation to be forms of genetic noise that are intrinsic problems for life, we see no need for natural selection to evolve mechanisms for generating further noise; i.e., there is no need to evolve mechanisms for generating the progressive impairments of function that define aging or for generating recombinational variation. We regard as unnecessary any implication that natural selection operates to reduce adaptive information to promote evolution of the group (species) or in anticipation of the problems of future generations.

We take the simple view that adaptations are generated by natural selection operating on each individual in its current circumstance. The DNA repair processes that promote longevity and the recombinational repair and complementation processes of sexual reproduction, we think, are adaptations achieved through the course of natural selection to cope with the intrinsic instability of DNA. Each human ages and dies, while her or his sexually transmitted genetic information may be passed on indefinitely. We think these human attributes reflect a compromise solution, arrived at by natural selection, to the problem of DNA damage.

References

Barlow, N. (1958). "The Autobiography of Charles Darwin 1809–1882." Norton, New York.

Culliton, B. J. (1990). Gene therapy clears first hurdle. *Science* **247,** 1287.

Cutler, R. G. (1975). Evolution of human longevity and the genetic complexity governing aging rate. *Proc. Natl. Acad. Sci. USA* **72,** 4664–4668.

Friedberg, E. C., and Hanawalt, P. C. (1988). "Mechanisms and Consequences of DNA Damage Processing." Alan R. Liss, New York.

Inouye, T., Sasaki, Y. F., Imanishi, H., Watanabe, M., Ohta, T., and Shirasu, Y. (1988). Suppression of mitomycin C-induced micronuclei in mice bone marrow cells by post-treatment with vanillin. *Mutat. Res.* **202,** 93–95.

Ludwig, F. C. (1980). What to expect from gerontological research? *Science* **209,** 1071.

Montgomery, G. (1990). The ultimate medicine. *Discover* **11,** 60–68.

Sasaki, Y. F., Imanishi, H., Ohta, T., and Shirasu, Y. (1987). Effects of vanillin on sister-chromatid exchanges and chromosome aberrations induced by mitomycin C in cultured Chinese hamster ovary cells. *Mutat. Res.* **191,** 193–200.

Sasaki, Y. F., Imanishi, H., Watanabe, M., Ohta, T., and Shirasu, Y. (1990a). Suppressing effect of antimutagenic flavorings on chromosome aberrations induced by UV-light or X-rays in cultured Chinese hamster cells. *Mutat. Res.* **229,** 1–10.

Sasaki, Y. F., Ohta, T., Imanishi, H., Watanabe, M., Matsumoto, K., Kato, T., and Shirasu, Y. (1990b). Suppressing effects of vanillin, cinnamaldehyde, and anisaldehyde on chromosome aberrations induced by X-rays in mice. *Mutat. Res.* **243,** 299–302.

Weindruch, R., and Walford, R. L. (1988). "The Retardation of Aging and Disease by Dietary Restriction." Charles C Thomas, Springfield, Illinois.

Subject Index

Adaptive benefit of sex, 2–4, 6–11, 272, 277, 287
Adenine methylase, 216
Adenosine deaminase deficiency, 363
Advantages of recombinational variation, 330–331
Aflatoxin, 28
Aging
  acceleration of, 116–121
  bacteria, 152–153
  biomolecules, 66
  cell division, 68
  cell physiology, 67–68
  cost effective, 67
  definition, 1, 347
  diet, 361–363
  DNA damage, 67
  DNA damage hypothesis of, 2, 4–6, 66–85, 108–139, 347–350
  evolutionary considerations, 67
  evolutionary strategies, 167–168
  filamentous fungi, 157–158, 165, 168
  genetic control of, 6, 315–317, 326–327
  insects, 160–164, 168
  nonadaptive, 5–6
  nonmammalian species, 152–168
  oxidative damage, 91–103
  plants, 163–164, 167–168
  programmed adaptation, 326–327
  protozoa, 7, 153–156, 165
  somatic line, 265
  somatic maintenance, 67
  theories of, 313–327
  wear and tear, 167, 319
  wild animals, 327
Albumin, 70, 74
Aldolase, 70, 74
Allelic recombination, 336

Allelic variation, 287–288
α-globulin, 70, 74
Alzheimer's disease, 122–123, 132–136, 138
  chromosome aberrations, 133
  genetic defect, 132
  MMS-sensitivity, 134–135, 138
  MNNG-sensitivity, 134–135, 138
  mRNA impairments, 135
  neuronal degeneration, 135–136
  X-ray sensitivity, 133–134, 138
Ameiotic apomixis, 283–285, 355
Amylase, 75
Animal viruses, 212, 296–299
Anisaldehyde, 362
Antioxidants, 97–98, 102, 160, 163, 361
  centrophenoxinol, 98
  DNA damage, 97–98
  increase life span, 97, 160, 163, 361
Apomixis, 273, 283–285
Apurinic–apyrimidinic sites, 30–31
Archaebacteria, 291
Asexual populations, 306
Ataxia telangiectasia, 122–123, 128–129, 138
Atherosclerosis, 166–167, 168
Autogamy, 153
Automixis, 281

B chromosomes, 225
Bacillus subtilis, 217
Bacillus subtilis
  sexual cycle, 217–219, 275–276
Bacterial decay of teeth, 319
Bacteriophage φ6, 298, 319
Basal metabolic rate, 96–97
Benzo(a)pyrene, 28
Bidirectional replication, 31–32
Binary fission, 153

Bracket fungi, 286
Brain
  aging, 76–79, 348
  carotenoid antioxidants, 98
  cell viability, 76–79
  cellular redundancy, 82
  cross-linked proteins, 320–321
  decline, cause and effect, 76–79
  low repair, 40–44
  mRNA synthesis, 98, 102
  oxidative damage, 98, 101–103
  oxygen utilization, 98, 102
  peroxide production, 98, 102
  protein synthesis, 69–72
  shrinkage, 77
  transcription, 68–71
Branch migration, 239
Bypass, 215

Caenorhabditis elegans, 158–160
  single-strand breaks, 159
Cardiac muscle cells, 38–39
Catalase, 70, 74, 95, 101, 161, 362
Cell death, 27, 32–34
Cell differentiation, theory of aging,
    317–318
Cell, self renewal, 320
Cell survival, 32–34
Cell types, 37–38
Cells, rapidly dividing, 83
Chiasmata, 258
Chromatids, 236–240
Chromosomal aberrations, 56–60, 94
  ataxia telangiectasia, 122, 128
  Down's syndrome, 101, 122
  Werner's syndrome, 122, 125
  xeroderma pigmentosum, 122
Chromosome abnormalities, 260
Cinnamaldehyde, 362
Clonal aging, 153
Closed system strategy, 277–278
Coadapted gene complexes, 283
Cockayne syndrome, 122–124, 138
Cognitive dysfunction, 76–78
Competence
  bacterial, 218, 276
Complementation, 3, 9, 272, 279, 282,
    354–355
  hybrid vigor, 279
  masking mutation, 272, 279, 282

Conjugation, 7
  bacterial, 215
Cost–benefit analysis, 295–296
Cost of rarity, 304–306
Costs of sex, 331
Coumarin, 362
Cross-link repair
  E. coli, 213–214
  fungi, 220
  humans, 224
Cross-linking theory of aging, 320–321
Cryptic recombination, 332–334
Cyanobacteria, 291
Cyclobutane, 183
Cytochrome P450, 70, 74

Damage tolerance, 185–187
Darwin's dilemma, 303–304
Death, 79
Death
  stages leading to, 84–85
Dendrite loss, 77
Developmental clock, 327–328, 358
Diabetes mellitus, 122–123, 130–132,
    138
Dichogamy, 273
Dictyotene (dictyate) stage, 251–252
Dietary restriction, 361
Dikaryons, 286
Dioecious, 273
Diploid organisms
  evolution of, 296, 300–303
Diploidy, 296, 356
  allows complementation, 301–303
Disadvantages of sex, 331
Displacement loop, 239
Disposable soma theory, 318
Disturbed habitats, 306
DNA damage, 15–26
  accumulation, 46–60, 367
  adducts, 17–18
  alkali labile sites, 55–56
  antioxidants, 97–98, 102
  blockage of transcription, 27–29
  bulky, 185, 187
  compared to mutation, 15–17, 347
  correlation with longevity, 96–97, 102
  cross-links, 17, 19, 22–23, 58, 196,
    213–214, 220, 224
  cytosine deamination, 19–20

depurinations, 17–20
depyrimidination, 18–20
diabetes, 131–132
4,6-diamino-5-formamido
    pyrimidine, 93
distinct from mutation, 15–18, 174
distribution of, 366–367
DNA-protein cross-links, 19, 21
double-strand breaks, 19, 22–23, 58,
    93–94, 96–97, 214–215, 220–221,
    223–224
efficiency of inactivation, 93–94
evolutionary strategies, 167–168,
    290–308
exogenous sources, 23–24
glucose adducts, 20–21, 131
glucose-6-phosphate, 19–21
heat, 22
housefly and fruitfly, 160–163
human muscle, 47–48
8-hydroxydeoxyguanosine, 19, 21, 95
5-hydroxymethyl uracil, 19, 21, 93–95
hypothesis of aging, 66–85, 108–139
hypoxanthine, 180
I compounds, 55, 59
imidazole-ring-opened form, 180
in nature, 18–24
lethal, 197–198
man-made chemicals, 24
3-methyladenine, 180–181
methyl adducts, 19, 21–22, 55, 56
methylated guanine, 47–48
5-methylcytosine, 159
7-methylguanine, 39
methyltartronylurea, 93
mitochondrial, 21
modified bases, 17–18
nematodes, 159–160
nuclear, 21
O^6-ethylguanine, 41
O^6-methyguanine, 19–20, 31, 41–42
O^4-methylthymine, 39
oxidative damage, 21, 91–103, 158,
    196–197
oxidized base(s), 54, 92–93, 95,
    96–97
plant chemicals, 23–24
postmitotic cells, 162
promotes sex, 299–300
pyrimidine dimers, 23, 29–30
rates of occurrence, 19, 24

reduced incidence, 361–362
single-strand, 175–189
single-strand breaks,17–20, 46–56,
    93–97, 102, 159
single-strand regions, 47, 53–54
skin cells, 23
somatic cells, 37–60
spontaneous, 18–20
spontaneous hydrolysis, 18–19, 24
strand breaks, 40, 46–55
sugar fragmentation, 91–92
sun, 23
theory of aging, 346–350
thymine dimers, 32
thymine glycol, 19, 21, 93–94, 96,
    179–180
transcription reduced, 68–76
universal problem, 2
uracil, 20, 180
urea, 93, 180
UV-irradiation, 23
DNA damaging chemicals,
    life-shortening, 120–121
DNA endonuclease, 249
DNA polymerase I, 30–31
DNA polymerase III, 30
DNA polymerase-α, 30, 44, 323
DNA polymerase-β, 31, 44, 250,
    323–324
DNA polymerase, T4, 30–31
DNA repair, 15–16, 173–198
    changes with age, 189–195
    cold-blooded vertebrates, 112
    coordinate expression, 114–115
    developmental decline, 42
    direct reversal, 183–185
    enzymes, 15
    error-prone repair, 185–187
    excision repair, 38–39, 42–43, 160,
        175–183, 190
    evolution, 45–46
    hepatocytes, 110–112
    humans, 41, 43–44
    increase level of, 362–365
    inefficiency of excision repair, 254,
        262
    lens epithelial cells, 110–111
    life span correlation, 108–116
    liver, 52–53
    luxury, 45–46
    lymphocytes, 109–110

mammalian cells, 178–198, 368
meiosis, 248–267
mitotic recombination, 255–259
muscle, 38–40
mutants, 92
neuronal tissue, 40–46
O^6-alkylguanine, 41–42, 184–185
O^6-methylguanine, 111, 113
photoreactivation, 111, 112–113,
   183–184
postmitotic cells, 38–46
postreplication repair, 187–189
primate species, 109–111
pyrimidine dimers, 114, 160
redundant information, 173–175
single-strand damages, 175–189
skin, 109
skin fibroblasts, 109–111
SOS response, 185–186
thymine glycol, 160
turn-off, 46
variety of organisms, 173
DNA repair capacity, correlation with
   life span, 108–113
DNA repair defect
   Alzheimer's disease, 133–136, 138
   ataxia telangiectasia, 128–129, 138
   Cockayne's syndrome, 124, 138
   Friedreich ataxia, 132, 138
   Huntington's disease, 130, 138
   Parkinson's disease, 136, 138
DNA repair enzymes
   AP endonucleases, 178–181
   apyrimidinic endonucleases, 176–178
   DNA polymerase I, 177–178, 182
   DNA polymerase III, 185
   exonucleases, 177–179, 185, 210
   expressing genes, 181–182
   glycosylases, 176–181
   methyl transferase, 184
   polynucleotide ligase, 177–178, 182
   uvrABC endonuclease, 182–183, 213
DNA repair genes
   lex, 185
   recA, 185
DNA repair glycosylases
   formamidopyrimidine, 93
   hydroxymethyl uracil, 93
   thymine glycol, 93
DNA repair rates
   O^6-methylguanine, 24

single-strand breaks, 24
DNA replication, 29–32
DNA stability, 2
DNA transmethylase, 327–328
Double-strand break repair, 196–198,
   214, 220–221, 223–224
Double-strand break repair model,
   237–240
Double-strand damages, 22–23, 219,
   226, 350
Double-strand gap, 238–240
Down's syndrome, 99–102, 138, 348
   catalase, 101
   chromosomal aberrations, 101
   DNA repair defect, 101–102
   glutathione peroxidase, 101
   increased leukemia, 101
   neuropathology, 99
   oxygen metabolism, 99–102
   premature aging, 99–103
   superoxide dismutase, 99–103
   trisomy 21, 99–100
Drosophila, 162–163
   fed amino acid analogs, 324–325
   half-tetrad analysis, 242
   initiation of recombination, 244–245
   mutation rate, 280, 283
   postmitotic cells, 317–318
   postreplication repair, 252
   recombination mutants, 252
   recombinational repair, 222–223
   stimulation of recombination, 257
Drosophila recombination
   absence in males, 252, 264

Endomitosis, 263–264, 281, 284
Epidermal growth factor, 70, 75
Epigenetic defect, 327–328
Epithelial cells, 37
Error catastrophe theory of aging,
   323–325, 358
Error-free repair, 211
Erythropoietic stem cells, 83
Escherichia coli, 152–153
   recombinational repair, 213–216
Eubacteria, 291
Eukaryotes, 291
Evolution, 6–7
   long-range consequences, 303
Evolution of aging, 67

Subject Index | 377

Evolution of sex, 3, 290–308, 356
  continuous history, 292–293
Evolutionary theory, of aging, 315–317
Excision repair, 109–112, 176–183
  pyrimidine dimers, 176–178, 182–183
Exonuclease, 5′ to 3′, 177, 238

Fanconi's anemia, 224
Ferns, 286
Fertilization, 273–274
Fertilized egg survival, 260
Fibroblasts, 224
Fission yeast, 221, 299–300
Flanking markers, 332–334
Flavorings, 362–363
Focal lesions, 165–167, 323, 349
Free radicals, 158, 91–94, 160–161
Free radical theory, of aging, 314–315
Friedreich ataxia, 132, 138
Fungi, gene conversion, 242
Future directions
  aging research, 306–368
  sex research, 368–369

Gamete production, 249–252, 261–262,
  265
Gametes, genetic engineering, 363–364
Gap repair, 215
Gaps in knowledge
  DNA damage theory of aging,
    366–368
  theory of sex, 368–369
Gene conversion, 241–245, 333
  mispair correction, 243
  polarity, 244–245
Gene therapy, 362–366
  danger of, 364, 366
  ethical issues, 365–366
  repair enzymes, 364
General implications
  of theories, 369–370
Genetic engineering, 362–365
Genetic noise, 370
Genetic variation, 255, 259
  benefit of sex, 4, 9–11, 331–332
  necessary for evolution, 9–10, 287,
    335–336, 360
Genome expansion, 301–303
Genome manipulation, 363–365
Genome redundancy, 295–296

Genome size, haploids versus diploids,
  303
Genotypes, earliest, 294
Germ cell formation, 235
Glial cells, 38
Glutathione, 97–98, 161
Glutathione peroxidase, 95, 101, 362
Glutathione reductase, 95, 97, 362
Glutathione S-transferase, 97
Group selection for sex, 10, 329–330,
  358

Habitats, newly created, 285
Hamster ovary cells, 223
Haplodiploid insects, 287
Haploids, multicellular, 286
Haploids, unicellular, 286
Hemopoietic cells, 37, 83
Hepatic cell decline, 80
Hepatocytes, 53, 73
Heteroduplex, 239, 243
Hiroshima subjects, 118
Historical background, 4–11
Holliday junction, 239, 243
Homeostatic regulation, 79
Human genetic syndromes, 121–139,
  280, 349, 367–368
Huntington's disease, 129–130, 138
Hutchinson–Guilford progeria,
  127–128, 138
Hybrid vigor, 9, 279, 354
Hydrogen peroxide, 21, 91–95, 300, 314
  double-strand breaks, 93–94
  induced damages, 93–94
  single-strand breaks, 93–94
Hydroxyl radical, 91–92, 94, 98, 100,
  102–103
Hypercycles, 294
Hypothalamus, 78

Idling, by DNA polymerase, 32, 187
Immortality, germ line, 265
Immune function, 80–81
Immune system, 59
Immunologic theory, of aging, 318–319
Independent assortment, 334–335
Influenza virus, 296–297
Information, 16–17
Inheritable variation, 8–11
Initiation of recombination, 244–245
Interleukin-2, 70, 75

Intestinal crypt cells, 83
Intrinsic pathenogenesis, 361
Ionizing radiation, life-shortening,
  117–120

Kidney
  damages, 59
  protein synthesis, 75
Klinefelter syndrome, 137
Kupffer cells, 53

Left hemisphere removed, 82
Life, basic properties, 293–294
Life-shortening
  effect of DNA damage, 349
  X-rays, 117–120
Life span, 56–57
  damage incidence, 96–97
  correlation with repair capacity,
    108–116
  proportional to brain size, 82
  versus repair capacity, 108–113
Lipofuscin, 78, 314
Liver
  chromosome aberrations, 56–58
  damages, 53–56
  DNA repair, 52–53
  function, 80
  polyploidy, 58–59
  protein synthesis, 70, 73–74
  transcription, 69–70, 73–74
Lymphocytes, 59–60, 318–319
  transcription, 70, 75
  protein synthesis, 70, 75
Lymphopoietic cells, 37

Macronucleus, 154–156
Mammalian cells, recombinational
  repair, 223–225
Mating success, density dependent,
  305–306
Medical implications, 360–366
Mei mutants
  Drosophila, 222–223
Meiosis, 3, 235–246, 248–267, 274–275,
  327–328, 352–353
  adaptation for repair, 248–267
  basic steps, 236–237
  compared to mitosis, 235–237
  timing of, 259–260

Meiosis and mitosis, relative
  recombination frequencies,
  258–259
Meiotic prophase, 248
Meiotic recombination, 235–246,
  248–267, 327–328
  mechanism, 237–240, 369
  repair process, 248–267, 368
Meiotic versus mitotic, recombination,
  359
Metabolic rate reduction, prolongs life
  span, 161
Metallothionein, 70, 74
Methylation of cytosine, 327–328
Micronucleus, 59, 154
Microsporocytes, Lily, DNA repair,
  249
Miscarriages, 260, 353
Mismatch repair, 216
Mitochondrial DNA, 157–158
Mitochondrial mutation, 157
Mitosis, 235–236
Mitotic recombination, 255–258
  DNA repair process, 255–258
  hamster cell mutant, 256–257
  recombination mutants, 256–257
  stimulation by DNA damage, 257
Mobile intron, 157–158
Models of recombination, 237–240
Molar odontoblasts, 38
Molecular drive, 337
Multiplicity reactivation, 209–212, 297
Muscle
  aging, 348
  damages, 46–48
  fiber loss, 79–80
  function, 79–80
  low repair, 38–40
  protein synthesis, 72–73
  transcription, 69, 72
Muscle decline, cause and effect, 81
Mushrooms, 286
Mutation, 15–16, 31–32, 57–58
Mutation
  as selective force, 272
  intrinsic, 279
  masking of, 281–282, 301–303
  most reduce fitness, 279–280
  removal by selection, 280
Mutational variation, 335–336

Myoblasts, 40
Myotubes, 40

Nematodes, 158–160
Neuritic plaques, 78
Neuroendocrine theory, of aging, 315
Neuron degeneration, *Drosophila*, 163
Neurons, reserve supply, 82
Neurospora crassa, 157, 286
Neurotransmitter systems, 315
Neutral modifier, hypothesis, 330–331
Niche, 304, 307
Neuroblastoma cells, 42–43
Neurofibrillary tangles, 78
Neurohormones, 78
Neuron loss, 76–77
Neuronal tissue
    damages, 48–52
    low repair, 40–44
Noise, 16
Nonreplication strategy, 168, 349

Oogenesis, DNA repair, 251–252
Open system strategy, 277
Origin
    of replication, 31–32
    of sex, 290–297
    of species, 303–304
Outcrossing, 1, 3–4, 9–11, 274–287,
    353–356
    benefit, 278
    costs, 278
    costs versus benefits, 278
    promotes complementation, 277
    selective advantage, 272–288
Overview, general, 346–360
Oxidative DNA damage, 91–103, 113,
    300, 348
    accumulation with age, 95–96
    brain, 98, 101
    causes, 91–95
    Down's syndrome, 99–102
    8-hydroxydeoxyguanosine, 95
    hydroxymethyluracil, 93–95
    nature of, 91–95
    single-strand breaks, 93–97, 102
    theory of aging, 314–315
    thymine glycol, 93–94, 96
    rate per cell per day, 94
Ozone, 23

Pachytene, 224–225, 249–251
Paramecium, 8
Paramecium tetraurelia, 153–156
Parasitic genes, 337–338
Parkinson's disease, 136, 138
Parotid gland, 75
Parthenogenesis, 263–264, 283–285, 353
    competition with outcrossers, 285
    lower reproductive rates, 285
Parthenogenetic species, distribution,
    306
Parthenogenic animals, frequency, 263
Parthenogens, 283–285
Phage λ, 211, 213
Phage T4, 210–212, 292–293
    sexual cycle, 276
Phages T2, T4, T6, 209
Phenotypes, earliest, 294
Physical recombination, 8, 332–333
Plant species
    apomicts, 273–274
    mostly sexual, 273
    self-fertilizing, 273
Plant toxins, 23–24
Pleiotropic genes, effect on aging,
    336–337
Podospera anserina, 157–158
Poeciliopsis, 277
Polarity of gene conversion, 244–245
Polynucleotide kinase, 249
Polynucleotide ligase, 249
Polyploidy, 58–59, 295
Population density, effect on mating,
    305–306
Postmitotic cells, 162
    low repair, 38–46
Postreplication recombinational repair,
    188–189, 212, 215, 217
Postreplicative gap, 188–189
Premature aging syndromes, 99,
    121–139, 349
    elevated DNA damage, 137–139
Premeiotic replication, 252–255, 332
    function in repair, 253–254
    common currency, 254
    sister chromatid exchange, 254–255
Primitive atmosphere, 23
Programmed aging, theory, 326–327,
    358
Promoters, 245

Proofreading, 32
Prophage reactivation, 213
Protein synthesis,
  brain, 69–72, 76
  decline with age, 76
  liver, 70, 73–74, 76
  lymphocytes, 70, 75
  muscle, 72–73, 76
Protocells, 295–296
Protozoans, 7
Punctuated evolution, 307
Pyrimidine dimers, 27, 29–30, 33

Rad 52, yeast, 220, 292–293
Radiologists, life-shortening, 117
Rapidly dividing cells
  E. coli, 152–153, 164
  fungi, 165
  P. tetraurelia, 165
  replacement strategy, 164
  trees, 163–164, 167
RecA gene, 216–217
RecA-like proteins, 249–250
RecA protein, 214–215, 293
RecN protein, 214
Recombination, 3, 7–9, 235–246, 272, 277, 287
  advantage of, 8–9
  cryptic, 332–334
  function of, 351–353
  higher in females, 264–265
  influenza virus, 296–297
  molecular model of, 352
  more basic than outcrossing, 277
  nonsister homologues, 260–261
  repair synthesis, 352–353
Recombination genes, functional homology, 292–293
Recombinational repair, 197–198, 208–226, 237–241, 245–246, 248–267, 351–353
  bacteria, 213–219
  eucaryotes, 219–226
  evolution of, 297–299
  fungi, 220–222
  gene products, 210–211, 213–214, 220–223
  higher organisms, 368
  plants, 225–226
  viruses, 209–213

Recombinational variation, 8–11, 335–337
Red blood cell, deterioration, 319–320
Redundancy, role in error correction, 173–175, 197
Redundant information, 17
Regulatory genes, control of aging, 364
Rejuvination, 7–8
Reovirus, 298
Replication accuracy, 301, 335
Replication impairment, 29–32
Replication strategy, 164–165, 168, 262–263, 348
Replicative machinery, 335–336
Replicon, 32
Reproductive isolation, 307
Reproductive systems, 281
Reproductive systems, shifts between them, 280–283
Reprogramming developmental clock, 327–328
Restriction enzymes, 338
Ribosomes, translational fidelity, 325
RNA replicators, 294
RNA synthesis, 27–29
RNA viruses, 296–299

S_1 endonuclease, 47–48, 51, 53–54
Saccharomyces cerevisiae, 220–221, 226, 256, 292–293
Schizosaccharomyces pombe, 221, 299–300
  sexual cycle, 275
Segmented genome, 295–299
Segmental progeroid syndromes, 116, 137
Self-fertilization, 273
Selfing, 281–283
Selfish DNA hypothesis of sex, 337–338
Senescence plasmid, 157–158
Senescent cell antigen, 319–320
Sertoli cells, 38
Sex
  advantage of, 6–11
  defining characteristics, 351
  definition, 1
  overview, 350–357
Sex, repair hypothesis, 2–4

Sex, repair (and complementation)
hypothesis, 6–11, 208, 226, 272,
313, 331–335
Sex, repair (and complementation)
theory, 346, 350–357, 368–369
Sex, short-term advantages, 10–11,
330–331, 358–359
Sex, short-term costs of, 330–331
Sex, theories of, 6–11, 327–338
Sexual cycle
common features, 274
fission yeast, 275
humans, 274–275
most primitive form of, 296
outcrossing, 274–277
recombination, 274–277
Sexual reproduction
dynamic of, 304–305
prevalence, 272–274
Silkworm moth, recombination reduced
in females, 264, 369
Single-strand damages, 350
Sister-chromatid exchange, 94,
224–226, 254–255, 359
Skeletal muscle cells, 38–39
Smooth muscle cells, 38
Smut fungus, recombinational repair,
221–222
Somatic mutation, 4–5, 321–323
aging, 165–167
atherosclerosis, 166–167
cancer, 166
focal lesions, 165
Somatic mutation theory of aging, 4–5,
321–323, 347, 357–358
Species competition, 329–330
Species concept, 307
Species
emergence of, 304–305, 356–357
morphological distinctions, 304
Spermatagonia, 224, 261
Spermatocytes, DNA repair, 224, 249
Spermatogenesis, 224, 249–250
Strand displacement, 239
Submandibular glands, 70, 75
Sugar phosphate backbone, 91–92
Suicide enzyme, 184
Superoxide, 21, 91–92, 94–95, 100
Superoxide dismutase, 70, 74, 95,
99–103, 161, 362

Superoxide dismutase gene, 100–102
SV40 virus, 30, 33–34
Sympatric distribution, 306
Synapse loss, 77
Syngamy, 237

Template, 17
Thelytoky, 263
Theories of aging, 314–327, 347–350,
357–358
Theories of sex, 327–338, 350–357,
358–360
Thymine dimer, 176, 188
Thymine glycol, 30
Transcription
brain, 69–72
muscle, 69, 72–73
liver, 69, 70, 73–74
lymphocytes, 70, 75
termination, 27–29, 33–34
Transfection, 33
Transformation
bacterial, 217–219, 275–276, 351
transgenic mice, 100–101
Trees, life span, 163–164
Trisomy 21, 99
Tryptophan oxygenase, 70, 74
Turbatrix aceti, 158, 160
Twin spots, 261
Tyrosine aminotransferase, 70, 74

Unicellular haploids, advantage of sex,
299–300
Universal cellular, ancestor, 295
Unscheduled DNA synthesis, 109
Urea, 30
Urea DNA glycosylase, 180
Urkaryotes, 291
Ustilago maydis, 221–222, 257
UvrA, uvrB endonuclease, 213–214

Vanillin, 362
Variation hypothesis of sex, 328–337,
358–360, 369–370
Vegetative death, 157–158
Vegetative reproduction, 262–263, 281,
283–285, 353

Wear and tear, 167
theory of aging, 319–321

Werner's syndrome, 125–126, 138

Xeroderma pigmentosum, 113–116,
   122–123, 137–138, 181
   accelerated aging, 115–116, 122–123
   AP endonuclease I, 115
   AP endonuclease defect, 115
   AP sites, 114–115
   excision repair defect, 114
   increased cancer, 114, 138

increased recombination, 257
neurological abnormalities, 115–116

Yeast, recombinational repair, 220–221,
   226

Zygotene–pachytene
   DNA repair, 249–250
   m-rec protein, 250